谨以此书祝贺

赵桢教授90寿辰！

赵桢文集

广义解析函数与积分方程

李仲来 / 主编

北京师范大学出版集团
BEIJING NORMAL UNIVERSITY PUBLISHING GROUP
北京师范大学出版社

2019 · 北京

图书在版编目(CIP)数据

广义解析函数与积分方程:赵桢文集/赵桢著,李仲来主编. 一北京:北京师范大学出版社,2019.12
(北京师范大学数学家文库)
ISBN 978-7-303-25235-0

Ⅰ. ①广… Ⅱ. ①赵… ②李… Ⅲ. ①广义解析函数一文集 ②积分方程一文集 Ⅳ. ①O174.55-53 ②O175.5-53

中国版本图书馆 CIP 数据核字(2019)第 235651 号

营销中心电话 010-58802181 58808006
北师大出版社高等教育与学术著作分社网 http://gaojiao.bnup.com
电子信箱 beishida168@126.com

出版发行:北京师范大学出版社 www.bnup.com
北京市西城区新街口外大街 12—3 号
邮政编码:100088

印 刷:鸿博昊天科技有限公司
经 销:全国新华书店
开 本:710 mm×1000 mm 1/16
印 张:14
插 页:4
字 数:215 千字
版 次:2019 年 12 月第 1 版
印 次:2019 年 12 月第 1 次印刷
定 价:52.00 元

策划编辑:岳昌庆 责任编辑:岳昌庆
美术编辑:李向昕 装帧设计:李向昕
责任校对:段立超 责任印制:马 洁

▲ 1955 年北京俄文专修学校留苏预备班，肩背手风琴者：赵桢。

▶ 20 世纪 70 年代末，分析教研室教师（左起，下同）：

张阳春、孙永生、赵桢、陆善镇。

◀ 20 世纪 70 年代全家福。

前排：张玉梅、方婧增（赵桢的母亲）、赵德麟（赵桢的父亲）。

后排：赵旭东、赵晓东、赵桢。

▶ 20 世纪 80 年代，赵桢等与项武义探讨数学教育。

王梓坤、项武义、赵桢、严士健、陈公宁。

◀ 赵桢与 1981 届研究生合影。

第 1 排: 赵桢、陈维曾、宋惠元、黄海洋。

第 2 排：李好好、赵达夫、林益。

第 3 排：陈永义、李正吾。

▲ 1982 年，奇异积分方程与边值问题讨论班。

第 1 排：黄海洋、蒋绍惠。

第 2 排：陈方权、李章泉、王泽义、宋惠元、刘来福。

第 3 排：李正吾、何致和、赵桢、赵达夫、沈乃禄。

第 4 排：张其友、陈永义、田长久、李好好、陈维曾、林益。

▶ 1995 年 10 月，赵桢在亚洲数学家大会上讲话。

◀ 1997 年 9 月 2 日，祝贺赵桢 66 寿辰。

第 1 排：谢宇、相登龙、朱戟。

第 2 排：陈方权、张朝晖、王荣良、赵桢、赵达夫、邓小琴、黄海洋。

第 3 排：郑神州。

▲ 2004 年 9 月 4 日全家福。

雷钧、赵明楷、赵晔、张帅、赵昊楷、赵桢、刘秀芳、赵旭东、赵晓东、宋华、赵钰、雷思雨、蒋卫红。

▲ 2004 年 1 月 27 日合影。
前排：赵桢、刘秀芳、柳藩、严士健。
后排：李希菲、唐守正、陈木法、罗丹。

▲ 2011 年全国积分方程与微分方程及其应用会议，会议中祝贺赵桢 80 寿辰。
第 1 排：蒋绍惠、谭学才、保继光、吴尽昭、赵桢、刘秀芳、刘来福、朱汝金。
第 2 排：何登旭、陈方权、张虹、杨宏奇、李美生、黄海洋、赵达夫、林益、陈永义、李正吾、楚泽甫、曾岳生、龙觉新、赵静、张超群、胡辉亚。
第 3 排：梁静、蓝师义、李杰权、宋惠元、沈乃禄、张立龙、文玉琼、刘春根、冼军、王子亭、刘焕文、刘振海、阿荣、岳淑泉、王福杰、殷文明、谷亚楠。
第 4 排：宋美玲、农丽娟、申贵成、周仕忠、朱兴龙、何崇南、郑神州、谭福锦、黄敬频、欧业林、陈德健、席东盟、莫愿斌、曹敦虔、闫沙沙。

序言

一

1952年院系调整后，北京师范大学数学系全职和兼职教师共有30人(含行政人员和工友共8人)，其中全职教授两人(傅种孙和张禾瑞)有出国经历．当时国家面临大规模经济建设，需要培养大批人才．重要措施之一就是选派青年教师留学苏联．时任数学系主任傅种孙教授对派出青年教师到苏联攻读副博士学位非常积极．1953年他让系里所有符合年龄条件的青年教师都去检查身体，由此可见他盼望青年教师成长的心情之殷．他的这一重要举措，使得数学系派到苏联攻读副博士学位的教师数量在全国高校数学系排第一名(全国共27人)，是数学系发展进程中的一大特色和一大亮点．从现在看，是一个卓有远见的做法，是改变数学系的师资结构、提高青年教师学术水平的一个重要办法，同时还为“文化大革命”后数学系在高等院校的崛起准备了相当强大的科研指导力量．

傅种孙教授对派出青年教师到苏联留学非常积极，数学系去的名额就会多一些；我们学校有的系对此事不积极，去的名额就少一些，甚至一个未去. 原校长方福康教授曾说，这是数学系办的一件明白事. 数学系与其他高校数学系相比，还有一个特点是：派出的青年教师全部到苏联攻读副博士学位，没有派出教师去苏联进修.

数学系派到苏联攻读副博士学位的教师数量在全国高校数学系排在第一位，还与体制有关：1952 年 11 月，增设高等教育部(1958 年 2 月，高等教育部和教育部合并为教育部). 当时教育部管理师范院校、中小学和学前教育，高等教育部管理普通大学. 教育部的出国名额分得少一些，北京师范大学是教育部部属的唯一一所大学，所以名额分配向北京师范大学倾斜，再照顾其他师范院校. 高等教育部出国名额分得多一些，但高校数量多，各校分配的名额相对就少一些.

20 世纪 50 年代，数学系陆续派出教师刘绍学(莫斯科大学，1953-09～1956-07，是中国派到苏联留学生中数学专业的第一个副博士学位获得者，研究方向：代数)、孙永生(莫斯科大学，1954-09～1958-02，研究方向：函数论)、袁兆鼎(莫斯科大学，1954-09～1958-06，他是数学系派到苏联攻读计算数学的研究生，是中国数学界到苏联学习计算数学的第一人)、丁尔陞(列宁格勒师范学院(圣彼得堡国立师范大学)，1956-11～1958-12，是中国数学界到苏联学习数学教学法唯一的一位研究生)、赵桢(莫斯科大学，1956-10～1960-05，研究方向：偏微分方程). 刘绍学、孙永生、丁尔陞、赵桢回国后，很快成为数学系教学和科研的重要骨干力量.

袁兆鼎是数学系派到苏联留学的青年教师. 我和袁老师有一次访谈(袁兆鼎教授访谈录. 见，李仲来主编：北京师范大学数学学科创建百年纪念文集. 北京师范大学出版社，2015：273-284)，当时我问他，您为什么不申请回到数学系？他说，上级分配我到哪里我就去哪里. 2005 年 12 月 25 日，在北京师范大学数学系成立 90 周年庆祝大会上，我在主持大会时，介绍袁兆鼎老师后说，如果袁老师在苏联学成回国之后，回到我们数学系，数学系的计算数学早就发展起来了. 紧接着，中国科学院石钟慈院士在自由发言中说："刚才介绍了袁兆鼎先生，袁先生是我们国家派到苏联学习计算数学的第一人. 1956 年中国科学院遵照华(罗庚)老的指示，派我出去学习计算数学. 当时在莫斯科大学我找到袁先生，他跟 Верезин 学习差分方法. 他答辩的时候我也去了，答辩完就在莫斯科的北京餐厅吃饭，吃北京烤鸭，这是我第一次吃烤鸭，我有机会和袁兆鼎先生在一起. 他的导师在苏联是非常有名的科学家，研究偏微分方程的解. 他当时是北京师范大学的人，回来一般会回到北京师范大学. 但回来后就到七机部(注：第七机械工业部的简称，是中华人民共和国曾经负责航天工业的机构，已撤销)，搞常微分方程稳定性，他一直领导七机部的一个小组. 如果袁先生当时回到北京师范大学，计算数学很早就会发展起来了. 当然袁先生到七机部以后，在导弹计算方面做了很重要的工作. 我认为，袁先生到七机部比回到北京师范大学对国家所作的贡献要大. 1956 年以后袁先生给我很多指导.”

1954～1955 年数学系同时还派出 1953 级傅德熏和吕乃刚两位本科生在北京俄语专科学校留苏预备部学习一年俄语，1955 年去苏联留学.

傅德熏 1960 年毕业于莫斯科大学数学力学系流体力学专业. 1960～1986 年分配到航空航天部 701 所工作，任研究员. 1989 年至今在中国科学院力学研究所工作，长期从事空气动力学、计算流体力学及湍流研究. 获国防科学技术工作委员会科技成果二等奖两项，中国科学院自然科学奖、中国科学院科技进步奖二等奖各一项.

吕乃刚在列宁格勒师范学院学习一年后，转到莫斯科大学数学力学系学习. 由于该校不承认在列宁格勒师范学院学习成绩，吕乃刚只能从一年级学习，比傅德熏晚一年毕业. 1961 年从苏联留学回国后，分配到华东师范大学数学系工作，曾任华东师范大学数理统计系教授、系副主任、图书馆副馆长.

傅德熏和吕乃刚两位本科生去莫斯科大学学习，毕业后没有分配到北京师范大学工作.

李占柄在 1954 年由北京第六中学推荐，经国家考试后被选送到俄语学院留苏预备部学习. 1956 年 6 月出国. 由于当年莫斯科大学正在装修、调整，于是被送往基辅大学，次年转入莫斯科大学学习直至毕业. 1961 年 7 月本科毕业后，11 月分配到北京师范大学数学系工作.

周美珂(1965-05～1966-10)到苏联科学院数学研究所攻读副博士学位，因中苏关系紧张提前同国.

以上是数学系对青年教师“送出去”的情况. 数学系没有“请进来”苏

联专家到数学系指导工作，时任北京师范大学副校长兼数学系主任傅种孙教授对此事持否定态度.

在全面学习苏联的大背景下，虽然没有“请进来”苏联专家到数学系指导工作，但数学系教师翻译了多种苏联教科书. 同时，在教学中还使用了兄弟院校翻译的苏联教科书，这些教科书体系严密，论证严谨，有效地提高了师范院校教师的教学法水平和在校学生的数学基础，培养了一大批优秀的教学人才. 数学系教师翻译且正式出版的教材名称如下.

20 世纪 50 年代：在人民教育出版社出版的有《中学数学教学法(第1册，通论；第 2 册，算术教学法；第 3 册，代数教学法；第 4 册，几何教学法；第 5 册，三角教学法)》《代数及三角习题汇编：中学八至十年级使用》《几何证题集》《青年工人学校算术教学法》《高中数学教学经验》《高中数学教学法：几何部分》，这些教材偏重于教材教法. 在教学形式和教学方法方面，学习苏联的经验，注意课堂讲授、辅导、讨论、作业、考试、考查和实习等多种教学形式的结合，收到了很好的教学效果，提高了教学质量.

在高等教育出版社出版的苏联师范学院用教学大纲 6 本《解析几何教学大纲》《高等代数教学大纲》《实变及复变函数论》《数论与群论教学大纲》《解析几何、数学分析、数学物理方法教学大纲》《射影几何及画法几何教学大纲》，还有非师范学院用的《高等代数教学大纲》；著作《算术》《代数与初等函数》《几何习题集》《数与多项式》《初等代数专门教程(上、下册)》《函数论与泛函分析初步(卷一)》. 翻译的两种高等代数教学大纲，与北京师范大学数学系负责编写师范院校的高等代数教材有关.

1953 年在商务印书馆出版的有《算术》.

20 世纪 60 年代：在人民教育出版社出版的有《群论(上册)》(第 2 版)；在高等教育出版社出版的有《数学分析专门教程》《算术》.

1952 年院系调整，数学系全面学习苏联经验. 新的课程中增加了俄语，不再开设公共英语课程，应是教学片面性的表现和一个全国性的大的失误.

二

赵桢老师 1931 年 9 月 2 日生于北京市，父亲赵德麟(1911—1975)是小商人，母亲方婧增(1912—1994)是家庭妇女. 1938 年 8 月至 1944

年7月，赵老师先后就读于北京市打磨厂小学初小和北京师范大学附属小学(创办于1912年9月5日，位置在北京琉璃厂古文化街北侧，前身是北京高等师范学校附属小学校．1955年10月改称北京第一实验小学)高小，1944年8月至1950年7月先后在北京师范大学附属中学和北京崇德中学读初中和高中，1950年9月至1953年7月在北京师范大学数学系本科学习．

赵桢老师本科提前一年毕业后在北京师范大学数学系任教．1955年2月加入中国共产党．1955年9月至1956年10月在北京俄语学院留苏预备部学习．1956年10月被派往苏联莫斯科大学数学力学系做研究生，师从苏联科学院依·涅·维库阿院士学习偏微分方程，1960年4月毕业并获数学物理副博士学位后，5月回国．1961～1966年和1977～1995年任数学系分析(二)教研室主任．1969年9月至1972年4月在学校机电厂劳动．1979年7月任副教授，1983年5月任教授．1985年3月至1989年11月任数学系主任，1989～1996年任北京数学会副理事长，1989～2007年任北京数学会主办的北京数学奥林匹克学校(现称北京数学培训学校)校长，曾任《北京数学》《偏微分方程》《数学研究与评论》等杂志编委．1995年10月退休．

1960年5月回国后，赵桢老师作为学术带头人，领导了数学系偏微分方程方向的研究工作．1962年暑假后，开始组织讨论班．数学系在1959级的本科生中试办代数、概率、(复)函数论共三个专门化．赵桢老师组织的专门化学生是以偏微分方程中的(复)函数论方法为研究方向，开设的专门化课程有奇异积分方程、广义解析函数、函数论边值问题等，讨论班上还研读一些有关的数学论文．参加专门化课程的学生约30人，参加讨论班的只有10多人．在1964年暑假毕业时完成了读书报告形式的毕业论文，这对数学系来说还是首次．

多年来，赵桢老师在国内外共发表过学术论文30多篇，专著1部，翻译专著1部，主编和编辑学术会议论文集10部(其中2部由新加坡科技出版社出版)，主编奥林匹克数学丛书3套．科研工作曾3次得到国家自然科学基金、教育部科学技术基金和北京市自然科学基金的支持，并两次获北京市科技进步奖三等奖．该方向到1997年共培养研究生36人，其中硕士生32人，接收国内、外访问学者15人．

在全国，北京大学、复旦大学、武汉大学、西南大学、北京科技大学等高校都有人从事这方面的研究工作，有闵嗣鹤、陈传璋、张孝礼、路见可等前辈学者．后来在我国形成一个学术集体，牵头的有路见可、赵桢、闻国椿、侯宗义、李明忠、林伟、容尔谦等．经过中国数学会批准，每两年召开一次全国性的学术会议．到 1999 年，先后在成都、厦门、西安、武汉、重庆、北京、广州、银川、郑州、承德等地举办了 10 次学术会议，其中北京和承德的是国际学术会议，有来自美国、德国、俄罗斯、乌克兰、塔吉克斯坦、伊朗、波兰、日本等国学者参加．北京师范大学数学系都是作为这些学术活动的主要组织者之一，在国内有一定的影响．每次学术会议都出版文集．

改革开放以来，赵桢老师作为访问教授访问过苏联、俄罗斯、乌克兰、捷克、德国、比利时、法国、波兰等国家；多次参加在美国、俄罗斯、德国、格鲁吉亚、泰国等国家举行的国际学术会议，从事国际交流活动．

华罗庚教授说："一个人最后余下的就是一本选集．"(龚昇论文选集．合肥：中国科学技术大学出版社，2008)这些选集的质量反映了我们学院某一学科，或几个学科，或学科群的整体学术水平．而将北京师范大学数学科学学院著名数学家、数学教育家和科学史专家论文进行整理和选编出版，是学院学科建设的一项重要的和基础性的工作，是学院的基本建设之一．它对提高学院的知名度和凝聚力，激励后人，有着重要的示范作用．当然，这项工作还在继续做下去，收集和积累数学科学学院各种资料的工作还在继续进行．

赵桢文集的出版，得到了数学科学学院党委书记唐仲伟教授，前院长李增沪教授和现任院长王恺顺教授的大力支持，同时得到了北京师范大学出版社的大力支持，在此表示衷心的感谢．

李仲来

2018-10-08

目　录

Contents

北京师范大学学报(自然科学版),
1962,(2):11—18.

用展级数法解二阶椭圆型方程的平面 Dirichlet 问题

Solving Plane Dirichlet Problem for Secondary Order Elliptic Differential Equation by Expanding Series

§ 1.

考虑二阶椭圆型方程①

$$\mathfrak{M}u \equiv \Delta u + a(x,y)\frac{\partial u}{\partial x} + b(x,y)\frac{\partial u}{\partial y} + c(x,y)u = 0, \qquad (1.1)$$

其中 Δ 是拉普拉斯算子,而系数 a,b,c 是在方程(1.1)的某一基本区域内解析的(实)函数.[2]

对方程(1.1)所提的狄利克雷(Dirichlet)问题是

问题 D:要求根据边界条件

$$u(t)\mid_{\Gamma} = \gamma(t), t \in \Gamma, \qquad (1.2)$$

来求方程(1.1)在区域 T 内正则,在 $T+\Gamma$ 上连续的解. 这里,T 是一个 $m+1$ 连通的有界区域,$\Gamma=\Gamma_0+\Gamma_1+\cdots+\Gamma_m$ 是它的边界,我们假设 Γ 的切线与某一固定方向之间的夹角是满足赫尔德条件的,另外,$C_1,C_2,\cdots,C_m$ 是分别由 $\Gamma_1,\Gamma_2,\cdots,\Gamma_m$ 所围出的有界区域,而 C_0 是由 Γ_0 外部的点所构成的无界区域(它包含无穷远点在内). 我们将假设 $T+C_1+C_2+\cdots+C_m$ 是位于方程(1.1)的某一个基本区域 $\mathfrak{D}$ 的内部;而且,不失一般性,可以规定坐标原点位于区域 T 的内部;$\gamma(t)$是给定的(实)连续函数.

问题 D 的可解性是熟知的结果.[2][4]本文的目的是给出问题 D 之解

① 一般的二阶椭圆型方程(两个自变量的)都可以化成标准型(1.1).

的一种表示式.[1]

我们知道,方程(1.1)的解可以表示成[2]

$$u(x,y)=\int_{\Gamma}\left[u(t)N\omega(x,y;\xi,\eta)-\frac{\mathrm{d}u(t)}{\mathrm{d}v}\omega(x,y;\xi,\eta)\right]\mathrm{d}s, \tag{1.3}$$

其中 $N\omega\equiv\frac{\mathrm{d}\omega}{\mathrm{d}v}-[a\cos(v,x)+b\cos(v,y)]\omega$;$v$ 是在点(ξ,η)处 Γ 之内法线,ω 是某一基本解.[2]我们知道,基本解 ω 是一定存在的,而且可以用逐次逼近法求出,特别是对于许多在数学物理中遇到的方程说来,它们的基本解 ω 是可以通过初等函数,或特殊函数而表成明显形式的[2],所以公式(1.3)给出了在区域 T 内方程(1.1)之解与其边界值 $u(t)$ 和导数$\frac{\mathrm{d}u(t)}{\mathrm{d}v}$之间的关系. 利用这个积分表示式时,不便之处在于被积函数中除了含有能在边界上给定的量以外,同时还含有不能事先给定的量,例如在问题 D 中 $u(t)$在 Γ 上是给定的,而$\frac{\mathrm{d}u}{\mathrm{d}v}$却不能事先任意给定. 因此,如果利用某种方法只要能通过已知条件来确定出$\frac{\mathrm{d}u}{\mathrm{d}v}$的值,那么,公式(1.3)就可以用来表示问题 D 的解,剩下的工作就是如何确定$\frac{\mathrm{d}u(t)}{\mathrm{d}v}$的值. 为此,我们将利用方程(1.1)的共轭方程

$$\mathfrak{N}v\equiv\Delta v-\frac{\partial av}{\partial x}-\frac{\partial bv}{\partial y}+cv=0 \tag{1.4}$$

的某一完备特解系$\{v_k(x,y)\}$. ①

§ 2.

为了简单,我们先来讨论单连通区域情形.

我们知道,[2]方程(1.4)的解可以表示成

$$v(x,y)=\mathrm{Re}\left[H_0(z)\varphi(z)+\int_0^z H(z,\tau)\varphi(\tau)\mathrm{d}\tau\right], \tag{2.1}$$

其中,$H_0(z)=G(z,0,z,\bar{z})$,$H_0(z)\neq 0$,$z\in\mathfrak{D}$,

$$H(z,\tau)=-\frac{\partial}{\partial\tau}(\tau,0;z,\bar{z}),$$

① 这里"完备"应该理解为:对方程(1.4)的任何在 T 内正则解和任何 T 内的闭区域 T^* 都可以用线性组合 $\xi_1v_1+\xi_2v_2+\cdots+\xi_nv_n$ 来一致逼近.

这里 $G(z,\zeta;t,\tau)$ 代表方程(1.4)对区域$(\mathfrak{D},\overline{\mathfrak{D}})$的黎曼函数,而 $\varphi(z)$ 是在区域 T 内的任意全纯函数.

如果在公式(2.1)中用函数组

$$\varphi_j(z)=\begin{cases} z^k, & j=2k+1, k\in\mathbf{N}, \\ \mathrm{i}z^k, & j=2k, \quad\ k\in\mathbf{N}^* \end{cases} \tag{2.2}$$

来代替 $\varphi(z)$,我们就可以得到方程(1.4)的一个完备特解系$\{v_j(x,y)\}, j\in\mathbf{N}^*$.

另外,利用格林公式,

$$\iint_T (v\mathfrak{M}u-u\mathfrak{N}v)\mathrm{d}x\mathrm{d}y=\int_\Gamma\left[u(t)Nv-\frac{\mathrm{d}u}{\mathrm{d}v}v(t)\right]\mathrm{d}s,$$

我们对方程(1.4)的任意解 v 和问题 D 的解 u,能够得到

$$\int_\Gamma\left[r(t)Nv(t)-\frac{\mathrm{d}u(t)}{\mathrm{d}v}v(t)\right]\mathrm{d}s=0.$$

显然,对于我们上面选定的$\{v_k(x,y)\}$来说,将有

$$\int_\Gamma \frac{\mathrm{d}u(t)}{\mathrm{d}v}v_k(t)\mathrm{d}s=\int_\Gamma r(t)Nv_k(t)\mathrm{d}s=C_k,\ k\in\mathbf{N}^*, \tag{2.3}$$

这里,C_k 是已知的(实)常数,这就给了我们通过已知条件来确定$\frac{\mathrm{d}u(t)}{\mathrm{d}v}$的可能.

§3. 齐次问题 $\mathbf{D}_0$ 没有非零解的情形

我们知道,在这种情形下,问题 D 对任何给定的连续函数 $r(t)$ 都是可解的,而且解是唯一的.[2][4] 这时候,对于所谓共轭问题 D^*①,也可以做出同样的结论,而且共轭齐次问题 $\mathrm{D}_0{}^*$ 也只有零解[4].

我们现在来证明以下定理.

定理 1 如果齐次问题 D_0 没有非零解,那么在边界 Γ 上函数组 $v_j(t), j\in\mathbf{N}^*$ 是线性无关的.

事实上,若对于某一个正整数 n,有 $\sum\limits_{j=1}^n \xi_j v_j(t)=0$,则根据定理条件,应该有 $\sum\limits_{j=1}^n \xi_j v_j(x,y)\equiv 0, (x,y)\in T_0$,但是,$v_1(x,y), v_2(x,y), \cdots, v_n(x,y)$是线性无关的,从而有 $\xi_j=0(j=1,2,\cdots,n)$. 这就证明了定理 1 成立.

---

① 问题 D^* 是指:要求根据边界条件

$$\{v(t)\mid_\Gamma=\delta(t), t\in\Gamma\}$$

来求方程(1.4)在区域 T 内正则,在 $T+\Gamma$ 上连续的解,其中 $\delta(t)$ 是给定的实连续函数.

这样，利用施密特方法总可以把函数组$\{v_j(t)\}(j\in\mathbf{N}^*)$标准正交化，为了简单起见，我们认为$\{v_j(t)\}(j\in\mathbf{N}^*)$已经就是标准正交函数系.

除此之外，我们知道[2]，如果齐次问题 D_0 没有非零解，那么，在边界 Γ 上函数组$\{v_j(t)\}(j\in\mathbf{N}^*)$关于空间 $L^2(\Gamma)$是封闭的.

这样一来，对于齐次问题 D_0 没有非零解的情形我们总可以认为，在边界 Γ 上，函数组$\{v_j(t)\}(j\in\mathbf{N}^*)$是完备的标准正交函数系（关于空间 $L^2(\Gamma)$）；如果再注意到等式(2.3)，就可以知道傅里叶级数 $\sum\limits_{j=1}^{+\infty} c_j v_j(t)$在边界 Γ 上平均收敛于函数$\dfrac{\mathrm{d}u(t)}{\mathrm{d}v}$. 也就是说，我们为函数$\dfrac{\mathrm{d}u(t)}{\mathrm{d}v}$找到了关于已知函数 $v_j(t)(j\in\mathbf{N}^*)$的傅里叶展开式.从而得到如下定理.

定理 2　如果齐次问题 D_0 没有非零解，那么在区域 T 内，对问题 D 的解来说，成立等式

$$u(x,y)=\int_\Gamma\Big[r(t)N\omega(x,y;\xi,\eta)-\sum_{j=1}^{+\infty}c_jv_j(t)\omega(x,y;\xi,\eta)\Big]\mathrm{d}s, \tag{3.1}$$

而且(3.1)的右端在闭区域 $T+\Gamma$ 上一致收敛于问题 D 的解 $u(x,y)$.

事实上，设

$$u_n(x,y)=\int_\Gamma\Big[r(t)N\omega(x,y;\xi,\eta)-\sum_{j=1}^{n}c_jv_j(t)\omega(x,y;\xi,\eta)\Big]\mathrm{d}s,$$

于是，很容易得到

$$|u(x,y)-u_n(x,y)|=\left|\int_\Gamma\Big[\frac{\mathrm{d}u(t)}{\mathrm{d}v}-\sum_{j=1}^{n}c_jv_j(t)\Big]\omega(x,y;\xi,\eta)\mathrm{d}s\right|$$
$$\leqslant\left\{\int_\Gamma|\omega(x,y;\xi,\eta)|^2\mathrm{d}s\right\}^{\frac12}\left\{\int_\Gamma\left|\frac{\mathrm{d}u(t)}{\mathrm{d}v}-\sum_{j=1}^{n}c_jv_j(t)\right|^2\mathrm{d}s\right\}^{\frac12}.$$

上面不等式右端之第一个因子是有界的，而第二个因子，当 $n\to+\infty$ 时，是一致趋于零的.

§4.　齐次问题 D_0 有非零解的情形

我们知道[2][4]，问题 D_0 之线性无关解的个数一定有限，而且，共轭齐次问题 D_0^* 与问题 D_0 有同样多个线性无关解[4]，假设 $u_1^0,u_2^0,\cdots,u_l^0$ 与 $v_1^0,v_2^0,\cdots,v_l^0$ 分别是问题 D_0 与 D_0^* 之线性无关解的完全组，这时候，问题 D 可解的充分必要条件是[4]

$$\int_{\Gamma} r(t)\,\frac{\mathrm{d}v_j^0(t)}{\mathrm{d}v}\mathrm{d}s = 0,\quad j = 1,2,\cdots,l. \tag{4.1}$$

定理 3 如果齐次问题 D_0 有 l 个线性无关解，那么，只要适当地选取方程(1.4)之完备特解系，总可以使得

(1)在边界 Γ 上有，$v_1(t)=v_2(t)=\cdots=v_l(t)=0$；

(2)函数组 $\{v_{l+j}(t)\}(j\in\mathbf{N}^*)$ 是线性无关的，从而，可以把它看作标准正交函数系.

事实上，我们考虑函数组

$$v_1^0(x,y),v_2^0(x,y),\cdots,v_l^0(x,y),v_1(x,y),v_2(x,y),\cdots,v_n(x,y),\cdots \tag{4.2}$$

其中 $v_j^0(x,y),j=1,2,\cdots,l$，是问题 $D_0{}^*$ 的线性无关解，而 $v_j(x,y),j\in\mathbf{N}^*$ 是方程(1.4)的完备特解系.

函数组(4.2)可能并不是线性无关的，亦即是说，可能会有 $v_j^0(x,y)=\sum\limits_{k=1}^{n_j}\xi_k^{(j)}v_k(x,y),j=1,2,\cdots,l$，成立，因为 $v_j^0(x,y)\neq 0$，所以系数 $\xi_1^{(j)},\xi_2^{(j)},\cdots,\xi_{n_j}^{(j)}$ 中至少有一个不等于零．我们假设 $\xi_{k_j}^{(j)}\neq 0$，那么，函数组

$$v_j^0(x,y),v_1(x,y),\cdots,v_{k_j-1}(x,y),v_{k_j+1}(x,y),\cdots,v_n(x,y),\cdots,$$

应该是线性无关的，不然的话就会推出 $\{v_j(x,y)\},j\in\mathbf{N}^*$ 是线性相关的结论来，而这与假设条件矛盾.

继续这样做 l 次(即令 $j=1,2,\cdots,l$)，并将得到的函数组重新进行编号，可以得到

$$\tilde{v}_1(x,y),\tilde{v}_2(x,y),\cdots,\tilde{v}_n(x,y),\cdots \tag{4.3}$$

其中 $\tilde{v}_j(x,y)=v_j^0(x,y),j=1,2,\cdots,l$，而其余的 $v_j(x,y)$ 就是由函数组 $\{v_j(x,y)\}$ 中抛掉有限个函数 $v_{k_j}(x,y)$ 以后，再重新编号所得到的函数组.

显然，函数组(4.3)可以作为方程(1.4)的完备特解系，因为，被我们抛掉的函数 $v_{k_j}(x,y)$ 都可以借助组(4.3)中函数的线性组合来表示，于是根据 $\{v_j(x,y)\}$ 的完备性，可以直接推得 $\{\tilde{v}_j(x,y)\}$ 的完备性.

因此，只要把函数组(4.3)取作方程(1.4)的完备特解系，就可以满足要求(1)，另外，要求(2)也是满足的；不然的话，将会有 $\sum\limits_{j=1}^{n}\xi_j\tilde{v}_{l+j}(x,y)=0$，其中 $\xi_1,\xi_2,\cdots,\xi_n$ 不全为零，而这与函数组(4.3)是线性无关的假设矛盾.

这样，为了简单起见，我们可以认为函数组 $\{v_j(x,y)\}(j\in\mathbf{N}^*)$ 本身

就满足定理 3 中的要求(1)(2);而且在边界 Γ 上它是标准正交函数系,根据(2.3),我们有

$$\int_{\Gamma} r(t)\,\frac{\mathrm{d}v_k(t)}{\mathrm{d}v}\mathrm{d}s = 0,\ k = 1,2,\cdots,l. \tag{4.4}$$

这刚好是问题 D 的可解条件.

我们考虑问题 D_0 之所有解构成的空间(是一个 l 维空间),对于这个空间中的任何非零元素 $\tilde{u}(x,y)\neq 0$,都有 $\dfrac{\mathrm{d}\tilde{u}(t)}{\mathrm{d}v}\neq 0, t\in\Gamma$,因为不然的话,根据公式(1.3)将有 $\tilde{u}(x,y)\equiv 0,(x,y)\in T$,而这与假设条件矛盾.边界值 $\dfrac{\mathrm{d}\tilde{u}(t)}{\mathrm{d}v}$ 构成连续函数空间 $L^2(\Gamma)$ 中一个 l 维子空间 R.容易看出,函数 $\dfrac{\mathrm{d}\mathring{u}_j(t)}{\mathrm{d}v}$ $(j=1,2,\cdots,l)$ 可以作为子空间 R 的基底.

我们来证明以下定理.

定理 4　如果齐次问题 D_0 有 l 个线性无关解 $\mathring{u}_1(x,y),\mathring{u}_2(x,y),\cdots,\mathring{u}_l(x,y)$,那么

$$\frac{\mathrm{d}\mathring{u}_1(t)}{\mathrm{d}v},\frac{\mathrm{d}\mathring{u}_2(t)}{\mathrm{d}v},\cdots,\frac{\mathrm{d}\mathring{u}_l(t)}{\mathrm{d}v},v_{l+1}(t),v_{l+2}(t),\cdots,v_{l+n}(t),\cdots \tag{4.5}$$

在空间 $L^2(\Gamma)$ 上是封闭的.

假设 $f(t)$ 是空间 $L^2(\Gamma)$ 中的任意函数,那么,对于任何 $\varepsilon>0$,都可以找到这样的连续函数 $g(t)$,使得 $\displaystyle\int_{\Gamma}|f(t)-g(t)|^2\mathrm{d}s<\frac{\varepsilon}{2}$,

另一方面,连续函数 $g(t)$ 可以唯一地分解成

$$g(t) = g_1(t) + g_2(t),$$

其中 $g_1(t)$ 属于子空间 R,即 $g_1(t)=\displaystyle\sum_{j=1}^{l}\eta_j\frac{\mathrm{d}\mathring{u}_j(t)}{\mathrm{d}v}$,而 $g_2(t)$ 属于 R 的正交余空间,即

$$\int_{\Gamma} g_2(t)\,\frac{\mathrm{d}\mathring{u}_j(t)}{\mathrm{d}v}\mathrm{d}s = 0,\ j = 1,2,\cdots,l, \tag{4.6}$$

这刚好是共轭问题 $\tilde{v}(t)|_{\Gamma}=g_2(t)$ 的可解条件,因此一定存在问题 D^* 的解,$\tilde{v}(x,y)$,它可以表成

$$\tilde{v}(x,y) = \mathrm{Re}\left[H_0(z)\tilde{\varphi}(z) + \int_O^Z H(z,\tau)\tilde{\varphi}(\tau)\mathrm{d}\tau\right],$$

其中 $\tilde{\varphi}(z)$ 是在 T 内的全纯函数,它在闭区域 $T+\Gamma$ 上(至少)是平方可积的.[3] 因此,在 Γ 上它可以用多项式来平均逼近,即一定存在这样的常数 $\xi_1,\xi_2,\cdots,\xi_n$,使得在 Γ 上成立不等式

$$\int_{\Gamma}\left|g_2(t)-\sum_{j=1}^{l}\xi_j v_j(t)\right|^2 \mathrm{d}s<\frac{\varepsilon}{2}.$$

从而得出

$$\int_{B}\left|f(t)-\sum_{j=1}^{l}\eta_j\frac{\mathrm{d}\mathring{u}_j(t)}{\mathrm{d}v}-\sum_{j=1}^{n}\xi_j v_j(t)\right|^2\mathrm{d}s$$

$$\leqslant\int_{\Gamma}|f(t)-g(t)|^2\mathrm{d}s+\int_{\Gamma}\left|g_1(t)-\sum_{j=1}^{l}\eta_j\frac{\mathrm{d}\mathring{u}_j(t)}{\mathrm{d}v}\right|^2\mathrm{d}s+$$

$$\int_{\Gamma}\left|g_2(t)-\sum_{j=1}^{n}\xi_j v_j(t)\right|^2\mathrm{d}s<\frac{\varepsilon}{2}+\frac{\varepsilon}{2}=\varepsilon.$$

于是定理 4 得证.

这样,函数组(4.5)可以看成关于 $L^2(\Gamma)$之完备标准正交函数系.

如果条件(4.6)满足,并且设 $a_j=\int_{\Gamma}\frac{\mathrm{d}u(t)}{\mathrm{d}v}+\int_{\Gamma}\frac{\mathrm{d}\mathring{u}_j(t)}{\mathrm{d}v}\mathrm{d}s, j=1,2,\cdots,l$,那么傅里叶级数 $\sum_{j=1}^{l}a_j\frac{\mathrm{d}\mathring{u}_j(t)}{\mathrm{d}v}+\sum_{j=1}^{+\infty}c_{l+j}v_{l+j}(t)$平均收敛于函数 $\frac{\mathrm{d}u(t)}{\mathrm{d}v}$. 特别是级数 $\sum_{j=1}^{+\infty}c_{l+j}v_{l+j}(t)$将平均收敛于函数 $\frac{\mathrm{d}}{\mathrm{d}v}\left(u(t)-\sum_{j=1}^{l}a_j\mathring{u}_j(t)\right)$. 如果记 $u^*(x,y)=u(x,y)-\sum_{j=1}^{+\infty}a_j\mathring{u}_j(x,y)$,那么显然,$u^*(x,y)$将是问题 D 的某一个特解.

定理 5 如果齐次问题 D_0 有 l 个线性无关解,那么,只要条件(4.6)满足,对于问题 D 的特解 $u^*(x,y)$有等式

$$u^*(x,y)=\int_{\Gamma}\left[r(t)N\omega(x,y;\xi,\eta)-\sum_{j=1}^{+\infty}c_{l+j}v_{l+j}(t)\omega(x,y;\xi,\eta)\right]\mathrm{d}s \tag{4.7}$$

成立,而且等式(4.7)之右端在闭区域 $T+\Gamma$ 上一致收敛于特解 $u^*(x,y)$,问题 D 的一般解可以由公式

$$u(x,y)=u^*(x,y)+\sum_{j=1}^{l}\lambda_j\mathring{u}_j(x,y) \tag{4.8}$$

得到,其中 $\lambda_j(j=1,2,\cdots,l)$是任意常数.

§ 5.

我们最后来讨论多连通区域情形. 对于 $m+1$ 连通区域情形,方程(1.4)的解可以表示成[2]

$$v(x,y)=\sum_{k=1}^{m}\beta_k\omega(x,y;x_k,y_k)+\mathrm{Re}\Big\{\mathfrak{H}[0,p(z)]+\sum_{k=1}^{m}\mathfrak{H}\Big[\bar{z}_k,q_k(z)-$$

$$\left(\frac{1}{2\pi i}\int_{\overline{\Gamma_k}} \overline{q_k(\bar{\tau})H(\bar{\tau},\bar{z}_k;z_k\bar{z})}\,d\tau\right)\lg(z-z_k)\Big]\Big\}, \tag{5.1}$$

其中 $p(z)$ 是在 $T+C_1+C_2+\cdots+C_m$ 内的全纯函数，$\beta_1,\beta_2,\cdots,\beta_m$ 是实常数；$q_1(z),q_2(z),\cdots,q_m(z)$ 分别是在 $S_1,S_2,\cdots,S_m$ 内的全纯函数，它们都在无穷远处等于零，其中 S_j 代表由 Γ_j 所围出的无界区域（$j=1,2,\cdots,m$），$z_1,z_2,\cdots,z_m$ 分别是在区域 $C_1,C_2,\cdots,C_m$ 中的定点；$\omega(x,y;x_k,y_k)$ 是以 $z_k=x_k+iy_k(k=1,2,\cdots,m)$ 为极点的基本解. 另外，

$$\mathfrak{H}[z_o,\varphi(z)]=G(z,\bar{z}_o;z,\bar{z})\varphi(z)-\int_o^z H(\tau,\bar{z}_0;z,\bar{z})\varphi(\tau)\,d\tau, \tag{5.2}$$

其中 $H(t,\tau;z,\zeta)=\dfrac{\partial}{\partial t}G(t,\tau;z,\zeta)$，而 $G(t,\tau;s,\zeta)$ 是方程(1.4)的黎曼函数.

考虑函数组

$$\begin{aligned}&\omega(x,y;x_k,y_k), \qquad k=1,2,\cdots,m,\\&\mathrm{Re}\{\mathfrak{H}[0,z^k]\},\mathrm{Re}\{\mathfrak{H}[0,iz^k]\},k\in\mathbf{N},\end{aligned} \tag{5.3}$$

$$\mathrm{Re}\left\{\mathfrak{H}[\bar{z}_t,(z-z_l)^{-k}]-\left[\frac{1}{2\pi i}\int_{\overline{\Gamma_k}}\overline{(\tau-z_l)^{-k}}\ \overline{H(\tau,\bar{z}_l;z_l,\bar{z})}\,d\tau\right]\lg(z-z_l)\right\}$$

$l=1,2,\cdots,m,\quad k\in\mathbf{N}$,

显然，只要把函数组(5.3)再进行重新编号，就可以得到关于多连通区域 T 的完备特解系 $\{v_j(x,y)\}$.

不言而喻，公式(1.3)对于多连通区域情形仍然是成立的. 此外，在 Γ 上，对边界值 $\dfrac{du(t)}{dv}$ 来说关系式

$$\int_\Gamma \frac{du(t)}{dv}v_j(t)\,ds=\int_\Gamma r(t)Nv_j(t)\,ds\equiv C_j,\quad j\in\mathbf{N}^*$$

也成立，其中 C_j 是已知常数.

有了这些结论以后，前两段中所讨论的结果就都可以无改变地推广到多连通区域情形上来，这里不再重复了.

参考文献

[1] Чжао Чжэн(赵桢). ДАН，1960，132：781-784.

[2] Векуа И Н. Новые методы решения эллилтического уравнения. ОТИЗ Гостехиздат М-Л，1948.

[3] Хведелидзе В В. Тр. Тбил. Мат. Ин-Та，1956：23.

[4] Миранла К. Уравнения с частными лроиволными эллилтического тила. Москва，1957.

北京师范大学学报(自然科学版),
1963,(3):1-26.

关于 n 重调和方程的基本边值问题①

On the Basic Boundary Value Problems for n-Harmonic Equations

本文主要结果是

(1) 给出对一类特殊的高阶椭圆型方程:n 重调和方程所提基本边值问题(问题 B_n)的一种等价边界条件.

(2) 给出基本边值问题解的一种级数表示式.

§1.

我们将要考虑的基本边值问题是[1]

问题 B_n:要求根据边界条件

$$u^{+}=f_0,\left(\frac{\mathrm{d}u}{\mathrm{d}v}\right)^{+}=f_1,\cdots,\left(\frac{\mathrm{d}^{n-1}u}{\mathrm{d}v^{n-1}}\right)^{+}=f_{n-1} \tag{1.1}$$

来确定方程

$$\Delta^{p}u(x,y)=0 \tag{1.2}$$

在区域 T 内的正则(实)解 $u(x,y)$;这里我们规定 T 是由足够光滑的边界 Γ 所围出的单连通区域,$f_0,f_1,\cdots,f_{n-1}$ 是在边界 Γ 上给定的实函数,而且规定 f_l 具有关于(弧参数)s 的一直到 $2n-k$ 阶连续导数,v 是边界 Γ 的外法线. 此外,我们还要求解 $u(x,y)=u\left(\frac{z+\bar{z}}{2},\frac{z+\bar{z}}{-2\mathrm{i}}\right)$ $(z=x+\mathrm{i}y)$之

① 本文与刘来福合作.

所有形如$\dfrac{\partial^{m+k}u}{\partial z^m\partial\bar{z}^k}$ ($k\leqslant n, m\leqslant n$)的导数在闭区域 $T+\Gamma$ 上是连续的.[1]

我们知道,问题 B_n 是永远有唯一解的.[1]

在往下讨论以前,我们先来重提一些已知结果:[1]首先有下列明显的关系式成立

$$\frac{\mathrm{d}}{\mathrm{d}s}\left(\frac{\partial^{k+m}u}{\partial z^k\partial\bar{z}^m}\right)^+=\left(\frac{\partial^{k+m+1}u}{\partial z^{k+1}\partial\bar{z}^m}\right)^+\frac{\mathrm{d}t}{\mathrm{d}s}+\left(\frac{\partial^{k+m+1}u}{\partial z^k\partial\bar{z}^{m+1}}\right)^+\frac{\mathrm{d}\bar{t}}{\mathrm{d}s},\tag{A}$$

$$m,k=0,1,2,\cdots;\ m+k<2n;$$

$$\left(\frac{\mathrm{d}^k u}{\mathrm{d}v^k}\right)^+=\mathrm{i}^k\sum_{l=0}^{k}(-1)^{k-l}\frac{k!}{l!(k-l)!}\left(\frac{\partial^k u}{\partial z^{k-l}\partial\bar{z}^l}\right)^+\left(\frac{\mathrm{d}t}{\mathrm{d}s}\right)^{k-2l},\tag{B}$$

$$k=0,1,2,\cdots,n-1.$$

其次如果引入记号

$$\left(\frac{\partial^{k+m}}{\partial z^k\partial\bar{z}^m}\right)^+=g_{t,m}(s)\text{(有时简记作 }g_{k,m}\text{)},\tag{C}$$

那么,边界条件(1.1)还等价于条件(参看[1])

$$u^+=g_{0,0},\left(\frac{\partial u}{\partial z}\right)^+=g_{1,0},\cdots,\left(\frac{\partial^{n-1}u}{\partial z^{n-1}}\right)^+=g_{n-1,0}.\tag{1.1$'$}$$

根据条件(1.1)和(1.1)′的等价性马上知道,函数 $f_0=f_1=\cdots=f_{n-1}\equiv 0$ 与 $g_{0,0}=g_{1,0}=\cdots=g_{n-1,0}\equiv 0$ 是等价的;此外,我们还能进一步证明如下的结果:

定理 1　如果在问题 B_n 的边界条件(1.1)′中出现的函数 $f_0,f_1,f_2,\cdots,f_{n-1}$ 里面有

$$f_0=f_1=\cdots=f_k\equiv 0,\quad 0\leqslant k\leqslant n-1,\tag{1.3}$$

那么,一定有

$$g_{0,0}=g_{1,0}=\cdots=g_{k,0}\equiv 0,\quad 0\leqslant k\leqslant n-1\tag{1.4}$$

成立;而且反过来也是成立的.

证　首先利用公式(A),可得

$$g_{k,m+1}(s)=\frac{\mathrm{d}}{\mathrm{d}s}g_{k,m}(s)\left(\frac{\mathrm{d}\bar{t}}{\mathrm{d}s}\right)-g_{k+1,m}(s)\left(\frac{\mathrm{d}t}{\mathrm{d}s}\right)^2.\tag{1.5}$$

然后考虑公式(B)中出现的函数 $g_{k-l,l}(s)$,如果对它重复应用 l 次公式(1.5),那么,我们将有

$$g_{k-l,l}(s)=\sum_{j_1=0}^{1}(-1)^{j_1}\frac{\mathrm{d}^{1-j_1}}{\mathrm{d}s^{1-j_1}}\Big\{\sum_{j_2=0}^{1}(-1)^{j_2}-$$

$$\frac{\mathrm{d}^{1-j_2}}{\mathrm{d}s^{1-j_2}}\cdots\left[\sum_{j_l=0}^{1}(-1)^{j_l}\frac{\mathrm{d}^{1-j_l}}{\mathrm{d}s^{1-j_l}}g_{k-l+j_1+\cdots+j_l,0}(s)\left(\frac{\mathrm{d}t}{\mathrm{d}s}\right)^{1+j_l}\right]\cdots\left(\frac{\mathrm{d}t}{\mathrm{d}s}\right)^{1+j_2}\Bigg\}\left(\frac{\mathrm{d}t}{\mathrm{d}s}\right)^{1+j_1}. \tag{1.6}$$

从而,根据公式(1.1)和(B)知道,函数 f_k 可以表示成函数 $g_{0,0},g_{1,0},\cdots,g_{k,0}$,和它们关于 s 的层数之(齐次)线性函数. 于是马上推得定理 1 的第二部分是成立的.

现在来证明定理 1 的另一部分. 为此,只要证明,如果 $f_0=f_1=\cdots=f_k\equiv0$,那么有 $g_{m,n}\equiv0,m+n\leqslant k,m,n\geqslant0$;那么,特别地,就更会有 $g_{0,0}=g_{1,0}=\cdots=g_{k,0}\equiv0$. 事实上,上面所说的结论,当 $k=0$ 时显然是成立的. 而当 $k=1$ 时,利用公式(A)(B)以及定理 1 的条件我们将有

$$\begin{cases}f_0=g_{0,0}\equiv0,\\ 0=\dfrac{\mathrm{d}f_0}{\mathrm{d}s}=g_{1,0}\left(\dfrac{\mathrm{d}t}{\mathrm{d}s}\right)+g_{0,1}\left(\dfrac{\mathrm{d}t}{\mathrm{d}s}\right)^{-1},\\ 0=f_1=\mathrm{i}\left[-g_{1,0}\left(\dfrac{\mathrm{d}t}{\mathrm{d}s}\right)+g_{0,1}\left(\dfrac{\mathrm{d}t}{\mathrm{d}s}\right)^{-1}\right].\end{cases} \tag{1.7}$$

显然,方程组(1.7)(对未知函数 $g_{1,0},g_{0,1}$)之系数行列式不等于零,从而一定有 $g_{1,0}=g_{0,1}\equiv0$.

现在假设这一结论对于 $k-1$ 的情况是成立的,也就是说,如果在问题 B_n 的边界条件(1.1)中出现的函数 $f_0,f_1,f_2,\cdots,f_{n-1}$ 里面,有

$$f_0=f_1=\cdots=f_{k-1}\equiv0,\quad 0\leqslant k\leqslant n-1,$$

那么一定有

$$g_{m,n}(s)\equiv0,\quad 0\leqslant m+n\leqslant k-1$$

成立. 我们用归纳法来证明这一结论对于 k 仍然是成立的.

考虑函数

$$\frac{\mathrm{d}^{i}f_{k-j}}{\mathrm{d}s^{j}}=i^{-j}\sum_{l=1}^{k-j}(-1)^{k-l-j}\mathrm{C}_{k-j}^{l}\left[\sum_{h=0}^{j}\mathrm{C}_j^h g_{k-l-h,l+h}\left(\frac{\mathrm{d}t}{\mathrm{d}s}\right)^{k-2(h+l)}\right]+F_{j,k}(g_{m,n}),$$

$$j=0,1,2,\cdots,k,$$

其中 $F_{j,k}(g_{m,n})$ 代表一个关于 $g_{m,n}(m+n\leqslant k-1)$ 及其导数的齐次线性函数. 根据归纳假设可知 $F_{j,k}(g_{m,n})\equiv0$,因此,为了证明我们的结论,只要证明齐次方程组

$$\sum_{l=0}^{k-j}(-1)^{k-j-l}\mathrm{C}_{k-j}^{l}\left[\sum_{h=0}^{j}\mathrm{C}_j^h g_{k-l-h,l+h}(s)\left(\frac{\mathrm{d}t}{\mathrm{d}s}\right)^{k-2(l+h)}\right]=0 \tag{1.8}$$

$$j=0,1,2,\cdots,k,$$

没有非零解就可以了. 事实上,如果把方程组(1.8)之左端展开,然后重新进行组合,并规定当 $m>n$ 时,有 $C_n^m=0$,那么,方程组(1.8)还可以写成

$$\sum_{l=0}^{k}\left[\sum_{h=0}^{l}(-1)^{k-j-h}C_{k-j}^{h}C_{j}^{l-h}\left(\frac{\mathrm{d}t}{\mathrm{d}s}\right)^{k-2t}\right]g_{k-l,l}(s)=0, \tag{1.9}$$

$$j=0,1,2,\cdots,k.$$

这时方程组(1.9)的系数矩阵是

$$\boldsymbol{A}=\begin{pmatrix} a_{00}\left(\frac{\mathrm{d}t}{\mathrm{d}s}\right)^{k} & a_{01}\left(\frac{\mathrm{d}t}{\mathrm{d}s}\right)^{k-2} & \cdots & a_{0,k-1}\left(\frac{\mathrm{d}t}{\mathrm{d}s}\right)^{2-k} & a_{0k}\left(\frac{\mathrm{d}t}{\mathrm{d}s}\right)^{-k} \\ a_{10}\left(\frac{\mathrm{d}t}{\mathrm{d}s}\right)^{k} & a_{11}\left(\frac{\mathrm{d}t}{\mathrm{d}s}\right)^{k-2} & \cdots & a_{1,k-1}\left(\frac{\mathrm{d}t}{\mathrm{d}s}\right)^{2-k} & a_{1k}\left(\frac{\mathrm{d}t}{\mathrm{d}s}\right)^{-k} \\ \vdots & \vdots & & \vdots & \vdots \\ a_{k0}\left(\frac{\mathrm{d}t}{\mathrm{d}s}\right)^{k} & a_{k1}\left(\frac{\mathrm{d}t}{\mathrm{d}s}\right)^{k-2} & \cdots & a_{k,k-1}\left(\frac{\mathrm{d}t}{\mathrm{d}s}\right)^{2-k} & a_{kk}\left(\frac{\mathrm{d}t}{\mathrm{d}s}\right)^{-k} \end{pmatrix},$$

其中

$$a_{jl}=\sum_{h=0}^{l}(-1)^{k-j-h}C_{k-j}^{h}C_{j}^{l-h}, l=0,1,2,\cdots,k, j=0,1,2,\cdots,k.$$

为了达到这目的,只要证明矩阵 $\boldsymbol{A}$ 是满秩的就够了. 如果考虑到矩阵 $\boldsymbol{A}$ 的特殊形式,那么,我们只要证明矩阵

$$\boldsymbol{A}^{*}=\begin{pmatrix} a_{00} & a_{01} & \cdots & a_{0k} \\ a_{10} & a_{11} & \cdots & a_{1k} \\ \vdots & \vdots & & \vdots \\ a_{k0} & a_{k1} & \cdots & a_{kk} \end{pmatrix}$$

满秩就可以了. 在证明这一结论以前,我们先研究矩阵 $\boldsymbol{A}^{*}$ 之元素的性质,如果设

$$b_{jl}=(-1)^{-k+j}a_{l}=\sum_{h=0}^{l}(-1)^{h}C_{k-j}^{h}C_{j}^{l-h}, \ 0\leqslant j,l\leqslant k,$$

那么,数列 $b_{1l},b_{2l},\cdots,b_{kl}$ $(l=0,1,2,\cdots,k)$ 就是一个 l 阶等差数列①. 为了证明,显然又只要证明数列

$$b_{jm}^{*}=\sum_{h=0}^{m}(-1)^{h}C_{k-j-l+m}^{h}C_{j}^{m-h}, j=1,2,\cdots,k,$$

是一个 m 阶等差数列就可以了. 这是因为,当 $m=l$ 时,将有 $b_{jl}^{*}=b_{jl}$ 成立.

① 这里指的是 l 阶差相等而异于零的数列.

事实上，当 $m=1$ 时，有

$$b_{jm}^{*}=\sum_{h=0}^{m}(-1)^{h}C_{k-j-l+m}^{h}C_{j}^{l-h}=2j-k+l-1,$$

所以 $b_{11}^{*},b_{21}^{*},\cdots,b_{k1}^{*}$ 是一个公差为 2 的等差数列（一阶等差数列）. 我们假设 $b_{1,m-1}^{*},b_{2,m-1}^{*},\cdots,b_{k,m-1}^{*}$ 是一个 $m-1$ 阶等差数列（$m\leqslant k$），现在来证明 $b_{1m}^{*},b_{2m}^{*},\cdots,b_{km}^{*}$ 将是一个 m 阶等差数列.

事实上，

$$\begin{aligned}
b_{j+1,m}^{*}-b_{jm}^{*}&=\sum_{h=0}^{m}(-1)^{h}C_{k-j-l+m-1}^{h}C_{j+1}^{m-h}-\sum_{h=0}^{m}(-1)^{h}C_{k-j-l+m}^{h}C_{j}^{m-h}\\
&=\sum_{h=0}^{m}(-1)^{h}\left[C_{k-j-l+m-1}^{h}C_{j+1}^{m-h}-(C_{k-j-l+m-1}^{h}+C_{k-j-l+m-1}^{h-1})C_{j}^{m-h}\right]\\
&=\sum_{h=0}^{m}(-1)^{h}\left[C_{k-j-l+m-1}^{h}(C_{j+1}^{m-h}-C_{j}^{m-h})-C_{k-j-l+m-1}^{h-1}C_{j}^{m-h}\right]\\
&=\sum_{h=0}^{m}(-1)^{h}\left[C_{k-j-l+m-1}^{h}C_{j+1}^{m-h-1}-C_{k-j-l+m-1}^{h-1}C_{j}^{m-h}\right]\\
&=\sum_{h=0}^{m-1}(-1)^{h}C_{k-j-l+m-1}^{h}C_{j+1}^{m-h-1}+\sum_{h=0}^{m}(-1)^{h-1}C_{k-j-l+m-1}^{h-1}C_{j}^{m-h}\\
&=2\sum_{h=0}^{m-1}(-1)^{h}C_{k-j-l+m-1}^{h}C_{j}^{m-h-1}\\
&=2b_{j,m-1}^{*},
\end{aligned}$$

应该指出的是，这里我们规定符号 $C_{m}^{n}=0$，如果 $n<0$；根据归纳假设知道它是一个 $m-1$ 阶等差数列，也就是说 b_{jm}^{*} 是一个 m 阶等差数列.

这样，我们得到，对应于矩阵 $\boldsymbol{A}^{*}$ 的行列式

$$|\boldsymbol{A}^{*}|=\begin{vmatrix}a_{00}&a_{01}&\cdots&a_{0k}\\a_{10}&a_{11}&\cdots&a_{1k}\\\vdots&\vdots&&\vdots\\a_{k0}&a_{k1}&\cdots&a_{kk}\end{vmatrix}=\begin{vmatrix}a_{00}&a_{01}&\cdots&a_{0k}\\0&a_{11}^{(1)}&\cdots&a_{1k}^{(1)}\\\vdots&\vdots&&\vdots\\0&a_{k1}^{(1)}&\cdots&a_{kk}^{(1)}\end{vmatrix},$$

其中，$a_{jl}^{(1)}=a_{j-1,l}+a_{jl}$，$j=1,2,\cdots,k$，$l=1,2,\cdots,k$.

继续利用这样的变换，并注意到上面刚刚讲过的数列 b_{jl} 的性质，最后可以得到

$$|\boldsymbol{A}^{*}|=\begin{vmatrix}a_{00}&a_{01}&a_{02}&\cdots&a_{0k}\\0&a_{11}^{(1)}&a_{12}^{(1)}&\cdots&a_{1k}^{(1)}\\0&0&a_{22}^{(2)}&\cdots&a_{2k}^{(2)}\\\vdots&\vdots&\vdots&&\vdots\\0&0&0&\cdots&a_{kk}^{(k)}\end{vmatrix}\neq0,$$

这是因为位于对角线上的元素 $a_{jj}^{(j)}$ 刚好是数列 $b_{mj}(m=0,1,2,\cdots,k)$ 的 j 阶差，所以有 $a_{jj}^{(j)}\neq 0$，$j=1,2,\cdots,k$. 于是齐次方程组(1.9)只能有零解，从而有 $g_{k-l,l}(s)=0$，$l=0,1,\cdots,k$，再综合归纳假设就得到 $g_{m,n}(s)=0(0\leqslant m+n\leqslant k)$. 特别地，当然更有 $g_{00}=g_{10}=\cdots=g_{k0}\equiv 0$，从而定理 1 得证.

§ 2.

为了以后应用，对基本边值问题 B_n，我们还将给出另一种等价的边界条件. 假设

$$h_0=u^+,h_1=\left(\frac{\mathrm{d}u}{\mathrm{d}v}\right)^+,h_2=(\Delta u)^+,h_3=\left(\frac{\mathrm{d}\Delta u}{\mathrm{d}v}\right)^+,\cdots,$$

$$h_{n-1}=\begin{cases}(\Delta^{\frac{n-1}{2}}u)^+, & n\text{ 是奇数},\\ \left(\dfrac{\mathrm{d}\Delta^{\frac{n}{2}-1}u}{\mathrm{d}v}\right)^+, & n\text{ 是偶数}.\end{cases}\tag{1.1}''$$

根据 § 1 中的关系式(B)以及关系式 $\Delta^m u=4^m\dfrac{\partial^{2m}u}{\partial z^m\partial\bar{z}^m}$，我们有

$$h_{2m}=4^m g_{mm}(s),\tag{2.1}$$

$$h_{2m+1}=4^m\mathrm{i}\left[-g_{m+1,m}(s)\left(\frac{\mathrm{d}t}{\mathrm{d}s}\right)+g_{m,m+1}(s)\left(\frac{\mathrm{d}t}{\mathrm{d}s}\right)^{-1}\right],\tag{2.2}$$

于是不难证明如下定理.

定理 2　对于问题 B_n 的解来说，函数 $h_0,h_1,h_2,\cdots,h_\lambda,0\leqslant\lambda\leqslant n-1$，可以通过边界条件(1.1)′中出现的函数 $g_{00},g_{10},g_{20},\cdots,g_{\lambda 0}$ 以及它们关于 s 的导数来表示；反过来，函数 $g_{00},g_{10},g_{20},\cdots,g_{\lambda 0}$ 也可以通过函数 $h_0,h_1,h_2,\cdots,h_\lambda$ 以及它们关于 s 的导数来表示.

事实上，定理的第一部分利用(2.1)(2.2)和(1.6)很容易得到证明. 现在来证明定理 2 的另一部分结论.

首先，当 $\lambda=0$ 时，显然有 $g_{00}=h_0$. 而当 $\lambda=1$ 时，除 $g_{00}=h_0$ 之外，还应该有

$$g_{10}(s)\left(\frac{\mathrm{d}t}{\mathrm{d}s}\right)+g_{01}(s)\left(\frac{\mathrm{d}t}{\mathrm{d}s}\right)^{-1}=\frac{\mathrm{d}h_0}{\mathrm{d}s},$$

$$\mathrm{i}\left[-g_{10}(s)\left(\frac{\mathrm{d}t}{\mathrm{d}s}\right)+g_{01}(s)\left(\frac{\mathrm{d}t}{\mathrm{d}s}\right)^{-1}\right]=h_1.\tag{2.3}$$

由于方程组(2.3)(对未知函数 g_{10} 和 g_{01})的系数行列式不等于零，所以函数 $g_{10}(s)$ 和 $g_{01}(s)$ 显然可以(唯一地)通过 h_0 关于 s 的导数和 h_1

来表示；从而，特别是边界条件(1.1)′中的函数 g_{00} 和 g_{10} 更可以通过给定的函数 h_0, h_1 以及 h_0 关于 s 的导数来表示.

现在假设 $\lambda=k-1$ 时 $(k\leqslant n-1)$ 函数 $g_{pq}(s)$，$0\leqslant p+q\leqslant k-1$，已经可以通过 $h_0, h_1, h_2, \cdots, h_{k-1}$ 以及它们关于 s 的导数来表示，我们来证明当 $\lambda=k$ 时，定理 2 的结论也是成立的.

为了清楚起见，我们分两种情形来讨论：

(1) k 是偶数的情形(无妨设 $k=2r$). 这时候，根据§1中的公式(A)，将有

$$\begin{cases} \dfrac{\mathrm{d}^l h_{k-l}}{\mathrm{d}s^l}=4^{r-\frac{l}{2}}\displaystyle\sum_{j=0}^{l}\mathrm{C}_l^j g_{\frac{k-l}{2}+l-j,\frac{k-l}{2}+j}(s)\left(\frac{\mathrm{d}t}{\mathrm{d}s}\right)^{l-2j}+F_{k,l}(g_{p,q}), \\ \qquad\qquad l=0,2,\cdots,2r, \\ \dfrac{\mathrm{d}^l h_{k-l}}{\mathrm{d}s^l}=4^{r-\frac{l+1}{2}}\mathrm{i}\displaystyle\sum_{j=0}^{l+1}\mathrm{A}_l^j g_{\frac{k-l-1}{2}+l-j+1,\frac{k-l-1}{2}+j}(s)\left(\frac{\mathrm{d}t}{\mathrm{d}s}\right)^{l-2j+1}+F_{k,l}(g_{p,l}), \\ \qquad\qquad l=1,3,5,\cdots,2r-1; \end{cases} \tag{2.4}$$

其中 $\mathrm{A}_l^j=\mathrm{C}_l^{j-1}-\mathrm{C}_l^j$(应该指出的是我们这里规定 $\mathrm{C}_m^n=0$，如果 $n<0$，或 $n>m$)，$F_{k,l}(g_{p,q})$ 是关于 $g_{pl}(s)$，$0\leqslant p+q\leqslant k-1$，及其关于 s 之导数的(齐次)函数；根据归纳假设知道，$F_{k,l}(g_{p,q})$ 是可以通过函数 $h_0, h_1, h_2, \cdots, h_{k-1}$ 以及它们关于 s 的导数来表示的. 于是方程组(2.4)还可以写成

$$\begin{cases} \displaystyle\sum_{j=0}^{l}\mathrm{C}_l^j\left(\frac{\mathrm{d}t}{\mathrm{d}s}\right)^{l-2j} g_{\frac{k-l}{2}+l-j,\frac{k-l}{2}+j}(s)=F_{k,t}^{*}, \\ \qquad\qquad l=0,2,\cdots,2r, \\ \displaystyle\sum_{j=0}^{l+1}\mathrm{A}_l^j\left(\frac{\mathrm{d}t}{\mathrm{d}s}\right)^{l-2j+1} g_{\frac{k-l-1}{2}+l-j+1,\frac{k-l-1}{2}+j}(s)=F_{k,t}^{*}, \\ \qquad\qquad l=1,3,\cdots,2r-1, \end{cases} \tag{2.5}$$

其中 $F_{k,t}^{*}$ 是可以通过给定的函数 $h_0, h_1, h_2, \cdots, h_k$ 以及它们关于 s 的导数求得的函数.

方程组(2.5)(按 $l=0,1,2,\cdots,2r$ 排列起来)的系数行列式是

$$
|\boldsymbol{A}|=\begin{vmatrix}
0 & 0 & \cdots & 0 & 0 & \mathrm{C}_0^0 & 0 & 0 & \cdots & 0 & 0\\
0 & 0 & \cdots & 0 & -\mathrm{C}_1^0\left(\frac{\mathrm{d}t}{\mathrm{d}s}\right)^2 & \mathrm{A}_1^1 & \mathrm{C}_1^1\left(\frac{\mathrm{d}t}{\mathrm{d}s}\right)^{-2} & 0 & \cdots & 0 & 0\\
0 & 0 & \cdots & 0 & -\mathrm{C}_2^1\left(\frac{\mathrm{d}t}{\mathrm{d}s}\right)^2 & \mathrm{C}_2^1 & \mathrm{C}_2^2\left(\frac{\mathrm{d}t}{\mathrm{d}s}\right)^{-2} & 0 & \cdots & 0 & 0\\
\vdots & \vdots & \cdots & \vdots & \vdots & \vdots & \vdots & \vdots & \cdots & \vdots & \vdots\\
\vdots & \vdots & \cdots & \vdots & \vdots & \vdots & \vdots & \vdots & \cdots & \vdots & \vdots\\
\vdots & \vdots & \cdots & \vdots & \vdots & \vdots & \vdots & \vdots & \cdots & \vdots & \vdots\\
\vdots & \vdots & \cdots & \vdots & \vdots & \vdots & \vdots & \vdots & \cdots & \vdots & \vdots\\
\vdots & \vdots & \cdots & \vdots & \vdots & \vdots & \vdots & \vdots & \cdots & \vdots & \vdots\\
\vdots & \vdots & \cdots & \vdots & \vdots & \vdots & \vdots & \vdots & \cdots & \vdots & \vdots\\
-\left(\frac{\mathrm{d}t}{\mathrm{d}s}\right)^k & \mathrm{A}_{k-1}^1\left(\frac{\mathrm{d}t}{\mathrm{d}s}\right)^{k-2} & \cdots & & \cdots & \mathrm{A}_{k-1}^{\frac{k}{2}} & \cdots & & \cdots & \mathrm{A}_{k-1}^{k-1}\left(\frac{\mathrm{d}t}{\mathrm{d}s}\right)^{2-k} & \left(\frac{\mathrm{d}t}{\mathrm{d}s}\right)^{-k}\\
\left(\frac{\mathrm{d}t}{\mathrm{d}s}\right)^k & \mathrm{C}_{k-1}^1\left(\frac{\mathrm{d}t}{\mathrm{d}s}\right)^{k-2} & \cdots & & \cdots & \mathrm{C}_k^{\frac{k}{2}} & \cdots & & \cdots & \mathrm{C}_k^{k-1}\left(\frac{\mathrm{d}t}{\mathrm{d}s}\right)^{2-k} & \left(\frac{\mathrm{d}t}{\mathrm{d}s}\right)^{-k}
\end{vmatrix}
$$

$$
=\begin{vmatrix}
0 & 0 & \cdots & 0 & 0 & 0 & 1 & 0 & 0 & 0 & \cdots & 0 & 0\\
0 & 0 & \cdots & 0 & 0 & -1 & 0 & 1 & 0 & 0 & \cdots & 0 & 0\\
0 & 0 & \cdots & 0 & 0 & 1 & 2 & 1 & 0 & 0 & \cdots & 0 & 0\\
0 & 0 & \cdots & 0 & -1 & -2 & 0 & 2 & 1 & 0 & \cdots & 0 & 0\\
0 & 0 & \cdots & 0 & 1 & 4 & 6 & 4 & 1 & 0 & \cdots & 0 & 0\\
\vdots & \vdots & \cdots & \vdots & \vdots & \vdots & \vdots & \vdots & \vdots & \vdots & \cdots & \vdots & \vdots\\
\vdots & \vdots & \cdots & \vdots & \vdots & \vdots & \vdots & \vdots & \vdots & \vdots & \cdots & \vdots & \vdots\\
\vdots & \vdots & \cdots & \vdots & \vdots & \vdots & \vdots & \vdots & \vdots & \vdots & \cdots & \vdots & \vdots\\
0 & -1 & \cdots & \cdots & \cdots & \cdots & 0 & \cdots & \cdots & \cdots & \cdots & 1 & 0\\
0 & 1 & \cdots & \cdots & \cdots & \cdots & \mathrm{C}_{k-2}^{\frac{k}{2}-1} & \cdots & \cdots & \cdots & \cdots & 1 & 0\\
-1 & 2-k & \cdots & \cdots & \cdots & \cdots & 0 & \cdots & \cdots & \cdots & \cdots & k-2 & 1\\
1 & k & \cdots & \cdots & \cdots & \cdots & \mathrm{C}_k^{\frac{k}{2}} & \cdots & \cdots & \cdots & \cdots & k & 1
\end{vmatrix}
$$

$$
=\begin{vmatrix}
0 & 0 & \cdots & 0 & 0 & 0 & 1 & 0 & 0 & 0 & \cdots & 0 & 0 \\
0 & 0 & \cdots & 0 & 0 & 0 & 0 & 1 & 0 & 0 & \cdots & 0 & 0 \\
0 & 0 & \cdots & 0 & 0 & 2 & 2 & 1 & 0 & 0 & \cdots & 0 & 0 \\
0 & 0 & \cdots & 0 & 0 & 0 & 0 & 2 & 1 & 0 & \cdots & 0 & 0 \\
0 & 0 & \cdots & 0 & 2 & 8 & 6 & 4 & 1 & 0 & \cdots & 0 & 0 \\
| & | & \cdots & | & | & | & | & | & | & | & \cdots & | & | \\
| & | & \cdots & | & | & | & | & | & | & | & \cdots & | & | \\
| & | & \cdots & | & | & | & | & | & | & | & \cdots & | & | \\
0 & 0 & \cdots & \cdots & \cdots & \cdots & 0 & \cdots & \cdots & \cdots & \cdots & 1 & 0 \\
0 & 2 & \cdots & \cdots & \cdots & \cdots & \mathrm{C}_{k-2}^{\frac{k}{2}-1} & \cdots & \cdots & \cdots & \cdots & 1 & 0 \\
0 & 0 & \cdots & \cdots & \cdots & \cdots & 0 & \cdots & \cdots & \cdots & \cdots & k-2 & 0 \\
2 & 2k & \cdots & \cdots & \cdots & \cdots & \mathrm{C}_{k}^{\frac{k}{2}} & \cdots & \cdots & \cdots & \cdots & k & 1
\end{vmatrix}
$$

$$
=(-1)^{2\left(1+2+\cdots+\frac{k}{2}\right)}\begin{vmatrix}
1 & 0 & 0 & 0 & 0 & \cdots & 0 & 0 & 0 & 0 \\
0 & 1 & 0 & 0 & 0 & \cdots & 0 & 0 & 0 & 0 \\
2 & 1 & 2 & 0 & 0 & \cdots & 0 & 0 & 0 & 0 \\
0 & 2 & 0 & 1 & 0 & \cdots & 0 & 0 & 0 & 0 \\
6 & 4 & 8 & 1 & 2 & \cdots & 0 & 0 & 0 & 0 \\
| & | & | & | & | & \cdots & | & | & | & | \\
| & | & | & | & | & \cdots & | & | & | & | \\
| & | & | & | & | & \cdots & | & | & | & | \\
0 & \cdots & \cdots & \cdots & \cdots & \cdots & 1 & 0 & 0 & 0 \\
\mathrm{C}_{k-2}^{\frac{k}{2}-1} & \cdots & \cdots & \cdots & \cdots & \cdots & 1 & 2 & 0 & 0 \\
0 & \cdots & \cdots & \cdots & \cdots & \cdots & k-2 & 0 & 1 & 0 \\
\mathrm{C}_{k}^{\frac{k}{2}} & \cdots & \cdots & \cdots & \cdots & \cdots & k & 2k & 1 & 2
\end{vmatrix}
$$

$$
=2^{\frac{k}{2}}\neq 0.
$$

从而方程组(2.5)是单值可解的；也就是说函数 $g_{p,g}(s)$，$p+q=k$ 可以通过 $h_0,h_1,h_2,\cdots,h_k$ 以及它们关于 s 的导数求得，再综合归纳假设就可以推出我们所要的结论.

(2) k 是奇数的情形(无妨假设 $k=2r+1$). 这时候，根据 §1 中公式(A)将有

$$
\begin{cases}
\dfrac{\mathrm{d}^l h_{k-l}}{\mathrm{d}s^l} = 4^{r-\frac{l}{2}}\mathrm{i}\displaystyle\sum_{j=0}^{l+1}\mathrm{A}_l^j g_{\frac{k-l-1}{2}+l-j+1,\frac{k-l-1}{2}+j}(s)\left(\frac{\mathrm{d}t}{\mathrm{d}s}\right)^{l-2j+1} + G_{k,l}(g_{p,l}), \\
\qquad\qquad l = 0,2,\cdots,2r, \\
\dfrac{\mathrm{d}^l h_{k-l}}{\mathrm{d}s^l} = 4^{r-\frac{l-1}{2}}\displaystyle\sum_{j=0}^{l}\mathrm{C}_l^j g_{\frac{k-l}{2}-j,\frac{k-l}{2}+j}(s)\left(\frac{\mathrm{d}t}{\mathrm{d}s}\right)^{l-2j} + G_{k,l}(g_{p,q}), \\
\qquad\qquad l = 1,3,\cdots,2r+1.
\end{cases}
\tag{2.6}
$$

其中 $\mathrm{A}_l^j=\mathrm{C}_l^{j-1}-\mathrm{C}_l^j$，而且我们规定 $\mathrm{C}_m^n=0$，如果 $n<0$ 或 $n>m$；$G_{k,l}(g_{p,q})$ 是关于 $g_{p,q}(s)$，$0\leqslant p+q\leqslant k-1$，及其关于 s 之导数的齐次函数；根据归纳假设知道，$G_{k,l}(g_{p,q})$ 是可以通过函数 $h_0,h_1,h_2,\cdots,h_{k-1}$ 以及它们关于 s 的导数来表示的. 于是方程组(2.6)还可以写成

$$
\begin{cases}
\displaystyle\sum_{j=0}^{l+1}\mathrm{A}_l^j\left(\frac{\mathrm{d}t}{\mathrm{d}s}\right)^{l-2j+1} g_{\frac{k-l-1}{2}+l-j+1,\frac{k-l-1}{2}+j}(s) = G_{k,l}^*, \\
\qquad\qquad l = 0,2,4,\cdots,2r, \\
\displaystyle\sum_{j=0}^{l}\mathrm{C}_l^j\left(\frac{\mathrm{d}t}{\mathrm{d}s}\right)^{l-2j} g_{\frac{k-l}{2}-j,\frac{k-l}{2}+j}(s) = G_{k,l}^*, \\
\qquad\qquad l = 1,3,\cdots,2r+1.
\end{cases}
\tag{2.7}
$$

其中 $G_{k,l}^*$ 是通过给定的函数 $h_0,h_1,h_2,\cdots,h_k$ 以及它们关于 s 的导数可以求得的函数.

方程组(2.7)(按 $l=0,1,2,\cdots,2r+1$ 排列起来)的系数行列式是

$$
|\boldsymbol{A}| = \begin{vmatrix}
0 & 0 & \cdots & 0 & -\left(\frac{\mathrm{d}t}{\mathrm{d}s}\right) & \left(\frac{\mathrm{d}t}{\mathrm{d}s}\right)^{-1} & 0 & \cdots & 0 & 0 \\
0 & 0 & \cdots & 0 & \mathrm{C}_1^0\left(\frac{\mathrm{d}t}{\mathrm{d}s}\right) & \mathrm{C}_1^1\left(\frac{\mathrm{d}t}{\mathrm{d}s}\right)^{-1} & 0 & \cdots & 0 & 0 \\
0 & 0 & \cdots & -\left(\frac{\mathrm{d}t}{\mathrm{d}s}\right)^3 & \mathrm{A}_2^1\left(\frac{\mathrm{d}t}{\mathrm{d}s}\right) & \mathrm{A}_2^2\left(\frac{\mathrm{d}t}{\mathrm{d}s}\right)^{-1} & \left(\frac{\mathrm{d}t}{\mathrm{d}s}\right)^{-3} & \cdots & 0 & 0 \\
0 & 0 & \cdots & \mathrm{C}_3^0\left(\frac{\mathrm{d}t}{\mathrm{d}s}\right)^3 & \mathrm{C}_3^1\left(\frac{\mathrm{d}t}{\mathrm{d}s}\right) & \mathrm{C}_3^2\left(\frac{\mathrm{d}t}{\mathrm{d}s}\right)^{-1} & \mathrm{C}_3^3\left(\frac{\mathrm{d}t}{\mathrm{d}s}\right)^{-3} & \cdots & 0 & 0 \\
\vdots & \vdots & & \vdots & \vdots & \vdots & \vdots & & \vdots & \vdots \\
-\left(\frac{\mathrm{d}t}{\mathrm{d}s}\right)^k & \mathrm{A}_{k-1}^1\left(\frac{\mathrm{d}t}{\mathrm{d}s}\right)^{k-2} & \cdots & \cdots & \mathrm{A}_{k-1}^r\left(\frac{\mathrm{d}t}{\mathrm{d}s}\right) & \mathrm{A}_{k-1}^{r+1}\left(\frac{\mathrm{d}t}{\mathrm{d}s}\right)^{-1} & \cdots & \cdots & \mathrm{A}_{k-1}^{k-1}\left(\frac{\mathrm{d}t}{\mathrm{d}s}\right)^{2-k} & \left(\frac{\mathrm{d}t}{\mathrm{d}s}\right)^{-k} \\
\left(\frac{\mathrm{d}t}{\mathrm{d}s}\right)^k & \mathrm{C}_k^1\left(\frac{\mathrm{d}t}{\mathrm{d}s}\right)^{k-2} & \cdots & \cdots & \mathrm{C}_k^r\left(\frac{\mathrm{d}t}{\mathrm{d}s}\right) & \mathrm{C}_k^{r+1}\left(\frac{\mathrm{d}t}{\mathrm{d}s}\right)^{-1} & \cdots & \cdots & \mathrm{C}_k^{k-1}\left(\frac{\mathrm{d}t}{\mathrm{d}s}\right)^{2-k} & \left(\frac{\mathrm{d}t}{\mathrm{d}s}\right)^{-k}
\end{vmatrix}
$$

$$
=\begin{vmatrix}
0 & 0 & \cdots & 0 & -1 & 1 & 0 & \cdots & 0 & 0 \\
0 & 0 & \cdots & 0 & 1 & 1 & 0 & \cdots & 0 & 0 \\
0 & 0 & \cdots & -1 & -1 & 1 & 1 & \cdots & 0 & 0 \\
0 & 0 & \cdots & 1 & 3 & 3 & 1 & \cdots & 0 & 0 \\
\vdots & \vdots & & \vdots & \vdots & \vdots & \vdots & & \vdots & \vdots \\
0 & -1 & \cdots & \cdots & \cdots & \cdots & \cdots & \cdots & 1 & 0 \\
0 & 1 & \cdots & \cdots & \cdots & \cdots & \cdots & \cdots & 1 & 0 \\
-1 & 2-k & \cdots & A_{k-1}^{r-1} & A_{k-1}^{r} & A_{k-1}^{r+1} & A_{k-1}^{r+2} & \cdots & k-2 & 1 \\
1 & k & \cdots & C_k^{r-1} & C_k^{r} & C_k^{r+1} & C_k^{r+2} & \cdots & k & 1
\end{vmatrix}
$$

$$
=\begin{vmatrix}
0 & 0 & \cdots & 0 & 0 & 1 & 0 & \cdots & 0 & 0 \\
0 & 0 & \cdots & 0 & 2 & 1 & 0 & \cdots & 0 & 0 \\
0 & 0 & \cdots & 0 & 0 & 1 & 1 & \cdots & 0 & 0 \\
0 & 0 & \cdots & 2 & 6 & 3 & 1 & \cdots & 0 & 0 \\
\vdots & \vdots & & \vdots & \vdots & \vdots & \vdots & & \vdots & \vdots \\
0 & 0 & \cdots & \cdots & \cdots & \cdots & \cdots & \cdots & 1 & 0 \\
0 & 2 & \cdots & \cdots & \cdots & \cdots & \cdots & \cdots & 1 & 0 \\
0 & 0 & \cdots & 0 & 0 & A_{k-1}^{r+1} & A_{k-1}^{r+2} & \cdots & k-2 & 1 \\
2 & 2k & \cdots & 2C_k^{r-1} & 2C_k^{r} & C_k^{r+1} & C_k^{r+2} & \cdots & k & 1
\end{vmatrix}
$$

$$
=(-1)^{r+1+2(1+2+\cdots+r)}\begin{vmatrix}
1 & 0 & 0 & 0 & \cdots & 0 & 0 & 0 & 0 \\
1 & 2 & 0 & 0 & \cdots & 0 & 0 & 0 & 0 \\
1 & 0 & 1 & 0 & \cdots & 0 & 0 & 0 & 0 \\
3 & 6 & 1 & 2 & \cdots & 0 & 0 & 0 & 0 \\
\vdots & \vdots & \vdots & \vdots & & \vdots & \vdots & \vdots & \vdots \\
\cdots & \cdots & \cdots & \cdots & \cdots & 1 & 0 & 0 & 0 \\
\cdots & \cdots & \cdots & \cdots & \cdots & 1 & 2 & 0 & 0 \\
A_{k-1}^{r+1} & 0 & A_{k-1}^{r+2} & 0 & \cdots & k-2 & 0 & 1 & 0 \\
C_k^{r+1} & 2C_k^{r} & C_k^{r+2} & 2C_k^{r+1} & \cdots & k & 2k & 1 & 2
\end{vmatrix}
$$

$$
=(-1)^{r+1}2^{r+1}\neq 0,\quad k=2r+1.
$$

从而方程组(2.7)是单值可解的;也就是说,通过给定的函数$h_0,h_1,h_2,\cdots,h_k$以及它们关于s的导数可以求得函数$g_{p,q}(s)$,$p+q=k$,再综合归纳假设就可以推出我们所要的结论.综合上述,定理2得到了证明.

推论 1　问题 B_n 的边界条件(1.1)′与边界条件(1.1)″是等价的. 从而边界条件(1.1)与边界条件(1.1)″也是等价的.

也就是说,在问题 B_n 的提法中可以用边界条件(1.1)″来代替边界条件(1.1). 问题 B_n 之这种提法的优越性在 §3 和 §5 中将可以看到.

推论 2　在问题 B_n 的边界条件(1.1)中出现的函数 $f_0,f_1,f_2,\cdots,f_{n-1}$ 里面,如果有

$$f_0=f_1=f_2=\cdots=f_k\equiv 0,\qquad 0\leqslant k\leqslant n-1,\tag{2.8}$$

或者在问题 B_n 的边界条件(1.1)′中出现的函数 $g_{0,0},g_{1,0},g_{2,0},\cdots,g_{n-1,0}$ 里面,如果有

$$g_{0,0}=g_{1,0}=g_{2,0}=\cdots=g_{k,0}\equiv 0,\quad 0\leqslant k\leqslant n-1,\tag{2.9}$$

那么,在边界条件(1.1)″中出现的函数 $h_0,h_1,h_2,\cdots,h_{n-1}$ 里面也一定有

$$h_0=h_1=h_2=\cdots=h_k\equiv 0,\ 0\leqslant k\leqslant n-1\tag{2.10}$$

成立,而且反过来说也是对的.

§3.

在文[2]中研究了对一般二阶椭圆型方程所提平面狄利克雷问题,该文中利用共轭方程的一组完备特解系,给出了平面狄利克雷问题解的一种级数表示式. 在文[3]中利用同样方法讨论了平面诺依曼问题. 现在我们将在文[2]的基础上,利用类似的方法来讨论在 §1 中提出的问题 B_n,并将给出解的一种级数表示式.

我们知道,文[1]方程(1.2)的解可以表示成

$$u(x,y)=\sum_{k=0}^{n-1}\int_{\Gamma}\left(\Delta^k u\,\frac{\mathrm{d}\Delta^{n-k-1}w}{\mathrm{d}v}-\frac{\mathrm{d}\Delta^k u}{\mathrm{d}v}\Delta^{n-k-1}w\right)\mathrm{d}s\tag{3.1}$$

的形式,其中 w 是方程(1.2)的一个已知的标准基本解,它可以表为

$$w=w(x,y;x_0,y_0)+w_0=\frac{r^{2n-2}}{2\pi 4^{n-1}(n-1)!^2}\ln\frac{1}{r}+w_0,\tag{3.2}$$

$$r=\sqrt{(x-x_0)^2+(y-y_0)^2},$$

而 w_0 是方程(1.2)的某一个在区域 T 内正则的解;容易知道,在公式(3.1)左端出现的函数

$$\text{或}\quad\begin{cases}u,\dfrac{\mathrm{d}u}{\mathrm{d}v},\Delta u,\dfrac{\mathrm{d}\Delta u}{\mathrm{d}v},\cdots,\Delta^{\frac{n-3}{2}}u,\dfrac{\mathrm{d}\Delta^{\frac{n-3}{2}}u}{\mathrm{d}v},\Delta^{\frac{n-1}{2}}u, & n\text{ 是奇数},\\ u,\dfrac{\mathrm{d}u}{\mathrm{d}v},\Delta u,\dfrac{\mathrm{d}\Delta u}{\mathrm{d}v},\cdots,\Delta^{\frac{n-2}{2}}u,\dfrac{\mathrm{d}\Delta^{\frac{n-2}{2}}u}{\mathrm{d}v}, & n\text{ 是偶数}\end{cases}\tag{3.3}$$

是完全可以通过给定的边界条件(1.1)来确定的(参看§2). 从而,可以把它们看作是已知的. 这里我们看出,如果问题 B_n 的边界条件用§2中(1.1)″给出的话,那么,在(3.3)中出现的函数本身就是事先可以给定的. 这样一来,就可以把问题 B_n 和在文[2]中讨论平面狄利克雷问题那样对照起来(参看下面),这也刚好表现在问题 B_n 的提法中利用边界条件(1.1)″的最大优越性.

应该指出,在公式(3.1)左端出现的另外 n 个关于 u 的函数(除去(3.3)中出现的函数之外)却是不能事先任意给定的. 因此若想利用公式(3.1)给出问题 B_n 的解,就应该设法通过已知条件确定出函数组

$$\frac{d\Delta^{n-1}u}{dv},\Delta^{n-1}u,\cdots,\frac{d\Delta^{\frac{n+1}{2}}u}{dv},\Delta^{\frac{n+1}{2}}u,\frac{d\Delta^{\frac{n-1}{2}}u}{dv},\quad n\text{ 是奇数},\tag{3.4a}$$

或

$$\frac{d\Delta^{n-1}u}{dv},\Delta^{n-1}u,\cdots,\frac{d\Delta^{\frac{n}{2}}u}{dv},\Delta^{\frac{n}{2}}u,\quad n\text{ 是偶数}\tag{3.4b}$$

在边界 Γ 上的值. 为此,需要设法做出方程(1.2)的某些完备特解系,然后通过这些完备特解系来达到我们要讨论解之表示式的目的.

§4.

为了简单,我们首先详细地讨论双调和方程的情形(亦即是 $n=2$ 的情形),这时候,将要考虑以下问题.

问题 B_2:要求根据边界条件

$$u^+=f_0,\ \left(\frac{du}{dv}\right)^+=f_1,\tag{4.1}$$

来确定方程

$$\Delta^2u(x,y)=0,\tag{4.2}$$

在区域 T 内的正则(实)解. 在这种情况下,(3.1)可以表示成

$$u(x,y)=\int_\Gamma\left(u\frac{d\Delta w}{dv}-\frac{du}{dv}\Delta w\right)ds+\int_\Gamma\left(\Delta u\frac{dw}{dv}-\frac{d\Delta u}{dv}w\right)ds,\tag{4.3}$$

其中 w 是方程(4.2)的一个已知的标准基本解. 为了能利用公式(4.3)给出问题 B_2 的解,只要能够再设法求出函数 Δu 和 $\frac{d\Delta u}{dv}$ 在边界 Γ 上的值就可以了.

如果对问题 B_2 的解 u 和方程(4.2)①的任意解 v 应用格林公式，那么，将成立等式

$$\int_{\Gamma}\left(\frac{\mathrm{d}\Delta u}{\mathrm{d}v}v-\Delta u\frac{\mathrm{d}u}{\mathrm{d}v}\right)\mathrm{d}s+\int_{\Gamma}\left(\Delta v\frac{\mathrm{d}u}{\mathrm{d}v}-\frac{\mathrm{d}\Delta u}{\mathrm{d}v}u\right)\mathrm{d}s=0. \tag{4.4}$$

据此可以看出，只要能适当地选取方程(4.2)的某一组完备特解系，使得我们通过公式(4.4)能够先确定出函数 Δu 在边界 Γ 上的值，然后再设法求出函数$\dfrac{\mathrm{d}\Delta u}{\mathrm{d}v}$在边界 Γ 上的值，我们的目的就可以达到了. 从公式(4.4)看出，如果我们能够选出一组完备特解系 $\{v_j^{(2,1)}(x,y)\}(j\in\mathbf{N}^*)$，要求它满足条件 $v_j^{(2,1)}(t)|_{\Gamma}=0$ 的话，那么，显然就有

$$\int_{\Gamma}\Delta u\frac{\mathrm{d}v_j^{(2,1)}}{\mathrm{d}v}\mathrm{d}s=\int_{\Gamma}\left(f_1\Delta v_j^{(2,1)}-f_0\frac{\mathrm{d}\Delta v_j^{(2,1)}}{\mathrm{d}v}\right)\mathrm{d}s=c_j^{(2,1)}, \tag{4.5}$$

$$j\in\mathbf{N}^*,$$

其中 $C_j^{(2,1)}$ 是已知的(实)常数. 这就提供了通过已知条件确定 $\Delta u(t)$的可能性.

为了做出适合我们要求的完备特解系，就必须首先找出方程(4.2)满足条件 $v(t)|_{\Gamma}=0$ 之解的一般表示式. 我们知道，方程(4.2)之(实)解的一般表示式是

$$v(x,y)=\mathrm{Re}[\psi_0(z)+\bar{z}\psi_1(z)], \tag{4.6}$$

其中 $\psi_0(z)$和 $\psi_1(z)$是在区域 T 内定义的任意全纯函数[1]；另外，如果要求所有形如$\dfrac{\partial^{m+k}v}{\partial z^m\partial\bar{z}^k}(m\leqslant 2,k\leqslant 2)$的导数在闭区域 $T+\Gamma$ 上连续的话，那么，$\psi_0(z)$和 $\psi_1(z)$的一阶导数在闭区域 $T+\Gamma$ 上还满足赫尔德条件[1].

根据表示式(4.6)和条件 $v(t)|_{\Gamma}=0$ 可以知道，全纯函数 $\psi_0(z)$和 $\psi_1(z)$在边界 Γ 上应该满足条件

$$\mathrm{Re}[\psi_0(t)]=-\mathrm{Re}[\bar{t}\psi_1(t)],\ t\in\Gamma. \tag{4.7}$$

这样，利用文[2]中的结果，马上可以得到

$$\mathrm{Re}[\psi_0(z)]=\int_{\Gamma}\left\{\mathrm{Re}[\bar{t}\psi_1(t)]\frac{\mathrm{d}w}{\mathrm{d}v}-w\sum_{j=1}^{+\infty}c_jv_j(t)\right\}\mathrm{d}s, \tag{4.8}$$

其中 c_j 是完全确定的(实)常数，它们依赖于函数 $\psi_1(t)$[2]. 以后我们把等式(4.8)之右端简记作 $\mathfrak{R}_1[\psi_1]$.

① 应该指出，方程(4.2)是自共轭的.

如果把式(4.8)代入式(4.6),那么将得到

$$v(x,y)=\operatorname{Re}[\bar{z}\psi_1(z)]+\mathfrak{K}_1[\psi_1]\equiv\mathfrak{H}_2[\psi_1(z)], \qquad (4.9)$$

这里 $\mathfrak{H}_2$ 代表一个(实的)齐次可加算子,于是得到以下定理.

定理 3 如果 $v(x,y)$是方程(4.2)在区域 T 内的正则解,它满足条件 $v(t)|_\Gamma=0$,那么,将有表示式(4.9)成立,其中 $\psi_1(z)$是在区域 T 内定义的任意全纯函数.

这就是说,方程(4.2)之满足条件 $v(t)|_\Gamma=0$ 的一般解可以只通过一个任意的全纯函数 $\psi_1(z)$来表示. 这样一来,假如用 z^k, $\mathrm{i}z^k$, $k\in\mathbf{N}$ 分别代替式(4.9)中的全纯函数 $\psi_1(t)$,然后再进行重新编号,就可以得到方程(4.2)的一组完备特解系$\{v_j^{(2,1)}(x,y)\}(j\in\mathbf{N}^*)$并且它满足条件

$$v_j^{(2,1)}(t)\,|_\Gamma=0. \qquad (4.10)$$

这样,就造出了一组我们所需要的完备特解系.

可以证明,对这组完备特解系来说,边界 Γ 上存在外法线方向导数 $\dfrac{\mathrm{d}v^{(2,1)}(t)}{\mathrm{d}v}$(我们把它理解为在区域 T 内点上 $v_j^{(2,1)}(x,y)$沿该方向导数的极限值). 根据公式(4.9)看出,$v_j^{(2,1)}(x,y)$的表示式中第一项对求外法线方向导数来说不会有任何困难,主要是需要讨论第二项.

为了简单,我们将限于讨论边界 Γ 是单位圆周的情形. 这时候,根据施瓦兹积分公式,我们知道

$$\mathfrak{K}_1[\psi_1]=\operatorname{Re}\left\{\frac{1}{2\pi\mathrm{i}}\int_\Gamma\operatorname{Re}[\tau\psi_1(\tau)]\frac{\tau+z}{\tau-z}\frac{\mathrm{d}\tau}{\tau}\right\}, ① \qquad (4.11)$$

如果考虑到我们的条件下 $\psi'_1(t)$在 Γ 上满足赫尔德条件以及柯西型积分的性质和边界值公式[2][4][5],那么,不难得到

$$\frac{\mathrm{d}}{\mathrm{d}v}\mathfrak{K}_1[\psi_1]=\operatorname{Re}\left\{\frac{1}{2}\mathrm{e}^{\mathrm{i}\varphi}[\bar{t}\psi'_1(t)+\overline{\psi_1}(\bar{t})]+\frac{\mathrm{e}^{\mathrm{i}\varphi}}{2\pi\mathrm{i}}\int_\Gamma\frac{\bar{\tau}\psi'_1(\tau)+\overline{\psi_1(\tau)}}{\tau-t}\mathrm{d}\tau\right\}, \qquad (4.12)$$

其中 φ 代表在点 t 处边界 Γ 之外法线 v 与某一确定方向(例如 Ox 轴方向)之间的夹角.

这样,我们得到

① 这里用了在区域 T 内的任意全纯函数 $\psi_1(z)$,但是为了讨论函数 $v_j^{(2,1)}(x,y)$只要分别用 z^k 和 $\mathrm{i}z^k$, $k\in\mathbf{N}$,来代替 $\psi_1(z)$就可以了.

$$\frac{\mathrm{d}v(t)}{\mathrm{d}v}=\frac{\mathrm{d}}{\mathrm{d}v}\mathrm{Re}[\bar{t}\psi_1(t)]+\frac{\mathrm{d}}{\mathrm{d}v}\mathfrak{K}[\psi_1]=\mathrm{Re}\{\mathrm{e}^{\mathrm{i}\varphi}\bar{t}\psi'_1(t)+\bar{\mathrm{e}}^{\mathrm{i}\varphi}\psi_1(t)\}+$$

$$\mathrm{Re}\left\{\frac{\mathrm{e}^{\mathrm{i}\varphi}}{2}[\bar{t}\psi'_1(t)+\overline{\psi_1(t)}]+\frac{\mathrm{e}^{\mathrm{i}\varphi}}{2\pi\mathrm{i}}\int_\Gamma\frac{\bar{\tau}\psi'_1(\tau)+\overline{\psi_1(\tau)}}{\tau-t}\mathrm{d}\tau\right\}.\quad(4.13)$$

除此之外,我们可以证明以下定理.

定理 4 函数$\left\{\frac{\mathrm{d}v_j^{(2,1)}(t)}{\mathrm{d}v}\right\}$($j\in\mathbf{N}^*$)关于空间 $L^2(\Gamma)$是封闭的;亦即对任意给定的 $\varepsilon>0$,和任何函数 $f(t)\in L^2(\Gamma)$,都可以找到这样的常数组 $\xi_1,\xi_2,\cdots,\xi_n$,使得不等式

$$\left\|f(t)-\sum_{j=1}^n\xi_j\frac{\mathrm{d}v_j^{(2,1)}(t)}{\mathrm{d}v}\right\|_{L^2}<\varepsilon$$

成立.

证 如果函数 $f(t)\in L^2(\Gamma)$,那么对任意给定的 $\varepsilon>0$,我们总可以找到这样的函数 $f^*(t)$,使得

$$\|f(t)-f^*(t)\|_{L^2}<\frac{\varepsilon}{2},$$

为了以后应用,我们无妨规定 $f^*(t)\in C^3(\Gamma)$.

然后,我们知道[1],对双调和方程所提的基本边值问题

$$u^+=0,\quad\left(\frac{\mathrm{d}u}{\mathrm{d}v}\right)^+=f^*(t),$$

一定有唯一解 $u(x,y)$,而且根据定理它可以表示成

$$u(x,y)=\mathrm{Re}[\bar{z}\psi^*(z)]+\mathfrak{K}_1[\psi^*]$$

其中 $\psi^*(z)$应该是某一个在区域 T 内安全确定的全纯函数,而且它在闭区域 $T+\Gamma$ 上具有满足赫尔德条件的导数.这样,我们有

$$f^*(t)=\left(\frac{\mathrm{d}u}{\mathrm{d}v}\right)^+=\mathrm{Re}\{\mathrm{e}^{\mathrm{i}\varphi}\bar{t}\psi^{*\prime}(t)+\mathrm{e}^{-\mathrm{i}\varphi}\psi^*(t)\}+$$

$$\mathrm{Re}\left\{\frac{\mathrm{e}^{\mathrm{i}\varphi}}{2}[\bar{t}\psi^{*\prime}(t)+\overline{\psi^*(t)}]+\frac{\mathrm{e}^{\mathrm{j}\varphi}}{2\pi\mathrm{j}}\int_\Gamma\frac{\bar{\tau}\psi^{*\prime}(\tau)+\overline{\psi^*(\tau)}}{\tau-t}\mathrm{d}\tau\right\}.$$

根据魏尔施特拉斯定理,以及带有柯西核的奇异积分是希尔伯特空间 $L^2(\Gamma)$中有界算子的性质[6],容易证明,应该有这样的常数组 ξ_1,$\xi_2,\cdots,\xi_n$ 存在,使得

$$\left\|f^*(t)-\sum_{j=1}^n\xi_j\frac{\mathrm{d}v_j^{(2,1)}(t)}{\mathrm{d}v}\right\|_{L^2}<\frac{\varepsilon}{2},$$

从而定理 4 得证.

不失一般性，我们可以认为函数组$\left\{\frac{\mathrm{d}v_j^{(2,1)}(t)}{\mathrm{d}v}\right\}$关于空间$L^2(\Gamma)$是标准正交完备系. 这样，根据(4.5)可以知道，傅里叶级数$\sum\limits_{j=1}^{+\infty} c_j^{(2,1)} \frac{\mathrm{d}v_j^{(2,1)}(t)}{\mathrm{d}v}$在$\Gamma$上将平均收敛于函数$\Delta u(t)$.

剩下的工作是还要设法确定函数$\frac{\mathrm{d}\Delta u}{\mathrm{d}v}$在边界$\Gamma$上的值. 为此，我们利用格林公式[1]

$$\iint_T [v\Delta^2 u - \Delta v\Delta u]\mathrm{d}T = \int_\Gamma \left[v\frac{\mathrm{d}\Delta u}{\mathrm{d}v} - \frac{\mathrm{d}v}{\mathrm{d}v}\Delta u\right]\mathrm{d}s.$$

如果令$\tilde{u}=\Delta u$，那么，显然，应该有$\Delta\tilde{u}=0$，而且在边界Γ上有$\tilde{u}(t)=\Delta u(t)$. 根据上面的讨论我们可以认为$\Delta u(t)$是已知的①.（这是对函数$\tilde{u}(x,y)$所提的狄利克雷问题）这样一来，把求函数$\frac{\mathrm{d}\Delta u}{\mathrm{d}v}$在边界$\Gamma$上之值的问题已经转化为对解狄利克雷问题时求函数$\frac{\mathrm{d}\tilde{u}(t)}{\mathrm{d}v}$，而如何求$\frac{\mathrm{d}\tilde{u}(t)}{\mathrm{d}v}$的问题刚好是文[2]中讨论过的，因此可以像文[2]中那样，求得傅里叶级数$\sum\limits_{j=1}^{+\infty} c_j^{(2,2)} v_j^{(2,2)}(t)$在边界$\Gamma$上平均收敛于函数$\frac{\mathrm{d}\Delta u(t)}{\mathrm{d}v}$. 最后，容易证明以下定理.

定理 5 对于问题B_2的解$u(x,y)$来说，成立等式

$$u(x,y) = \int_\Gamma \left[f_0\frac{\mathrm{d}\Delta w}{\mathrm{d}v} - f_1\Delta w\right]\mathrm{d}s + \int_\Gamma \left[\frac{\mathrm{d}w}{\mathrm{d}v}\sum_{j=1}^{+\infty} c_j^{(2,1)} \frac{\mathrm{d}v_j^{(2,1)}}{\mathrm{d}v} - w\sum_{j=1}^{+\infty} c_j^{(2,2)} v_j^{(2,2)}\right]\mathrm{d}s,$$

而且上式右端的级数任何闭区域$T^* + \Gamma^* \subset T$上都一致收敛于问题B_2的解$u(x,y)$②.

如果设

$$u_n(x,y) = \int_\Gamma \left[f_0\frac{\mathrm{d}\Delta w}{\mathrm{d}v} - f_1\Delta w\right]\mathrm{d}s + \int_\Gamma \left[\frac{\mathrm{d}w}{\mathrm{d}v}\sum_{j=1}^{n} c_j^{(2,1)} \frac{\mathrm{d}v_j^{(2,1)}}{\mathrm{d}v} - w\sum_{j=1}^{n} c_j^{(2,2)} v_j^{(2,2)}\right]\mathrm{d}s;$$

---

① 上面求得的只是$\Delta u(t)$的傅里叶表示式，但这对我们应用来说并不受影响.

② 在闭区域$T+\Gamma$上，可以证明右端的级数是在空间$L^2(\Gamma)$的范数意义下一致收敛于问题B_2的解$u(x,y)$.

那么,将有

$$|u(x,y)-u_n(x,y)|=\left|\int_\Gamma\begin{bmatrix}\frac{\mathrm{d}w}{\mathrm{d}v}\left(\Delta u-\sum_{j=1}^{n}c_j^{(2,1)}\frac{\mathrm{d}v_j^{(2,1)}}{\mathrm{d}v}\right)-\\ w\left(\frac{\mathrm{d}\Delta u}{\mathrm{d}v}-\sum_{j=1}^{n}c_j^{(2,2)}v_j^{(2,2)}\right)\end{bmatrix}\mathrm{d}s\right|$$

$$\leqslant\left\{\int_\Gamma\left|\frac{\mathrm{d}w}{\mathrm{d}v}\right|^2\mathrm{d}s\right\}^{\frac{1}{2}}\left\{\int_\Gamma\left|\Delta u-\sum_{j=1}^{n}c_j^{(2,1)}\frac{\mathrm{d}v_j^{(2,1)}}{\mathrm{d}v}\right|^2\mathrm{d}s\right\}^{\frac{1}{2}}+$$

$$\left\{\int_\Gamma|w|^2\mathrm{d}s\right\}^{\frac{1}{2}}\left\{\int_\Gamma\left|\frac{\mathrm{d}\Delta u}{\mathrm{d}v}-\sum_{j=1}^{n}c_j^{(2,2)}v_j^{(2,2)}\right|^2\mathrm{d}s\right\}^{\frac{1}{2}}.$$

但是上式右端第一项的第一个因子对于$(x,y)\in T^*+\Gamma^*\subset T$是一致有界的,而当$n$足够大以后,第二个因子却可以任意小.另外上式右端第二项的第一个因子对于$(x,y)\in T+\Gamma$是一致有界的,而当n足够大以后,第二个因子也可以任意小,从而定理5得证.

§5.

现在转来讨论n重调和方程的情形,n是任意整数.这时候,为了方便,我们可以讨论与问题B_n等价的问题B_n''(参看§2)亦即

问题B_n'':要求根据边界条件

$$u^+=h_0,\ \left(\frac{\mathrm{d}u}{\mathrm{d}v}\right)^+=h_1,\ (\Delta u)^+=h_2,\ \left(\frac{\mathrm{d}\Delta u}{\mathrm{d}v}\right)^+=h_3,\cdots,$$

$$\begin{cases}(\Delta^{\frac{n-1}{2}}u)^+=h_{n-1}, & n\text{ 是奇数},\\ \left(\dfrac{\mathrm{d}\Delta^{\frac{n}{2}-1}u}{\mathrm{d}v}\right)^+=h_{n-1}, & n\text{ 是偶数}\end{cases}\tag{5.1}$$

来确定方程

$$\Delta^n u(x,y)=0\tag{5.2}$$

在区域T内的正则(实)解$u(x,y)$.对于问题B_n''的解$u(x,y)$来说,在公式(3.1)右端出现的前n项,亦即含有函数

$$u,\frac{\mathrm{d}u}{\mathrm{d}v},\Delta u,\frac{\mathrm{d}\Delta u}{\mathrm{d}v},\cdots,\Delta^{\frac{n-3}{2}}u,\frac{\mathrm{d}\Delta^{\frac{n-3}{2}}u}{\mathrm{d}v},\Delta^{\frac{n-1}{2}}u,\quad n\text{ 是奇数},$$

或

$$u,\frac{\mathrm{d}u}{\mathrm{d}v},\Delta u,\frac{\mathrm{d}\Delta u}{\mathrm{d}v},\cdots,\Delta^{\frac{n-2}{2}}u,\frac{\mathrm{d}\Delta^{\frac{n-2}{2}}u}{\mathrm{d}v},\qquad n\text{ 是偶数}$$

的项是已经给定的.为了利用公式(3.1)给出问题B_n的解,只要能够再设

法通过已知条件确定出函数组(3.4a)或(3.4b)——在边界Γ上的值就可以了.

我们下面将要利用数学归纳法进行证明. 因此,如果假设问题B''_{n-1}(或者说与之等价的问题B_{n-1})的解能够利用§4中所讲的方法表示,这就是说,假设能够利用已知条件确定出平均收敛的傅里叶级数,

$$\begin{cases} \sum_{j=1}^{+\infty} c_j^{(n-1,1)} \dfrac{\mathrm{d}\Delta^{\frac{n-3}{2}} v_j^{(n-1,1)}(t)}{\mathrm{d}v} \sim \Delta^{\frac{n-1}{2}} u(t), \\ \sum_{j=1}^{+\infty} c_j^{(n-1,2)} \Delta^{\frac{n-3}{2}} v_j^{(n-1,2)}(t) \sim \dfrac{\mathrm{d}\Delta^{\frac{n-1}{2}} u(t)}{\mathrm{d}v}, \\ \cdots\cdots \\ \sum_{j=1}^{+\infty} c_j^{(n-1,n-2)} \dfrac{\mathrm{d}v_j^{(n-1,n-2)}(t)}{\mathrm{d}v} \sim \Delta^{n-2} u(t), \\ \sum_{j=1}^{+\infty} c_j^{(n-1,n-1)} v_j^{(n-1,n-1)}(t) \sim \dfrac{\mathrm{d}\Delta^{n-2} u(t)}{\mathrm{d}v}, \end{cases} \quad n\text{是奇数}, \tag{5.3a}$$

或

$$\begin{cases} \sum_{j=1}^{+\infty} c_j^{(n-1,1)} \Delta^{\frac{n}{2}-1} v_j^{(n-1,1)}(t) \sim \dfrac{\mathrm{d}\Delta^{\frac{n}{2}-1} u(t)}{\mathrm{d}v}, \\ \sum_{j=1}^{+\infty} c_j^{(n-1,2)} \dfrac{\mathrm{d}\Delta^{\frac{n}{2}-2} v_j^{(n-1,2)}(t)}{\mathrm{d}v} \sim \Delta^{\frac{n}{2}} u(t), \\ \sum_{j=1}^{+\infty} c_j^{(n-1,3)} \Delta^{\frac{n}{2}-2} v_j^{(n-1,3)}(t) \sim \dfrac{\mathrm{d}\Delta^{\frac{n}{2}} u(t)}{\mathrm{d}v}, \\ \cdots\cdots \\ \sum_{j=1}^{+\infty} c_j^{(n-1,n-2)} \dfrac{\mathrm{d}v_j^{(n-1,n-2)}(t)}{\mathrm{d}v} \sim \Delta^{n-2} u(t), \\ \sum_{j=1}^{+\infty} c_j^{(n-1,n-1)} v_j^{(n-1,n-1)}(t) \sim \dfrac{\mathrm{d}\Delta^{n-2} u(t)}{\mathrm{d}v}, \end{cases} \quad n\text{是偶数}. \tag{5.3b}$$

为了说话确定起见,我们将规定n是奇数(n是偶数的情形完全可以类似地讨论).

如果对问题B''_n的解u和方程(5.2)的任意解v应用格林公式,那么成立等式

$$\sum_{k=0}^{n-1} \int_\Gamma \left[\Delta^k v \frac{\mathrm{d}\Delta^{n-k-1} u}{\mathrm{d}v} - \frac{\mathrm{d}\Delta^k v}{\mathrm{d}v} \Delta^{n-k-1} u \right] \mathrm{d}s = 0. \tag{5.4}$$

根据这个公式可以看出,只要我们能够适当地选取方程(5.2)的一组

完备特解系$\{v_j^{(n,1)}(x,y)\}(j\in\mathbf{N}^*)$要求它在边界$\Gamma$上满足条件

$$v^+=0,\left(\frac{\mathrm{d}u}{\mathrm{d}v}\right)^+=0,(\Delta v^+)=0,\left(\frac{\mathrm{d}\Delta u}{\mathrm{d}v}\right)^+=0,\cdots,$$

$$\left(\Delta^{\frac{n-3}{2}}v\right)^+=0,\left(\frac{\mathrm{d}\Delta^{\frac{n-3}{2}}v}{\mathrm{d}v}\right)^+=0. \tag{5.5}$$

那么,显然将有

$$\int_\Gamma \frac{\mathrm{d}\Delta^{\frac{n-1}{2}}u}{\mathrm{d}v}\Delta^{\frac{n-1}{2}}v_j^{(n,1)}\mathrm{d}s=\int_\Gamma h_{n-1}\frac{\mathrm{d}\Delta^{\frac{n-1}{2}}v_j^{(n,1)}}{\mathrm{d}v}\mathrm{d}s+\cdots+$$

$$\int_\Gamma\left[h_0\frac{\mathrm{d}\Delta^{n-1}v_j^{(n,1)}}{\mathrm{d}v}-h_1\Delta^{n-1}v_j^{(n,1)}\right]\mathrm{d}s\equiv c_j^{(n,1)}, \tag{5.6}$$

$$j\in\mathbf{N}^*,$$

其中$c_j^{(n,1)}$是已知的(实)常数．这就提供了通过已知条件来确定函数$\frac{\mathrm{d}^{\frac{n-1}{2}}u(t)}{\mathrm{d}v}$的可能性.

仿照§4中所做,我们首先必须找出方程(5.2)满足条件(5.5)之解的一般表示式.

我们知道[1],方程(5.2)之一般(实)解可以表示成

$$v(x,y)=\mathrm{Re}[\psi_0(z)+\bar{z}\psi_1(z)+\cdots+\bar{z}^{n-1}\psi_{n-1}(z)], \tag{5.7}$$

其中$\psi_k(z)$在区域T内定义的任意全纯函数,$k=0,1,2,\cdots,n-1$.另外,如果要求所有形如$\frac{\partial^{m+k}v}{\partial z^m\partial \bar{z}^k}$ $(m\leqslant n,k\leqslant n)$的导数在闭区域$T+\Gamma$上连续,那么$\psi_k(z)$的$n-1$阶导数在闭区域$T+\Gamma$上还满足赫尔德条件[1].

如果令

$$\tilde{v}(x,y)=\mathrm{Re}[\psi_0(z)+\bar{z}\psi_1(z)+\cdots+\bar{z}^{n-2}\psi_{n-2}(z)],$$

那么,显然,$\tilde{v}(x,y)$将满足方程$\Delta^{n-1}u(x,y)=0$[1],而且根据(5.7)和(5.5)还应该满足边界条件

$$\begin{cases}\tilde{v}(t)=-\mathrm{Re}[\bar{t}^{n-1}\psi_{n-1}(t)],\\ \dfrac{\mathrm{d}\tilde{v}(t)}{\mathrm{d}v}=-\mathrm{Re}\left[\dfrac{\mathrm{d}}{\mathrm{d}v}(\bar{t})^{n-1}\psi_{n-1}(t)\right],\\ \cdots\cdots\\ \dfrac{\mathrm{d}\Delta^{\frac{n-3}{2}}\tilde{v}(t)}{\mathrm{d}v}=-\mathrm{Re}\left[\dfrac{\mathrm{d}}{\mathrm{d}v}\Delta^{\frac{n-3}{2}}(\bar{t})^{n-1}\psi_{n-1}(t)\right],\end{cases} \tag{5.8}$$

也就是说,$\tilde{v}(x,y)$刚好是一个问题B''_{n-1}的解．根据归纳假设知道,它的

解一定可以表示成级数的形式．为了简单，我们把它记作

$$\tilde{v}(x,y)=\mathfrak{K}_{n-1}[\psi_{n-1}], \tag{5.9}$$

这里 $\mathfrak{K}_{n-1}$ 是一个（实的）齐次可加算子．

如果把式(5.9)代入式(5.7)，那么我们将得到表示式

$$v(x,y)=\mathfrak{H}_n[\Psi_{n-1}(z)]\equiv \mathrm{Re}[\bar{z}^{n-1}\psi_{n-1}(z)]+\mathfrak{K}_{n-1}[\psi_{n-1}]. \tag{5.10}$$

显然，这里 $\mathfrak{H}_n$ 也是一个（实的）齐次可加算子．最后我们有以下定理

定理 6 如果 $v(x,y)$ 是方程(5.2)在区域 T 内的正则解，它满足条件(5.5)，那么，将有式(5.10)成立，其中 $\psi_{n-1}(z)$ 是区域 T 内定义的任意全纯函数．

也就是说，方程(5.2)满足条件(5.5)之一般解可以只通过一个任意的全纯函数 $\psi_{n-1}(z)$ 来表示．例如用 z^k，$\mathrm{i}z^k$，$k\in\mathbf{N}$ 分别代替式(5.10)中的全纯函数 $\psi_{n-1}(t)$，然后再进行重新编号，就可以得到方程(5.2)之一组完备特解系 $\{v_j^{(n,1)}(x,y)\}(j\in\mathbf{N}^*)$ 并且它的每一个函数都满足条件(5.5)．这样，我们就造出了一组我们所需要的完备特解系．

类似于§4中那样，利用 $\psi_n(z)$ 的 $n-1$ 阶导数满足赫尔德条件，以及格林函数的性质[1]，不难证明，对这组完备特解系来说，在边界 Γ 上存在导数 $\Delta^{\frac{n-1}{2}}v_j^{(n,1)}(t)$．另外，类似于§4中那样，还可以证明以下定理．

定理 7 函数组 $\{\Delta^{\frac{n-1}{2}}v_j^{(n,1)}(t)\}(j\in\mathbf{N}^*)$ 关于空间 $L^2(\Gamma)$ 是封闭的．即任意给定的 $\varepsilon>0$ 和任何函数 $f(t)\in L^2(\Gamma)$，都可以找到这样的常数组 $\xi_1,\xi_2,\cdots,\xi_n$，使得不等式

$$\left\| f(t)-\sum_{j=1}^{n}\xi_j\Delta^{\frac{n-1}{2}}v_j^{(n,1)}(t)\right\|_{L^2}<\varepsilon$$

成立．

不失一般性，我们可以认为函数组 $\left\{\Delta^{\frac{n-1}{2}}v_j^{(n,1)}(t)\right\}$ 关于空间 $L^2(\Gamma)$ 是标准正交完备系．于是根据(5.6)知道，傅里叶级数 $\sum\limits_{j=1}^{+\infty}c_j^{(n,1)}\Delta^{\frac{n-1}{2}}v_j^{(n,1)}(t)$ 平均收敛于函数 $\dfrac{\mathrm{d}\Delta^{\frac{n-1}{2}}u(t)}{\mathrm{d}v}$．

剩下的工作是设法确定(3.4a)中的其他函数在边界 Γ 上的值．如果令 $\tilde{u}=\Delta u$，那么，显然应该有 $\Delta^{n-1}\tilde{u}=0$．这时候，(3.4a)中需要确定的其他函数可以写作

$$\begin{cases} \dfrac{\mathrm{d}\Delta^{n-1}u}{\mathrm{d}v} = \dfrac{\mathrm{d}\Delta^{n-2}\tilde{u}}{\mathrm{d}v}, \\ \Delta^{n-1}u = \Delta^{n-2}\tilde{u}, \\ \cdots\cdots \\ \dfrac{\mathrm{d}\Delta^{\frac{n+1}{2}}u}{\mathrm{d}v} = \dfrac{\mathrm{d}\Delta^{\frac{n-1}{2}}\tilde{u}}{\mathrm{d}v}, \\ \Delta^{\frac{n+1}{2}}u = \Delta^{\frac{n-1}{2}}\tilde{u}. \end{cases} \tag{5.11}$$

容易看出，这些函数刚好是在解问题 B''_{n-1} 时需要确定的函数，根据归纳假设函数(5.11)在边界上的值是可以通过已知条件确定的，无妨把它们的傅里叶级数分别记作

$$\sum_{j=1}^{+\infty} c_j^{(n,2)} \frac{\mathrm{d}\Delta^{\frac{n-3}{2}} v_j^{(n,2)}(t)}{\mathrm{d}v} \sim \Delta^{\frac{n-1}{2}}\tilde{u}(t) = \Delta^{\frac{n+1}{2}}u(t)$$

$$\sum_{j=1}^{+\infty} c_j^{(n,3)} \Delta^{\frac{n-3}{2}} v_j^{(n,3)}(t) \sim \frac{\mathrm{d}\Delta^{\frac{n-1}{2}}\tilde{u}(t)}{\mathrm{d}v} = \frac{\mathrm{d}\Delta^{\frac{n+1}{2}}u(t)}{\mathrm{d}v},$$

$$\cdots\cdots$$

$$\sum_{j=1}^{+\infty} c_j^{(n,n-1)} \frac{\mathrm{d}v_j^{(n,n-1)}(t)}{\mathrm{d}v} \sim \Delta^{n-2}\tilde{u}(t) = \Delta^{n-1}u(t),$$

$$\sum_{j=1}^{+\infty} c_j^{(n,n)} v_j^{(n,n)}(t) \sim \frac{\mathrm{d}\Delta^{n-2}\tilde{u}(t)}{\mathrm{d}v} = \frac{\mathrm{d}\Delta^{n-1}u(t)}{\mathrm{d}v}.$$

于是不难得到以下定理．

定理 8　对于问题 B''_n 的解 $u(x,y)$ 来说，成立等式

$$\begin{aligned} u(x,y) = {} & \sum_{k=1}^{\frac{n-3}{2}} \int_\Gamma \left[h_{2k} \frac{\mathrm{d}\Delta^{n-k-1}w}{\mathrm{d}v} - h_{2k+1}\Delta^{n-k-1}w \right] \mathrm{d}s + \\ & \int_\Gamma h_{n-1} \frac{\mathrm{d}\Delta^{\frac{n-1}{2}}w'}{\mathrm{d}v} \mathrm{d}s - \int_\Gamma \Delta^{\frac{n-1}{2}}w \sum_{j=1}^{+\infty} c_j^{(n,1)} \Delta^{\frac{n-1}{2}} v_j^{(n,1)} \mathrm{d}s + \\ & \sum_{k=0}^{\frac{n-3}{2}} \int_\Gamma \left[\frac{\mathrm{d}\Delta^{\frac{n-3}{2}-k}w}{\mathrm{d}v} \sum_{j=1}^{+\infty} c_j^{(n,2k+3)} \frac{\mathrm{d}\Delta^{\frac{n-3}{2}-k} v_j^{(n,2k+2)}}{\mathrm{d}v} - \right. \\ & \left. \Delta^{\frac{n-3}{2}-k}w \sum_{j=1}^{+\infty} c_j^{(n,2k+3)} \Delta^{\frac{n-3}{2}-k} v_j^{(n,2k+3)} \right] \mathrm{d}s, \end{aligned}$$

而且上式右端的级数在任何闭区域 $T^* + \Gamma^* \subset T$ 上都是一致收敛于问题 B_n 的解 $u(x,y)$．

对于 n 是偶数的情形有完全类似的结果，不再详细写出．

参考文献

[1] Векуа И Н. Новые методы решения эллиптических уравнений. ОГИЗ Гостехиздат, 1948.

[2] 赵桢．北京师范大学学报(自然科学版),1962,(2):11-18.

[3] 刘来福．北京师范大学学报(自然科学版),1962,(2):19-26.

[4] Мусхелишвили Н И. Сингулярные интеяральные уравнения. Гостехиздат. Москва, 1962.

[5] Векуа И Н. 北京大学数力组,译．广义解析函数．北京:人民教育出版社,1960.

[6] Михлин С Г. Успехи матем. Наук, 3 Вып 3 Стр. 1948,3(25):29-112.

应用数学与计算数学，
1979,(6):53-62.

带位移的奇异积分方程的 Noether 理论

Noether's Theory of Singular Integral Equations with Shifts

§1.　引言

奇异积分方程理论对于很多实际问题来说有着很大意义. 古典的 Fredholm 积分方程理论产生以后，不久 Poincare 和 Hilbert 就开始了对带 Cauchy 核的奇异积分方程的研究，但是这在很长时间内并未引起数学家们应有的重视.

在 20 世纪四五十年代苏联格鲁吉亚学派大大发展了一维奇异积分方程理论. Н. И. Мусхелишвили，И. Н.　Векуа，Н. П. Векуа，В. В. Хведелидзе 等人都做了大量的工作，专著[50]系统地总结了这方面的工作. 关于奇异积分方程和方程组的理论还可以参考专著[5][48]. 关于多维奇异积分方程理论可以参考专著[49].

借助于奇异积分方程这一有力工具，解析函数与广义解析函数边值问题的理论得到了极大地推进[3][7].

早在 20 世纪初 Hilbert，Haseman 以及晚些时候 Carleman 都研究过带位移的边值问题. 对这类边值问题系统地进行研究是由 Н И. Мусхелишвили 在 20 世纪 40 年代开始的. Д. А. Квеселава 对这类边值问题理论做了具有奠基意义的工作[19]~[21]. 由于这类边值问题理论的需要，近十多年来带位移的奇异积分方程理论①也得到了相应的发展. 在苏联第比利斯、罗斯托夫、喀山、明斯克、杜尚别、敖德萨等地都有人做了不少工作. 专著[33]系统地总结了有关带位移的奇异积分方程边值问题的工作，并且给出了大量参考文献.

本文目的主要是综述有关带位移的奇异积分方程理论的某些结果.

假设 Γ 是简单的封闭 Ляпунов 曲线，$a(t)$ 是曲线 Γ 到它自身的同胚变换，以后我们将把 $a(t):\Gamma\to\Gamma$ 叫作位移. 函数 $a(t)$ 实现的变换可以保持或者改变 Γ 的原来方向，前者又叫作正位移，而后者又叫作反位移. 为了容易区别，我们有时候把正、反位移可以分别用 $a_+(t)$，$a_-(t)$ 来表示.

对于某一整数 $n\geqslant 2$，如果满足条件

$$\begin{cases} a_n(l)\equiv l, & t\in\Gamma, \\ a_k(t)\neq t, & k=1,2,\cdots,n-1, \end{cases} \tag{K_n}$$

那么我们将把这样的位移 $\alpha(t)$ 叫作 Carleman 位移，或者说 $a(t)$ 满足条件 (K_n). 除此以外，我们以后还要求 $a'(t)\in H_\mu(\Gamma)$，$a'(t)\neq 0$. 下面将主要介绍带 Carleman 位移的奇异积分方程理论.

讨论奇异积分方程

$$(\mathscr{K}\varphi)(t)=\sum_{k=0}^{n-1}\left\{a_k(t)\varphi[a_k(t)]+\frac{c_k(t)}{\pi \mathrm{i}}\int_\Gamma\frac{\varphi(\tau)}{\tau-a_k(t)}\mathrm{d}\tau\right\}+$$

$$\int_\Gamma K(t,\tau)\varphi(\tau)\mathrm{d}\tau=g(t), \tag{1.1}$$

其中 $a(t)$ 满足条件 (K_n)，$a_k(t)=a[a_{k-1}(t)]$，$1\leqslant k\leqslant n-1$，$a_0(t)\equiv t$. 假设核 $K(t,\tau)$ 在 Γ 上只具有弱奇性，另外方程(1.1)的系数 $a_k(l)$，$c_k(t)\in C(\Gamma)$，而右端 $g(t)\in L_p(\Gamma)$. 有时候为了方便，还假设系数和右端都属于空间 $H_\mu(\Gamma)$.

如果在 Γ 上，$K(t,\tau)\equiv 0$，那么方程(1.1)叫作特征方程. 借助于密度是 $\varphi(t)$ 的 Cauchy 型积分，可以把特征方程归结为与它等价的边值问题

$$\sum_{k=0}^{n-1}\{\tilde{a}_k(t)\Phi^+[a_k(t)]+\tilde{c}_k(t)\Phi^-[a_k(t)]\}=g(t), \tag{1.2}$$

其中 $\tilde{a}_k(t)=a_k(t)+b_k(t)$，$\tilde{c}_k(t)=b_k(t)-a_k(t)$，$\Phi^-(\infty)=0$.

① 这一理论对具有实际意义的邻近理论有着大量的应用，例如，混合型偏微分方程边值问题、正曲率曲面的无穷小变形问题、理想流体的空泡流动理论以及各向异性的弹性理论等.

如果引入算子符号还可以把算子 $\mathscr{K}$ 和边值问题(1.2)改写成如下形式

$$\mathscr{K}\equiv\sum_{k=0}^{n-1}\{a_k(t)\mathscr{W}^k+c_k(t)\mathscr{W}^k\mathscr{S}\}+\mathscr{D},\tag{1.3}$$

$$\sum_{k=0}^{n-1}[\tilde{a}_k(t)\mathscr{W}^k\mathscr{P}-\tilde{c}_k(t)\mathscr{W}^k\mathscr{Q}]\varphi=g,\tag{1.4}$$

这里，
$$(\mathscr{W}^k\varphi)(t)=\varphi[a_k(t)],$$
$$(\mathscr{S}\varphi)(t)=\frac{1}{\pi\mathrm{i}}\int_\Gamma\frac{\varphi(\tau)}{\tau-t}\mathrm{d}\tau,$$
$$(\mathscr{D}\varphi)(t)=\int_\Gamma K(t,\tau)\varphi(\tau)\mathrm{d}\tau,$$
$$\mathscr{P}=\frac{1}{2}(\mathscr{T}+\mathscr{S}),\ \mathscr{Q}=\frac{1}{2}(\mathscr{T}-\mathscr{S}),$$
$\mathscr{T}$ 是恒等算子.

为了讲述直观和简单起见，我们将首先讨论 $n=2$ 的情形，主要原因：这种情况几乎包括了一般情形的所有基本特点，但研究方法却是非常简单的. 然后我们只对一般情形做些必要的说明.

§2. 带 Carleman 位移的奇异积分方程的 Noether 理论

如果没有特殊说明，我们规定的本节中出现的位移 $a(t)$，都是满足条件(K_2)的，即满足条件 $a[a(t)]=t$.

讨论奇异积分方程

$$(\mathscr{K}\varphi)(t)\equiv a(t)\varphi(t)+b(t)\varphi[a(t)]+\frac{c(t)}{\pi\mathrm{i}}\int_\Gamma\frac{\varphi(\tau)}{\tau-t}\mathrm{d}\tau+$$
$$\frac{d(t)}{\pi\mathrm{i}}\int_\Gamma\frac{\varphi(\tau)}{\tau-a(t)}\mathrm{d}\tau+\int_\Gamma K(t,\tau)\varphi(\tau)\mathrm{d}\tau=g(t).\tag{2.1}$$

我们可以在空间 $H_\mu(\Gamma)$ 或者 $L_p(\Gamma)$ 中来讨论算子 $\mathscr{K}$. 为了确定起见，我们先假设算子 $\mathscr{K}$ 是作用在 $H_\mu(\Gamma)$ 中. 利用 §1 中的记号，算子 $\mathscr{K}$ 还可以写成

$$\mathscr{K}\equiv a(t)\mathscr{T}+b(t)\mathscr{W}+c(t)\mathscr{S}+d(t)\mathscr{W}\mathscr{S}+\mathscr{D},\tag{2.2}$$

其中 $\mathscr{D}$ 是完全连续算子. 为了研究带位移的奇异积分方程(2.1)，需要引入由两个不带位移的积分方程组成的方程组，我们将把它叫作对应方程

组[33]. 为此,只需要在方程(2.1)中把 z 换成 $a(t)$,然后再把得到的方程与原来的方程联立,并引入新的未知函数 $\rho_1(t)=\varphi(t),\rho_2(t)=\varphi[a(t)]$,最后就可以得到关于未知函数向量 $\rho(t)=\{\rho_1(t),\rho_2(t)\}$ 的对应方程组①

$$a(t)\rho_1(t)+b(t)\rho_2(t)+\frac{c(t)}{\pi i}\int_\Gamma\frac{\rho_1(\tau)}{\tau-t}d\tau+\frac{vd(t)}{\pi i}\int_\Gamma\frac{a'(\tau)\rho_2(t)}{a(\tau)-a(t)}d\tau+\int_\Gamma K(t,\tau)\rho_1(\tau)d\tau=g(t),$$

$$b[a(t)]\rho_1(t)+a[a(t)]\rho_2(t)+\frac{d[a(t)]}{\pi i}\int_\Gamma\frac{\rho_1(\tau)}{\tau-t}d\tau+\frac{vc[a(t)]}{\pi i}\cdot\int_\Gamma\frac{a'(\tau)\rho_2(t)}{a(\tau)-a(t)}d\tau+v\int_\Gamma K[a(t),a(\tau)]a'(\tau)\rho_2(\tau)d\tau=g[a(t)]\tag{2.3}$$

与 $a(t)$ 对应的是正位移或反位移. 这里的系数 v 分别取值+1 或-1.

为了阐明存在于方程(2.1)和对应方程组(2.3)之间的联系,我们还需要引入奇异积分方程

$$(\mathscr{T}x)(t)\equiv a(t)x(t)-b(t)x[a(t)]+\frac{c(t)}{\pi i}\int_\Gamma\frac{x(\tau)}{\tau-t}d\tau-\frac{d(t)}{\pi i}\int_\Gamma\frac{x(\tau)}{\tau-a(t)}d\tau+\int_\Gamma K(t,\tau)x(\tau)d\tau=0,\tag{2.4}$$

方程(2.4)叫作方程(2.1)的伴随方程. 借助上面讲过的方法,如果假设 $\rho_1(t)=x(t),\rho_2(t)=-x[a(t)]$,由方程(2.4)也可以得到同样的对应方程组(2.3).

最后,还要分别引入方程(2.1)和方程(2.4)的相联方程

$$(\mathscr{K}\psi)(t)\equiv a(t)\psi(t)+va'(t)b[a'(t)]\psi[a(t)]-\frac{1}{\pi i}\int_\Gamma\frac{c(\tau)\psi(\tau)}{\tau-t}d\tau-\frac{v}{\pi i}\int_\Gamma\frac{d[a(\tau)]a'(\tau)\psi[a(\tau)]}{\tau-t}d\tau+\int_\Gamma m(t,\tau)\psi(\tau)d\tau=0,\tag{2.5}$$

$$(\mathscr{T}\omega)(t)\equiv a(t)\omega(t)-va'(t)b[a(t)]\omega[a(t)]-\frac{1}{\pi i}\int_\Gamma\frac{c(\tau)\omega(\tau)}{\tau-t}d\tau+\frac{v}{\pi i}\int_\Gamma\frac{d[a(\tau)]a'(\tau)\omega[a(\tau)]}{\tau-t}d\tau+\int_\Gamma m(t,\tau)\omega(\tau)d\tau=0.\tag{2.6}$$

① 这实际上是带 Cauchy 核的两个奇异积分方程.

容易看出，方程(2.6)刚好是方程(2.5)的伴随方程，而方程(2.5)和(2.6)的对应方程组将是

$$a(t)\omega_1(t)+b[\alpha(t)]\omega_2(t)-\frac{1}{\pi i}\int_\Gamma\frac{c(\tau)\omega_1(\tau)}{\tau-t}d\tau-$$
$$\frac{1}{\pi i}\int_\Gamma\frac{d[\alpha(\tau)]\omega_2(\tau)}{\tau-t}d\tau+\int_\Gamma m(t,\tau)\omega(\tau)d\tau=0,$$
$$b(t)\omega_1(t)+a[\alpha(t)]\omega_2(t)-\frac{v\alpha'(t)}{\pi i}\int_\Gamma\frac{\alpha'(\tau)\omega_1(\tau)}{\alpha(\tau)-\alpha(t)}d\tau-\frac{v\alpha'(t)}{\pi i}\times$$
$$\int_\Gamma\frac{c[\alpha(\tau)]\omega_2(\tau)}{\alpha(\tau)-\alpha(t)}d\tau+v\alpha'(t)\int_\Gamma m[\alpha(\tau),\alpha(t)]\omega_2(\tau)d\tau=0.\tag{2.7}$$

可以验证，方程组(2.7)是对应方程组(2.3)的相联方程组.

方程(2.1)和(2.4)的解与对应方程组(2.3)的解之间存在着密切的联系.首先，如果齐次方程(2.1)有非零解 $\varphi(t)$，那么向量

$$\boldsymbol{\rho}(t)=\{\boldsymbol{\rho}_1(t),\boldsymbol{\rho}_2(t)\},\tag{2.8}$$

其中 $\boldsymbol{\rho}_1(t)=\varphi(t)$，$\boldsymbol{\rho}_2(t)=\varphi[\alpha(t)]$，将是齐次对应方程组(2.3)的解.显然这组解满足条件

$$\boldsymbol{\rho}_2[\alpha(t)]=\boldsymbol{\rho}_1(t).\tag{2.9}$$

同理，如果齐次方程(2.4)有非零解 $x(t)$，那么向量(2.8)也将是齐次对应方程组(2.3)的解，这里 $\boldsymbol{\rho}_1(t)=x(t)$，$\boldsymbol{\rho}_2(t)=-x[\alpha(t)]$.显然，这组解满足条件

$$\boldsymbol{\rho}_2[\alpha(t)]=\boldsymbol{\rho}_1(t).\tag{2.10}$$

反过来，如果向量(2.8)是齐次对应方程组(2.3)的非零解，并且满足条件(2.9)或(2.10)中的一个，那么其第一个分量 $\boldsymbol{\rho}_1(t)$将相应地是齐次方程(2.1)或(2.4)的非零解.于是有以下定理成立：[33]

定理 1　齐次对应方程组(2.3)的线性无关解个数 l 等于齐次方程(2.1)和齐次伴随方程(2.4)的线性无关解个数 l_1 与 l_2 之和，亦即

$$l=l_1+l_2.$$

同理，齐次对应方程组(2.7)与齐次方程(2.5)和(2.6)的线性无关解的个数之间将有

定理 2　l^*，$l_1{}^*$，$l_2{}^*$之间有等式

$$l^*=l_1{}^*+l_2{}^*$$

成立.应该指出的只是代替条件(2.9)和(2.10)，这里要利用条件

$$\omega_2(t) - va'(t)\omega_1[a(t)] = 0, \tag{2.11}$$

和

$$\omega_2(t) + va'(t)\omega_1[a(t)] = 0. \tag{2.12}$$

定理 3 带 Carleman 位移的非齐次方程(2.1)可解的充分必要条件是非齐次对应方程组(2.3)可解.

实际上,假设方程(2.1)有解 $\varphi(t)$,于是向量 $\boldsymbol{\rho}_1(t)=\varphi(t)$, $\boldsymbol{\rho}_2(t)=\varphi[a(t)]$就是对应方程组(2.3)的解. 反过来,假设向量 $\boldsymbol{\rho}(t)=\{\boldsymbol{\rho}_1(t), \boldsymbol{\rho}_2(t)\}$是方程组(2.3)的解,可以验证向量$\{\boldsymbol{\rho}_2[a(t)], \boldsymbol{\rho}_1[a(t)]\}$也是方程组(2.3)的解,从而向量 $\tilde{\boldsymbol{\rho}}(t)=\{\tilde{\boldsymbol{\rho}}_1(t), \tilde{\boldsymbol{\rho}}_2(t)\}$就是方程组(2.3)满足条件(2.9)的解. 这里

$$\tilde{\boldsymbol{\rho}}_1(t)=\frac{1}{2}\{\boldsymbol{\rho}_1(t)+\boldsymbol{\rho}_2[a(t)]\}, \quad \tilde{\boldsymbol{\rho}}_2(t)=\frac{1}{2}\{\boldsymbol{\rho}_2(t)+\boldsymbol{\rho}_1[a(t)]\},$$

从而 $\varphi(t)=\tilde{\boldsymbol{\rho}}_1(t)$就是原方程(2.1)的解.

通过直接验证,还不难得到算子 $\mathscr{K}$ 和 $\mathscr{T}$ 之间具有以下形式的联系:

定理 4 如果 $a(t)$是正位移,那么有

$$u\mathscr{K} - \mathscr{T}u = \mathscr{D}, \tag{2.13}$$

其中 $u=a(t)-t$, $\mathscr{D}$ 是完全连续算子.

定理 5 如果 $a(t)$是反位移,那么有

$$\mathscr{S}\mathscr{K} - \mathscr{T}\mathscr{S} = \mathscr{D},$$

其中 $\mathscr{S}$ 是奇异积分算子,而 $\mathscr{D}$ 是完全连续算子.

根据定理 4 和定理 5 马上得到

定理 6 如果 $\mathscr{K}$ 或 $\mathscr{T}$ 之一是 Noether 算子,那么另一个也必是 Noether 算子,而且 $\operatorname{lnd}\mathscr{K}=\operatorname{lnd}\mathscr{T}$.

为了方便,我们还可以把(2.3)左端写成算子形式

$$\mathscr{M} \equiv \boldsymbol{p}(t)\mathscr{T} + \boldsymbol{q}(t)\mathscr{S} + \mathscr{D}_1, \tag{2.14}$$

这里 $\boldsymbol{p}(t)=\begin{pmatrix} a(t) & b(t) \\ b[a(t)] & a[a(t)] \end{pmatrix}$, $\boldsymbol{q}(t)=\begin{pmatrix} c(t) & vd(t) \\ d[a(t)] & vc[a(t)] \end{pmatrix}$,

而 $\mathscr{D}_1$ 是完全连续算子,或者写成

$$\mathscr{M} \equiv M_2(t,1)\mathscr{P} + M_2(t,-1)\mathscr{Q} + \mathscr{D}_1, \tag{2.15}$$

这里 $M_2(t,j)=\boldsymbol{p}(t)+j\boldsymbol{q}(t)$叫作算子 $\mathscr{M}$ 的标符, $j=\pm 1$,而算子 $\mathscr{M}$ 也叫作对应算子.

我们还引入以下记号

$$\left.\begin{aligned}\Delta_1(t) &= \det[\boldsymbol{p}(t)-\boldsymbol{q}(t)]\\ &= c_1(t)c_1[a(t)]-d_1(t)d_1[a(t)],\\ \Delta_2(t) &= \det[\boldsymbol{p}(t)+\boldsymbol{q}(t)]\\ &= a_1(t)a_1[a(t)]-b_1(t)b_1[a(t)],\end{aligned}\right\} a(t)\text{ 是正位移.}$$

$$\left.\begin{aligned}\Delta(t) &= \det[\boldsymbol{p}(t)-\boldsymbol{q}(t)]\\ &= c_1(t)a_1[a(t)]-b_1(t)d_1[a(t)],\\ \Delta[a(t)] &= \det[\boldsymbol{p}(t)+\boldsymbol{q}(t)]\\ &= a_1(t)c_1[a(t)]-d_1(t)b_1[a(t)],\end{aligned}\right\} a(t)\text{ 是反位移,}$$

其中 $a_1(t)=a(t)+c(t)$，$b_1(t)=b(t)+d(t)$，$c_1(t)=c(t)-a(t)$，$d_1(t)=d(t)-b(t)$.

定理 7　带 Carleman 位移的奇异积分方程(2.1)是 Noether 方程的充分条件是

(1) $\Delta_1(t)\neq 0$，$\Delta_2(t)\neq 0$，如果 $a(t)$是正位移；

(2) $\Delta(t)\neq 0$，如果 $a(t)$是反位移.

事实上，当满足条件 A 或 B 时不难得到 $l_1<+\infty$，$l_1^*<+\infty$. 另外方程组(2.3)可解的充分必要条件是

$$\int_\Gamma G(t)W(t)\mathrm{d}t=0, \tag{2.16}$$

其中 $G(t)=\{g(t),g[a(t)]\}$，而 $W(t)=\{\omega_1(t),\omega_2(t)\}$是相联方程组(2.7)的任意解[5]. 如果 $W(t)$满足条件(2.12)将得到 $2\int_\Gamma g(t)\omega_1(t)\mathrm{d}t=0$，这里 $\omega_1(t)\equiv\psi(t)$是相联方程(2.5)的任意解，如果 $W(t)$满足条件(2.13)可以看出，条件(2.16)将自动满足. 于是，方程(2.1)可解的充分必要条件是

$$\int_\Gamma g(t)\psi_k(t)\mathrm{d}t=0,\quad k=1,2,\cdots,l_1{}^*,$$

其中 $\psi_k(t)(k=1,2,\cdots,l_1{}^*)$是方程(2.5)的基本解系. 这就说明方程(2.1)是 Noether 方程.

定理 8　当满足 Noether 条件时，方程(2.1)的指数可按以下公式来计算：

(1) 如果 $a(t)$是正位移，那么

$$\operatorname{Ind}\mathscr{K}=\frac{1}{4\pi}\left\{\arg\frac{\Delta_1(t)}{\Delta_2(t)}\right\}_\Gamma; \tag{2.17}$$

(2) 如果 $a(t)$是反位移,那么

$$\operatorname{Ind}\mathscr{K}=\frac{1}{2\pi}\{\arg\Delta(t)\}\Gamma. \tag{2.18}$$

事实上,根据定理 1 和定理 2 我们有

$$\operatorname{Ind}\mathscr{M}=\operatorname{Ind}\mathscr{K}+\operatorname{Ind}\mathscr{T},$$

再根据定理 6 马上得到 $\operatorname{Ind}\mathscr{K}=\frac{1}{2}\operatorname{Ind}\mathscr{M}$,从而不难得到定理 8 的结论.

§3. 一般情形下带 Carleman 位移的奇异积分方程的 Noether 理论

如果讨论奇异分方程(1.1),那么对于 $n>2$ 只需要讨论 $a(t)$是正位移情形[33].

可以用与 $n=2$ 情形完全相同的原则建立方程(1.1)的对应方程组.只需要在方程(1.1)中分别用 $a_k(t)$代替 $t,k=1,2,\cdots,n-1$,然后把得到的方程与(1.1)联立,并引入未知向量 $\boldsymbol{\rho}(t)=\{\rho_1(t),\rho_2(t),\cdots,\rho_n(t)\}$,这里 $\rho_k(t)=\varphi[a_{k-1}(t)],k=1,2,\cdots,n$,于是得到对应方程组

$$\sum_{k=0}^{n-1}\left\{a_k[a_{s-1}(t)]\rho_{k+s}(t)+\frac{c_k[a_{s-1}(t)]}{\pi\mathrm{i}}\times\int_\Gamma\frac{a'_{k+s-1}(\tau)\rho_{k+s}(\tau)}{a_{k+s-1}(\tau)-a_{k+s-1}(t)}\mathrm{d}\tau\right\}+$$
$$\int_\Gamma K[a_{s-1}(t),\tau]\rho_1(\tau)\mathrm{d}\tau=g[a_{s-1}(t)],s=1,2,\cdots,n, \tag{3.1}$$

其中 $\rho_{n+v}(t)=\rho_v(t),v=1,2,\cdots,n.$

类似地还要引入 $n-1$ 个伴随方程

$$(\mathscr{T}_jx)(t)\equiv\sum_{k=0}^{n-1}\omega_k^j\left\{a_k(t)x[a_k(t)]+\frac{c_k(t)}{\pi\mathrm{i}}\int_\Gamma\frac{x(\tau)}{\tau-a_k(t)}\mathrm{d}\tau\right\}+$$
$$\int_\Gamma K(t,\tau)x(\tau)\mathrm{d}\tau=0,\ j=1,2,\cdots,n-1, \tag{3.2}$$

其中,$\omega_k=\mathrm{e}^{2\pi\mathrm{i}\frac{k}{n}},k=0,1,2,\cdots,n-1$,以及相联方程

$$(\mathscr{K}'\psi)(t)\equiv\sum_{k=0}^{n-1}\{a_k[a_{n-k}(t)]a'_{n-k}(t)\times\psi[a_{n-k}(t)]-$$
$$\frac{1}{\pi\mathrm{i}}\int_\Gamma\times\frac{c_k[a_{n-k}(\tau)]a'_{n-k}(\tau)\psi[a_{n-k}(\tau)]}{\tau-t}\mathrm{d}\tau\right\}+\int_\Gamma K(t,\tau)\psi(\tau)\mathrm{d}\tau=0 \tag{3.3}$$

就可以得到与§2 中完全相似的一些定理,从而得到相应的 Noether 理论[33].这里就不详细介绍了.

除此以外,还可以讨论更一般的奇异积分方程

$$(\mathscr{K}\psi)(t) \equiv \sum_{k=0}^{n-1}\Big\{a_k(t)\varphi[a_k(t)] + b_k(t)\,\overline{\varphi[a_k(t)]} + \frac{c_k(t)}{\pi i}\times\int_\Gamma \frac{\varphi(\tau)}{\tau - a_k(t)}d\tau + d_k(t)\times\overline{\frac{1}{\pi i}\int_\Gamma \frac{\varphi(\tau)}{\tau - a_k(t)}d\tau}\Big\} + \int_\Gamma K_1(t,\tau)\varphi(\tau)d\tau + \overline{\int_\Gamma K_2(t,\tau)\varphi(\tau)d\tau} = g(t), \quad (3.4)$$

通过建立对应方程组的方法也可以得到相应的 Noether 理论[33].

§4.　奇异积分算子 $\mathscr{K}$ 的 Noether 充分必要条件

类似于 §2,对奇异积分方程(1.1)也引入相应的标符概念,我们把 n 阶函数矩阵

$$\boldsymbol{\sigma}\mathscr{K}(t,j) = \begin{pmatrix} a_0(t) & a_1(t) & \cdots & a_{n-1}(t) \\ a_{n-1}[a(t)] & a_0[a(t)] & \cdots & a_{n-2}[a(t)] \\ \vdots & \vdots & & \vdots \\ a_1[a_{n-1}(t)] & a_2[a_{n-1}(t)] & \cdots & a_0[a_{n-1}(t)] \end{pmatrix} + j\begin{pmatrix} c_0(t) & c_1(t) & \cdots & c_{n-1}(t) \\ c_{n-1}[a(t)] & c_0[a(t)] & \cdots & c_{n-2}[a(t)] \\ \vdots & \vdots & & \vdots \\ c_1[a_{n-1}(t)] & c_2[a_{n-1}(t)] & \cdots & c_0[a_{n-1}(t)] \end{pmatrix} \quad (4.1)$$

叫作算子 $\mathscr{K}$ 的标符,这里 $t\in\Gamma$, $j=\pm 1$.

为了讲述直观和简单起见,我们将只讨论 $n=2$ 的情形.

为此我们定义方程(2.1)的标符

$$\boldsymbol{\sigma}\mathscr{K}(t,j) = \begin{pmatrix} a(t) & b(t) \\ b[a(t)] & a[a(t)] \end{pmatrix} + j\begin{pmatrix} c(t) & v d(t) \\ d[a(t)] & vc[a(t)] \end{pmatrix}, \quad (4.2)$$

$t\in\Gamma$, $j=\pm 1$,而 $v=+1$ 或 $v=-1$ 要依赖于位移 $a(t)$ 是正位移还是反位移.

应该指出,在 §2 中得到的结果只是 Noether 充分条件,现在来证明 Noether 条件也是必要的. 为此,我们在 $L_p^2(\Gamma)$, $p>1$ 中讨论带 Cauchy 核的对应奇异积分算子

$$\mathscr{M} \equiv \boldsymbol{p}_2(t)\mathscr{T}_2 + \boldsymbol{q}_2(t)\mathscr{S}_2 + \mathscr{D}_2, \quad (4.3)$$

其中　$\boldsymbol{p}_2(t) = \begin{pmatrix} a(t) & b(t) \\ b[a(t)] & a[a(t)] \end{pmatrix}$,

$$\boldsymbol{q}_2(t)=\begin{pmatrix} c(t) & vd(t) \\ d[a(t)] & vc[a(t)] \end{pmatrix},$$

$$\mathscr{T}_2(\rho_1,\rho_2)=(\mathscr{T}\rho_1,\mathscr{T}\rho_2),$$

$$\mathscr{S}_2(\rho_1,\rho_2)=(\mathscr{S}\rho_1,\mathscr{S}\rho_2),$$

$$\mathscr{D}_2(\rho_1,\rho_2)=(d(t)[\mathscr{W}\mathscr{S}\mathscr{W}-v\mathscr{S}]\rho_2,\ c[a(t)][\mathscr{W}\mathscr{S}\mathscr{W}-v\mathscr{S}]\rho_2).$$

首先证明两个引理：

引理 4.1 二维向量$\{\boldsymbol{\varphi}_1(t),\boldsymbol{\varphi}_2(t)\}$的空间 $L_p^2(\Gamma)$可以分解成分别由二维向量$\{\boldsymbol{\rho}(t),\boldsymbol{\rho}[a(t)]\}$和$\{\boldsymbol{\rho}(t),-\boldsymbol{\rho}[a(t)]\}$组成的子空间 $L_p^{2,0}(\Gamma)$和 $L_p^{2,1}(\Gamma)$的直接和.

显然 $L_p^{2,0}(\Gamma)\cap L_p^{2,1}(\Gamma)=\{0\}$，而每一个元素 $\varphi=\{\varphi_1,\varphi_2\}\in L_p^2(\Gamma)$都可以表示成 $\varphi=\psi_1+\psi_2$ 的形式，其中

$$\psi_1(t)=\left\{\frac{\varphi_1(t)+\varphi_2[a(t)]}{2},\frac{\varphi_1[a(t)]+\varphi_2(t)}{2}\right\},$$

$$\psi_2(t)=\left\{\frac{\varphi_1(t)-\varphi_2[a(t)]}{2},\frac{\varphi_2(t)-\varphi_1[a(t)]}{2}\right\}.$$

明显看出，$\psi_1\in L_p^{2,0}(\Gamma)$，而 $\psi_2\in L_p^{2,1}(\Gamma)$.

引理 4.2 子空间 $L_p^{2,0}(\Gamma)$和 $L_p^{2,1}(\Gamma)$关于算子 $\mathscr{M}$ 是不变的，算子 $\mathscr{M}$ 到子空间 $L_p^{2,0}(\Gamma)$的收缩 $\mathscr{M}_0=\mathscr{M}|_{L_p^{2,0}(\Gamma)}$ 在 Noether 意义下与算子 $\mathscr{K}$ 等价，而算子 $\mathscr{M}$ 到子空间 $L_p^{2,1}(\Gamma)$的收缩 $\mathscr{M}_1=\mathscr{M}|_{L_p^{2,1}(\Gamma)}$ 在 Noether 意义下与伴随算子 $\mathscr{T}$ 等价.

实际上，不失一般性，可以把 $\mathscr{K}$ 和 $\mathscr{T}$ 都看成特征算子(即 $\mathscr{D}\equiv0$). 于是只要证明：

$$\mathscr{M}_0(\rho,\rho(a))=\mathscr{M}(\rho,\rho(a))=(g_1,g_2)=(\mathscr{K}\rho,\mathscr{W}\mathscr{K}\rho),\tag{4.4}$$

$$\mathscr{M}_1(\rho,-\rho(a))=\mathscr{M}(\rho,-\rho(a))=(g_1{}^*,g_2{}^*)=(\mathscr{T}\rho,\ -\mathscr{W}\mathscr{T}\rho).\tag{4.5}$$

事实上，我们有

$$\begin{aligned} g_1 &= a\rho+b\rho(a)+c\mathscr{S}\rho+vd\,\mathscr{S}\rho(a)+d(\mathscr{W}\mathscr{S}\mathscr{W}-v\mathscr{S})\rho(a) \\ &= a\rho+b\mathscr{W}\rho+c\mathscr{S}\rho+d\mathscr{W}\mathscr{S}\rho=\mathscr{K}\rho, \\ g_2 &= b(a)\rho+a(a)\rho(a)+d(a)\mathscr{S}\rho+vc(a)\mathscr{S}\rho(a)+c(a)(\mathscr{W}\mathscr{S}\mathscr{W}-v\mathscr{S})\rho(a) \\ &= \mathscr{W}a\rho+\mathscr{W}b\mathscr{W}\rho+\mathscr{W}c\mathscr{S}\rho+\mathscr{W}d\mathscr{W}\mathscr{S}\rho=\mathscr{W}\mathscr{K}\rho, \\ g_1{}^* &= a\rho-b\rho(a)+c\mathscr{S}\rho-vd\,\mathscr{S}\rho(a)-d(\mathscr{W}\mathscr{S}\mathscr{W}-v\mathscr{S})\rho(a) \\ &= a\rho-b\mathscr{W}\rho+c\mathscr{S}\rho-d\mathscr{W}\mathscr{S}\rho=\mathscr{T}\rho, \end{aligned}$$

$$g_2{}^* = b(a)\rho - a(a)\rho(a) + d(a)\mathscr{S}\rho - v c(a)\mathscr{S}\rho(a) - c(a)(\mathscr{W}\mathscr{S}\mathscr{W} - v\mathscr{S})\rho(a)$$
$$= -\mathscr{W}a\rho + \mathscr{W}b\mathscr{W}\rho - \mathscr{W}c\mathscr{S}\rho + \mathscr{W}d\mathscr{W}\mathscr{S}\rho = -\mathscr{W}\mathscr{T}\rho.$$

这样,我们证明了对应算子 $\mathscr{M}$ 能够表成分别作用在空间 $L_p^{2,0}(\Gamma)$ 和 $L_p^{2,1}(\Gamma)$ 上的算子 $\mathscr{M}_0$ 和 $\mathscr{M}_1$ 的直接和,并且在 Noether 意义下,它们分别等价于算子 $\mathscr{K}$ 和 $\mathscr{T}$.

定理 9　为了使算子

$$\mathscr{K} \equiv a(t)\mathscr{T} + b(t)\mathscr{W} + c(t)\mathscr{S} + d(t)\mathscr{W}\mathscr{S} + \mathscr{D}$$

是 Noether 算子的充分必要条件是它的标符为非退化的,即

$$\det\boldsymbol{\sigma}\mathscr{K}(t,j) \neq 0, \quad t \in \Gamma, \quad j = \pm 1, \tag{4.6}$$

而算子 $\mathscr{K}$ 的指数是

$$\operatorname{lnd}\mathscr{K} = \frac{1}{4\pi}\left\{\arg\frac{\det\boldsymbol{\sigma}\mathscr{K}(t,-1)}{\det\boldsymbol{\sigma}\mathscr{K}(t,1)}\right\}_\Gamma \tag{4.7}$$

证　首先证明充分性. 根据定义显然有

$$\boldsymbol{\sigma}\mathscr{K}(t,j) = \boldsymbol{\sigma}\mathscr{M}(t,j).$$

由条件(4.6)得到对应算子 $\mathscr{M}$ 是 Noether 算子,再根据引理 4.1 和引理 4.2 知道,算子 $\mathscr{M}$ 是算子 $\mathscr{M}_0$ 和 $\mathscr{M}_1$ 的直接和,从而算子 $\mathscr{M}_0 = \mathscr{M}|_{L_\rho^{2,0}}(\Gamma)$ 也是 Noether 算子,即 $\mathscr{K}$ 是 Noether 算子.

下面再来证明必要性. 假设算子 $\mathscr{K}$ 是 Noether 算子,容易知道,这时 $\mathscr{T}$ 也是 Noether 算子. 再根据引理 4.1 和引理 4.2 知道,算子 $\mathscr{M}_0$ 和 $\mathscr{M}_1$ 从而还有算子 $\mathscr{M}$ 都是 Noether 算子. 于是有(4.6)成立.

再利用 $\operatorname{lnd}\mathscr{M} = 2\operatorname{lnd}\mathscr{K}$ 容易推得公式(4.7).

可以证明带 Carleman 位移的奇异积分算子 $\mathscr{K}$ 构成一个代数 $\mathscr{U}$,而标符 $\boldsymbol{\sigma}\mathscr{K}(t,j)$ 作为 n 阶函数矩阵也构成一个代数 M. 如果用 D 代表所有完全连续算子 $\mathscr{D}$ 所组成的集合(它可以作为代数 $\mathscr{U}$ 的双侧理想),于是有以下结论:

商代数 $\mathscr{U}/D$ 与标符代数 M 是同构的.

最后指出,作者还讨论了带两个 Carleman 位移的奇异积分方程,并利用类似的方法建立了相应的 Noether 理论.

关于带非 Carleman 位移的奇异积分方程理论,目前还很不完整,可以参看[33](第九章).

参考文献

[1] Вашкарев П Г, Карлович Ю И, Нечаев А П. ДАН СССР, 1974, 219(2): 272-274.

[2] Василевский Н Л, Шапиро М В, Укр матем. ж., 1975, 27(2): 216-223.

[3] Векуа И Н. Обобшенные аналитические функции. 1959.（有中译本，人民教育出版社）

[4] Векуа И Н. Сообщ. Груа ССР, 1948, 3: 153-160.

[5] Векуа И Н. Системы сингулярных интеяралъных уравнений и некоторые граничные задачи. 1970.（1950年版有中译本，上海科技出版社）

[6] Гахов ф Д. Дифференп УР-НИЯ, 1966, 2(4): 533-543.

[7] Гахов ф Д. Краевые задачи, 1977.

[8] Гохберг И Ц, Крупник Н Я. Изв. АН СССР сер матем., 1971, 35(4): 940-964.

[9] Гохберг И Ц, Крупник Н Я. функл әнәп из, 1970, 4(3): 27-38.

[10] Гохберг И Ц, Крупник Н Я. Введение в теорию одномерных сингуральвых интеяральных операторов, 1973.

[11] Гохберя И Ц, Крупник Н Я. Изв АН Арм. ССР матем., 1973, 8(1): 3-12.

[12] Гохберя И Ц, Крупник Н Я. Матемтические исследования (Кишииев), 1973, 8(2): 170-175.

[13] Исахавов Р С. Сообщ. АН Груз. ССР, 1958, 20(1): 9-12.

[14] Карапетянп Н К, Самко С Г. ДАН СССР, 1972, 202(2): 273-276.

[15] Карапетянп Н К, Самко С Г. ДАН СССР, 1972, 204(3): 536-539.

[16] Карапетянп Н К, Самко С Г. ДАН СССР, 1973, 211(2): 281-284.

[17] Карлович Ю И. ТР. матем. ин-та АН Груз. ССР, 1977, 44: 113-124.

[18] Карлович Ю И. ДАН СССР, 1974, 216(1): 32-35.

[19] Квеселава Д А. ДАН СССР, 1946, 53(8): 683-686.

[20] Квеселава Д А. Сообщ. АН Груз. ССР, 1946, 7(9—10): 609-614.

[21] Квеселава Д А. ТР. матем. ин-та АН Груз. ССР, 1948, 16: 39-80.

[22] Кордзадза Р А. ДАН СССР, 1964, 154(6): 1 250-1 253.

[23] Кордзадза Р А. ДАН СССР, 1965, 160(6): 1 242-1 243.

[24] Кордзадза Р А. ДАН СССР, 1966, 168(6): 1 245-1 248.

[25] Кордзадза Р А. ДАН СССР, 1969, 185(4): 753-756.

[26] Кравченко В Г. ДАН СССР, 1971, 201(6): 1 275-1 278.

[27] Кравчеко В Г. Укр. матем. ж., 1972, 24(6): 752-762.

[28] Кравченко В Г. ДАН СССР,1974,215(6):1 301-1 304.

[29] Кравченко В Г,Литвинчук Г С. Труды симпозиума по механике сплошной среды и родственным проблехам анализа,Том1,142-152.

[30] Кравченко В Г,Литвинчук Г С. Укр. матем. ж. ,1973,25(4):541-545.

[31] Крейн С Г. Линейвые уравненияө банаховом пространстве.

[32] Крулвик Н Я,Няга В И. Сообш. АН Гртуз. ССР,1974,76(1):25-28.

[33] Литвинчук Г С. Краевые задачии сингулярные интегральные уравнения со сдвигом,1977.

[34] Литвинчук Г С. V Всесоюзная конференция по теории функций,тезисы докладов,Изд-во АН Арм. ССР,Ереван,1960:65-66.

[35] Литвинчук Г С. ДАН СССР,1960,134(6):1 295-1 298.

[36] Литвинчук Г С. Научн. сообщ. РГУ за 1963г. ,Ростов-Надону,1964:11-13.

[37] Литвинчук Г С. ДАН СССР,1965,162(1):26-29.

[38] Литвинчук Г С. Тезисы научн. конф. ,посвящ. 100-Петию Одесск. ун-та,Одесса,1965:22-24.

[39] Литвинчук Г С. Тезисы докәплов 21 научн. конф. Одесск. ун-та, Одесса,1966:10-11.

[40] Литвинчук Г С. Изв. АН СССР сер. матем. ,1967,31(3):563-586.

[41] Литвинчук Г С. Изв. АН СССР сер. матем. ,1968,32(6):1 414-1 417.

[42] Литвинчук Г С. Хасабов Э Г. ДАН СССР,1961,140(1):48-51.

[43] Литвинчук Г С. Хасабов Э Г. ДАН СССР,1962,145(4):731-734.

[44] Литвинчук Г С. Хасабов Э Г. Сибир. матем. ж. ,1964,5(3):608-625.

[45] Литвинчук Г С. Хасабов Э Г. Сибир. матем. ж. ,1964,5(4):858-880.

[46] Мавлжавидзе Г Ф. Сообщ. АН Груз ССР,1950,11(6):351-356.

[47] Мавлжавидзе Г Ф. Тр. международв. симпозиума в Тбилиси 17-23 сент 1963. т. ,1965,1:237-247.

[48] Михлин С Г. Сигулярные интеяралвые уравнения,УМНЭ,1948,3(25).(有中译本,数学进展,1958,4(1))

[49] Михлив С Г. Многомерные сингулярвые интсгралы и интеральные уравнения,1962.(有中译本,上海科技出版社)

[50] Мусхелишвили Н И. Сингулярные интеяральные уравнения,1968.(1962 年版有中译本,上海科技出版社)

[51] Сазонов Л Н. Матем. заметки,1973,13(3):385-393.

[52] Сосунов А С. Нзв. вузов,матем. ,1967,4:103-111.

[53] Сосунов А С. Материалы республ конф матем. Велоруссии,Минск,1967.

[54] Сосунов А С. Сообщ. на конф. Рост. матем. об-ва, Ростов-на-дону, 1968: 84-90.

[55] Сосунов А С. Материалы всесоюзн. конф. по каревым задазам,Казань,1970: 249-253.

[56] Шапиро М В. Сообщ. АН Груз. ССР,1973,71(1):37-40.

[57] Шапиро М В. Сибир. матем. ж. ,1975,16(1):158-168.

[58] Kuczma M. Functional equations in a singie variabie. Polish Scientific publishers,Warszawa,1968.

[59] Orth D. Singular integral equations with shifts, Journ. Math and Mech. , 1968,18(6):491-515.

[60] Orth D. singular integral equations with shifts,11,Indiana Univ. Math. Journ. , 1971,20(7):603-621.

北京师范大学学报(自然科学版),
1980,(2):1-18.

带两个 Carleman 位移的奇异积分方程的可解性问题①

On the Problem of Solvability of Singular Integral Equations with Two Carleman's Shifts

Abstract In this paper the problem of solvability of singular integral equations with two Carleman's shifts (equation(1.1)) is considered using the method of corresponding system of equations.

Suppose that Γ is a closed simple Lyapunoff's curve and $\alpha(t)$, $\beta(t)$, which both satisfy Carleman's conditions and $\alpha[\beta(t)]=\beta[\alpha(t)]$, are two different homeomorphisms of Γ onto itself. Coefficients $a_k(t)$, $b_k(t)$, $k=0,1,2,3$ and $g(t)$ of equation (1.1) all belong to the space $H_\mu(\Gamma)$.

The following main results are obtained.

(1) Singular integral equation(1.1) is solvable, if the Noether's conditions

$$\det(\boldsymbol{p}(t)\pm\boldsymbol{q}(t))\neq 0$$

are satisfied.

(2) The index of the singular integral equation (1.1) will be calculated by the formula

$$\operatorname{Ind}\mathcal{K}=\frac{1}{8\pi}\left\{\arg\frac{\det(\boldsymbol{p}(t)-\boldsymbol{q}(t))}{\det(\boldsymbol{p}(t)+\boldsymbol{q}(t))}\right\}_\Gamma$$

where $\boldsymbol{p}(t)$ and $\boldsymbol{q}(t)$ are matrices of coefficients of the corresponding system of equations.

① 收稿日期:1979-09-04.

带 Carleman 位移的奇异积分方程理论近年来得到了很大发展. 在[1]～[3]中完整地建立了这种奇异积分方程的 Noether 理论. 所用的基本方法是建立所谓的对应方程组(是不带位移的奇异积分方程组,它的理论是已知的,参看[5][6]). 本文目的是利用类似方法解决带两个位移的奇异积分方程的可解性问题.

§1.

假设 $\Gamma=\Gamma_0+\Gamma_1+\Gamma_2+\cdots+\Gamma_m$ 是由 $m+1$ 条简单闭 Ляпунов 曲线组成,它围出一个连通区域 D^+. 我们用 D^- 表示闭区域 $D^+\cup\Gamma$ 到全平面的补集. 假设 $\alpha(t),\beta(t)$ 是把 Γ 仍映射成它自身的两个不同的同胚,它们可以是正位移或反位移,并规定它们都满足 Carleman 条件 K_2,即满足 $\alpha[\alpha(t)]=t,\beta[\beta(t)]=t$,此外还假设 $\alpha[\beta(t)]=\beta[\alpha(t)]$,为了方便,我们记

$$\gamma(t)=\alpha[\beta(t)]=\beta[\alpha(t)].$$

显然,$\gamma(t)$ 也是满足 Carleman 条件 K_2 的位移.

关于位移 $\alpha(t),\beta(t)$ 我们还要求 $\alpha'(t),\beta'(t)\in H_\mu(\Gamma)$,而且

$$\alpha'(t)\cdot\beta'(t)\neq 0.$$

我们将讨论带两个 Carleman 位移的奇异积分方程

$$(\mathscr{K}\varphi)(t)\equiv a_0(t)\varphi(t)+a_1(t)\varphi[a(t)]+a_2(t)\varphi[\beta(t)]+a_3(t)\varphi[\gamma(t)]+$$
$$\frac{b_0(t)}{\mathrm{i}\pi}\int_\Gamma\frac{\varphi(\tau)}{\tau-t}\mathrm{d}\tau+\frac{b_1(t)}{\mathrm{i}\pi}\int_\Gamma\frac{\varphi(\tau)}{\tau-a(t)}\mathrm{d}\tau+\frac{b_2(t)}{\mathrm{i}\pi}\int_\Gamma\frac{\varphi(\tau)}{\tau-\beta(t)}\mathrm{d}\tau+$$
$$\frac{b_3(t)}{\mathrm{i}\pi}\int_\Gamma\frac{\varphi(\tau)}{\tau-\alpha(t)}\mathrm{d}\tau+\int_\Gamma K(t,\tau)\varphi(\tau)\mathrm{d}\tau=g(t),\tag{1.1}$$

其中 $a_k(t),b_k(t),k=0,1,2,3$ 及 $g(t)$ 都属于空间 $H_\mu(\Gamma)$,而 $K(t,\tau)$ 最多只具有弱奇性. 我们还可以把(1.1) 写成算子形式

$$\mathscr{K}\equiv a_0(t)\mathscr{I}+a_1(t)\mathscr{W}+a_2(t)_2\mathscr{W}_2+a_3(t)\mathscr{W}_3+b_0(t)\mathscr{I}+$$
$$b_1(t)\mathscr{W}_1\mathscr{I}+b_2(t)\mathscr{W}_2\mathscr{I}+b_3(t)\mathscr{W}_3\mathscr{I}+\mathscr{D},\tag{1.2}$$

其中 $\mathscr{I}$ 是恒等算子,$\mathscr{D}$ 是完全连续算子,

$$(\mathscr{W}_1\varphi)(t)\equiv\varphi[\alpha(t)],\ (\mathscr{W}_2\varphi)(t)\equiv\varphi[\beta(t)],\ (\mathscr{W}_3\varphi)(t)\equiv\varphi[\gamma(t)],$$

$$(\mathscr{S}\varphi)(t)\equiv\frac{1}{\mathrm{i}\pi}\int_\Gamma\frac{\varphi(t)}{\tau-t}\mathrm{d}\tau,\ (\mathscr{W}_1\mathscr{S}\varphi)(t)\equiv\frac{1}{\mathrm{i}\pi}\int_\Gamma\frac{\varphi(t)}{\tau-a(t)}\mathrm{d}\tau,$$

$$(\mathscr{W}_2\mathscr{S}\varphi)(t)\equiv\frac{1}{\mathrm{i}\pi}\int_\Gamma\frac{\varphi(t)}{\tau-\beta(t)}\mathrm{d}\tau,\ (\mathscr{W}_3\mathscr{S}\varphi)(t)\equiv\frac{1}{\mathrm{i}\pi}\int_\Gamma\frac{\varphi(t)}{\tau-\gamma(t)}\mathrm{d}\tau.$$

如果 $K(t,\tau)\equiv 0$，我们将得到特征方程

$$(\mathscr{K}^{\circ}\varphi)(t)=g(t). \tag{1.3}$$

我们把方程(1.1)中的 t 分别换成 $\alpha(t)$，$\beta(t)$ 和 $\gamma(t)$，并把这样得到的三个方程与(1.1)联立，再假设

$\varphi(t)=\rho_1(t)$，$\varphi[\alpha(t)]=\rho_2(t)$，$\varphi[\beta(t)]=\rho_3(t)$，$\varphi[\gamma(t)]=\rho_4(t)$，

最后得到方程组

$$\begin{aligned}
&a_0(t)\rho_1(t)+a_1(t)\rho_2(t)+a_2(t)\rho_3(t)+a_3(t)\rho_4(t)+\frac{b_0(t)}{\mathrm{i}\pi}\int_\Gamma\frac{\rho_1(\tau)}{\tau-t}\mathrm{d}\tau+\\
&\frac{\nu_1 b_1(t)}{\mathrm{i}\pi}\int_\Gamma\frac{\alpha'(\tau)}{\alpha(\tau)-\alpha(t)}\rho_2(\tau)\mathrm{d}\tau+\frac{\nu_2 b_2(t)}{\mathrm{i}\pi}\int_\Gamma\frac{\beta'(\tau)}{\beta(\tau)-\beta(t)}\rho_3(\tau)\mathrm{d}\tau+\\
&\frac{\nu_1\nu_2 b_3(t)}{\mathrm{i}\pi}\int_\Gamma\frac{\gamma'(\tau)}{\gamma(\tau)-\gamma(t)}\rho_4(t)\mathrm{d}\tau+\int_\Gamma K(t,\tau)\rho_1(\tau)\mathrm{d}\tau=g(t),\\
&a_1[\alpha(t)]\rho_1(t)+a_0[\alpha(t)]\rho_2(t)+a_3[\alpha(t)]\rho_3(t)+a_2[\alpha(t)]\rho_4(t)+\\
&\frac{b_1[\alpha(t)]}{\mathrm{i}\pi}\int_\Gamma\frac{\rho_1(\tau)}{\tau-t}\mathrm{d}\tau+\frac{\nu_1 b_0[\alpha(t)]}{\mathrm{i}\pi}\int_\Gamma\frac{\alpha'(\tau)}{\alpha(\tau)-\alpha(t)}\rho_2(\tau)\mathrm{d}\tau+\\
&\frac{\nu_2 b_3[\alpha(t)]}{\mathrm{i}\pi}\int_\Gamma\frac{\beta'(\tau)}{\beta(\tau)-\beta(t)}\rho_3(\tau)\mathrm{d}\tau+\frac{\nu_1\nu_2 b_2[\alpha(t)]}{\mathrm{i}\pi}\int_\Gamma\frac{\gamma'(\tau)}{\gamma(\tau)-\gamma(t)}\rho_4(\tau)\mathrm{d}\tau+\\
&\int_\Gamma K[\alpha(t),\tau]\rho_1(\tau)\mathrm{d}\tau=g[\alpha(t)], \qquad\qquad (1.4)\\
&a_2[\beta(t)]\rho_1(t)+a_3[\beta(t)]\rho_2(t)+a_0[\beta(t)]\rho_3(t)+a_1[\beta(t)]\rho_4(t)+\\
&\frac{b_2[\beta(t)]}{\mathrm{i}\pi}\int_\Gamma\frac{\rho_1(\tau)}{\tau-t}\mathrm{d}\tau+\frac{\nu_1 b_3[\beta(t)]}{\mathrm{i}\pi}\int_\Gamma\frac{\alpha'(\tau)}{\alpha(\tau)-\alpha(t)}\rho_2(\tau)\mathrm{d}\tau+\\
&\frac{\nu_2 b_0[\beta(t)]}{\mathrm{i}\pi}\int_\Gamma\frac{\beta'(\tau)}{\beta(\tau)-\beta(t)}\rho_3(\tau)\mathrm{d}\tau+\frac{\nu_1\nu_2 b_1[\beta(t)]}{\mathrm{i}\pi}\int_\Gamma\frac{\gamma'(\tau)}{\gamma(\tau)-\gamma(t)}\rho_4(\tau)\mathrm{d}\tau+\\
&\int_\Gamma K[\beta(t),\tau]\rho_1(\tau)\mathrm{d}\tau=g[\beta(t)],\\
&a_3[\gamma(t)]\rho_1(t)+a_2[\gamma(t)]\rho_2(t)+a_1[\gamma(t)]\rho_3(t)+a_0[\gamma(t)]\rho_4(t)+\\
&\frac{b_3[\gamma(t)]}{\mathrm{i}\pi}\int_\Gamma\frac{\rho_1(\tau)}{\tau-t}\mathrm{d}\tau+\frac{\nu_1 b_2[\gamma(t)]}{\mathrm{i}\pi}\int_\Gamma\frac{\alpha'(\tau)}{\alpha(\tau)-\alpha(t)}\rho_2(\tau)\mathrm{d}\tau+\\
&\frac{\nu_2 b_1[\gamma(t)]}{\mathrm{i}\pi}\int_\Gamma\frac{\beta'(\tau)}{\beta(\tau)-\beta(t)}\rho_3(\tau)\mathrm{d}\tau+\frac{\nu_1\nu_2 b_0[\gamma(t)]}{\mathrm{i}\pi}\int_\Gamma\frac{\gamma'(\tau)}{\gamma(\tau)-\gamma(t)}\rho_4(\tau)\mathrm{d}\tau+\\
&\int_\Gamma K[\gamma(t),\tau]\rho_1(\tau)\mathrm{d}\tau=g[\gamma(t)],
\end{aligned}$$

这里 $\nu_1=+1$ 或 -1 依赖于 $\alpha(t)$ 是正位移或反位移，而 $\nu_2=+1$ 或 -1 依赖于 $\beta(t)$ 是正位移或反位移. 这个方程组实际上是不带位移的奇异积分

方程组[1]，如果写成算子形式，将有

$$\mathscr{M} \equiv \boldsymbol{p}(t)\mathscr{I} + \boldsymbol{q}(t)\mathscr{S} + \mathscr{D}^* \tag{1.5}$$

其中

$$\boldsymbol{p}(t) = \begin{pmatrix} a_0(t) & a_1(t) & a_2(t) & a_3(t) \\ a_1[\alpha(t)] & a_0[\alpha(t)] & a_3[\alpha(t)] & a_2[\alpha(t)] \\ a_2[\beta(t)] & a_3[\beta(t)] & a_0[\beta(t)] & a_1[\beta(t)] \\ a_3[\gamma(t)] & a_2[\gamma(t)] & a_1[\gamma(t)] & a_0[\gamma(t)] \end{pmatrix},$$

$$\boldsymbol{q}(t) = \begin{pmatrix} b_0(t) & \nu_1 b_1(t) & \nu_2 b_2(t) & \nu_1\nu_2 b_3(t) \\ b_1[\alpha(t)] & \nu_1 b_0[\alpha(t)] & \nu_2 b_3[\alpha(t)] & \nu_1\nu_2 b_2[\alpha(t)] \\ b_2[\beta(t)] & \nu_1 b_3[\beta(t)] & \nu_2 b_0[\beta(t)] & \nu_1\nu_2 b_1[\beta(t)] \\ b_3[\gamma(t)] & \nu_1 b_2[\gamma(t)] & \nu_2 b_1[\gamma(t)] & \nu_1\nu_2 b_0[\gamma(t)] \end{pmatrix},$$

$\mathscr{D}^*$ 是完全连续算子，或者有

$$\mathscr{M} \equiv M(t,1)\mathscr{P} + M(t,-1)\mathscr{Q} + \mathscr{D}^*,$$

其中 $M(t,j) = \boldsymbol{p}(t) + j\boldsymbol{q}(t)$，$j = \pm 1$，而

$$\mathscr{P} = \frac{1}{2}(\mathscr{I} + \mathscr{S}),\ \mathscr{Q} = \frac{1}{2}(\mathscr{I} - \mathscr{S}).$$

以后我们把方程组(1.4)叫作方程(1.1)的对应方程组，(1.5)叫作对应算子，而 $M(t,j)$ 叫作算子 $\mathscr{M}$ 的标符.

为了方便，我们还引入方程(1.1)的三个伴随方程

$$(\mathscr{T}_1 x)(t) \equiv a_0(t)x(t) - a_1(t)x[\alpha(t)] - a_2(t)x[\beta(t)] + a_3(t)x[\gamma(t)] + \frac{b_0(t)}{\mathrm{i}\pi}\int_\Gamma \frac{x(\tau)}{\tau - t}\mathrm{d}\tau - \frac{b_1(t)}{\mathrm{i}\pi}\int_\Gamma \frac{x(\tau)}{\tau - \alpha(t)}\mathrm{d}\tau - \frac{b_2(t)}{\mathrm{i}\pi}\int_\Gamma \frac{x(\tau)}{\tau - \beta(t)}\mathrm{d}\tau + \frac{b_3(t)}{\mathrm{i}\pi}\int_\Gamma \frac{x(\tau)}{\tau - \gamma(t)}\mathrm{d}\tau + \int_\Gamma K(t,\tau)x(\tau)\mathrm{d}\tau = 0, \tag{1.6}$$

$$(\mathscr{T}_2 x)(t) \equiv a_0(t)x(t) - a_1(t)x[\alpha(t)] + a_2(t)x[\beta(t)] - a_3(t)x[\gamma(t)] + \frac{b_0(t)}{\mathrm{i}\pi}\int_\Gamma \frac{x(\tau)}{\tau - t}\mathrm{d}\tau - \frac{b_1(t)}{\mathrm{i}\pi}\int_\Gamma \frac{x(\tau)}{\tau - \alpha(t)}\mathrm{d}\tau + \frac{b_2(t)}{\mathrm{i}\pi}\int_\Gamma \frac{x(\tau)}{\tau - \beta(t)}\mathrm{d}\tau - \frac{b_3(t)}{\mathrm{i}\pi}\int_\Gamma \frac{x(\tau)}{\tau - \gamma(t)}\mathrm{d}\tau + \int_\Gamma K(t,\tau)x(\tau)\mathrm{d}\tau = 0, \tag{1.7}$$

$$(\mathscr{T}_3 x)(t) \equiv a_0(t)x(t) + a_1(t)x[\alpha(t)] - a_2(t)x[\beta(t)] - a_3(t)x[\gamma(t)] + \frac{b_0(t)}{\mathrm{i}\pi}\int_\Gamma \frac{x(\tau)}{\tau - t}\mathrm{d}\tau + \frac{b_1(t)}{\mathrm{i}\pi}\int_\Gamma \frac{x(\tau)}{\tau - \alpha(t)}\mathrm{d}\tau - \frac{b_2(t)}{\mathrm{i}\pi}\int_\Gamma \frac{x(\tau)}{\tau - \beta(t)}\mathrm{d}\tau - \frac{b_3(t)}{\mathrm{i}\pi}\int_\Gamma \frac{x(\tau)}{\tau - \gamma(t)}\mathrm{d}\tau + \int_\Gamma K(t,\tau)x(\tau)\mathrm{d}\tau = 0. \tag{1.8}$$

如果对于方程(1.6) ～ (1.8) 分别假设

$$\rho_1(t)=x(t),\ \rho_2(t)=-x[\alpha(t)],\ \rho_3(t)=-x[\beta(t)],$$
$$\rho_4(t)=x[\gamma(t)];$$
$$\rho_1(t)=x(t),\ \rho_2(t)=-x[\alpha(t)],\ \rho_3(t)=x[\beta(t)],$$
$$\rho_4(t)=-x[\gamma(t)];$$
$$\rho_1(t)=x(t),\ \rho_2(t)=x[\alpha(t)],\ \rho_3(t)=-x[\beta(t)],$$
$$\rho_4(t)=-x[\gamma(t)].$$

可以验证它们的对应方程组仍然是方程组(1.4). 这也就是说，方程(1.1)(1.6) ～ (1.8) 具有相同的对应方程组(1.4).

最后，还要引入方程(1.1) 的相联方程和方程(1.6) ～ (1.8) 的相联方程

$$(\mathscr{K}\varphi')(t)\equiv a_0(t)\varphi(t)+\nu_1\alpha'(t)a_1[\alpha(t)]\varphi[\alpha(t)]+$$
$$\nu_2\beta'(t)a_2[\alpha(t)]\varphi[\beta(t)]+\nu_1\nu_2\gamma'(t)a_3[\gamma(t)]\varphi[\gamma(t)]-\frac{1}{\mathrm{i}\pi}\int_\Gamma\frac{b_0(\tau)\varphi(\tau)}{\tau-t}\mathrm{d}\tau-$$
$$\frac{\nu_1}{\mathrm{i}\pi}\int_\Gamma\frac{b_1[\alpha(\tau)]\alpha'(\tau)\varphi[\alpha(\tau)]}{\tau-t}\mathrm{d}\tau-\frac{\nu_2}{\mathrm{i}\pi}\int_\Gamma\frac{b_2[\beta(\tau)]\beta'(\tau)\varphi[\beta(\tau)]}{\tau-t}\mathrm{d}\tau-$$
$$\frac{\nu_1\nu_2}{\mathrm{i}\pi}\int_\Gamma\frac{b_3[\gamma(\tau)]\gamma'(\tau)\varphi[\gamma(\tau)]}{\tau-t}\mathrm{d}\tau+\int_\Gamma m(\tau,t)\varphi(\tau)\mathrm{d}\tau=0,\tag{1.9}$$

$$(\mathscr{T}'_1\omega)(t)\equiv a_0(t)\omega(t)-\nu_1\alpha'(t)a_1[\alpha(t)]\omega[\alpha(t)]-$$
$$\nu_2\beta'(t)a_2[\beta(t)]\omega[\beta(t)]+\nu_1\nu_2\gamma'(t)a_3[\gamma(t)]\omega[\gamma(t)]-\frac{1}{\mathrm{i}\pi}\int_\Gamma\frac{b_0(\tau)\omega(\tau)}{\tau-t}\mathrm{d}\tau+$$
$$\frac{\nu_1}{\mathrm{i}\pi}\int_\Gamma\frac{b_1[\alpha(\tau)]\alpha'(\tau)\omega[\alpha(\tau)]}{\tau-t}\mathrm{d}\tau+\frac{\nu_2}{\mathrm{i}\pi}\int_\Gamma\frac{b_2[\beta(\tau)]\beta'(\tau)\omega[\beta(\tau)]}{\tau-t}\mathrm{d}\tau-$$
$$\frac{\nu_1\nu_2}{\mathrm{i}\pi}\int_\Gamma\frac{b_3[\gamma(\tau)]\gamma'(\tau)\omega[\gamma(\tau)]}{\tau-t}\mathrm{d}\tau+\int_\Gamma m(\tau,t)\varphi(\tau)\mathrm{d}\tau=0,\tag{1.10}$$

$$(\mathscr{T}'_2\omega)(t)\equiv a_0(t)\omega(t)-\nu_1\alpha'(t)a_1[\alpha(t)]\omega[\alpha(t)]+$$
$$\nu_2\beta'(t)a_2[\beta(t)]\omega[\beta(t)]-\nu_1\nu_2\gamma'(t)a_3[\gamma(t)]\omega[\gamma(t)]-\frac{1}{\mathrm{i}\pi}\int_\Gamma\frac{b_0(\tau)\omega(\tau)}{\tau-t}\mathrm{d}\tau+$$
$$\frac{\nu_1}{\mathrm{i}\pi}\int_\Gamma\frac{b_1[\alpha(\tau)]\alpha'(\tau)\omega[\alpha(\tau)]}{\tau-t}\mathrm{d}\tau-\frac{\nu_2}{\mathrm{i}\pi}\int_\Gamma\frac{b_2[\beta(\tau)]\beta'(\tau)\omega[\beta(\tau)]}{\tau-t}\mathrm{d}\tau+$$
$$\frac{\nu_1\nu_2}{\mathrm{i}\pi}\int_\Gamma\frac{b_3[\gamma(\tau)]\gamma'(\tau)\omega[\gamma(\tau)]}{\tau-t}\mathrm{d}\tau+\int_\Gamma m(\tau,t)\varphi(\tau)\mathrm{d}\tau=0,\tag{1.11}$$

$$(\mathscr{T}'_3\omega)(t)\equiv a_0(t)\omega(t)+\nu_1\alpha'(t)a_1[\alpha(t)]\omega[\alpha(t)]-$$
$$\nu_2\beta'(t)a_2[\beta(t)]\omega[\beta(t)]-\nu_1\nu_2\gamma'(t)a_3[\gamma(t)]\omega[\gamma(t)]-\frac{1}{\mathrm{i}\pi}\int_\Gamma\frac{b_0(\tau)\omega(\tau)}{\tau-t}\mathrm{d}\tau-$$

$$\frac{\nu_1}{\mathrm{i}\pi}\int_\Gamma \frac{b_1[\alpha(\tau)]\alpha'(\tau)\omega[\alpha(\tau)]}{\tau-t}\mathrm{d}\tau+\frac{\nu_2}{\mathrm{i}\pi}\int_\Gamma \frac{b_2[\beta(\tau)]\beta'(\tau)\omega[\beta(\tau)]}{\tau-t}\mathrm{d}\tau+$$

$$\frac{\nu_1\nu_2}{\mathrm{i}\pi}\int_\Gamma \frac{b_3[\gamma(\tau)]\gamma'(\tau)\omega[\gamma(\tau)]}{\tau-t}\mathrm{d}\tau+\int_\Gamma m(\tau,t)\varphi(\tau)\mathrm{d}\tau=0. \tag{1.12}$$

显然方程(1.10)～(1.12)是方程(1.9)的伴随方程.方程(1.9)的对应方程组是

$$a_0(t)\mu_1(t)+a_1[\alpha(t)]\mu_2(t)+a_2[\beta(t)]\mu_3(t)+a_3[\gamma(t)]\mu_4(t)-$$

$$\frac{1}{\mathrm{i}\pi}\int_\Gamma \frac{b_0(\tau)\mu_1(\tau)}{\tau-t}\mathrm{d}\tau-\frac{1}{\mathrm{i}\pi}\int_\Gamma \frac{b_1[\alpha(\tau)]\mu_2(\tau)}{\tau-t}\mathrm{d}\tau-\frac{1}{\mathrm{i}\pi}\int_\Gamma \frac{b_2[\beta(\tau)]\mu_3(\tau)}{\tau-t}\mathrm{d}\tau-$$

$$\frac{1}{\mathrm{i}\pi}\int_\Gamma \frac{b_3[\gamma(\tau)]\mu_4(\tau)}{\tau-t}\mathrm{d}\tau+\int_\Gamma m(\tau,t)\mu_1(\tau)\mathrm{d}\tau=0,$$

$$a_1(t)\mu_1(t)+a_0[\alpha(t)]\mu_2(t)+a_3[\beta(t)]\mu_3(t)+a_2[\gamma(t)]\mu_4(t)-$$

$$\frac{\nu_1\alpha'(t)}{\mathrm{i}\pi}\int_\Gamma \frac{b_1(\tau)\mu_1(\tau)}{\alpha(\tau)-\alpha(t)}\mathrm{d}\tau-\frac{\nu_1\alpha'(t)}{\mathrm{i}\pi}\int_\Gamma \frac{b_0[\alpha(\tau)]\mu_2(\tau)}{\alpha(\tau)-\alpha(t)}\mathrm{d}\tau-$$

$$\frac{\nu_1\alpha'(t)}{\mathrm{i}\pi}\int_\Gamma \frac{b_3[\beta(\tau)]\mu_3(\tau)}{\alpha(\tau)-\alpha(t)}\mathrm{d}\tau-\frac{\nu_1\alpha'(t)}{\mathrm{i}\pi}\int_\Gamma \frac{b_2[\gamma(\tau)]\mu_4(\tau)}{\alpha(\tau)-\alpha(t)}\mathrm{d}\tau+$$

$$\nu_1\alpha'(t)\int_\Gamma m[\alpha(\tau),\alpha(t)]\mu_2(\tau)\mathrm{d}\tau=0, \tag{1.13}$$

$$a_2(t)\mu_1(t)+a_3[\alpha(t)]\mu_2(t)+a_0[\beta(t)]\mu_3(t)+a_1[\gamma(t)]\mu_4(t)-$$

$$\frac{\nu_2\beta'(t)}{\mathrm{i}\pi}\int_\Gamma \frac{b_2(\tau)\mu_1(\tau)}{\beta(\tau)-\beta(t)}\mathrm{d}\tau-\frac{\nu_2\beta'(t)}{\mathrm{i}\pi}\int_\Gamma \frac{b_3[\alpha(\tau)]\mu_2(\tau)}{\beta(\tau)-\beta(t)}\mathrm{d}\tau-$$

$$\frac{\nu_2\beta'(t)}{\mathrm{i}\pi}\int_\Gamma \frac{b_0[\beta(\tau)]\mu_3(\tau)}{\beta(\tau)-\beta(t)}\mathrm{d}\tau-\frac{\nu_2\beta'(t)}{\mathrm{i}\pi}\int_\Gamma \frac{b_1[\gamma(\tau)]\mu_4(\tau)}{\beta(\tau)-\beta(t)}\mathrm{d}\tau+$$

$$\nu_2\beta'(t)\int_\Gamma m[\beta(\tau),\beta(t)]\mu_3(\tau)\mathrm{d}\tau=0,$$

$$a_3(t)\mu_1(t)+a_2[\alpha(t)]\mu_2(t)+a_1[\beta(t)]\mu_3(t)+a_0[\gamma(t)]\mu_4(t)-$$

$$\frac{\nu_1\nu_2\gamma'(t)}{\mathrm{i}\pi}\int_\Gamma \frac{b_3(\tau)\mu_1(\tau)}{\gamma(\tau)-\gamma(t)}\mathrm{d}\tau-\frac{\nu_1\nu_2\gamma'(t)}{\mathrm{i}\pi}\int_\Gamma \frac{b_3[\alpha(\tau)]\mu_2(\tau)}{\gamma(\tau)-\gamma(t)}\mathrm{d}\tau-$$

$$\frac{\nu_1\nu_2\gamma'(t)}{\mathrm{i}\pi}\int_\Gamma \frac{b_1[\beta(\tau)]\mu_3(\tau)}{\gamma(\tau)-\gamma(t)}\mathrm{d}\tau-\frac{\nu_1\nu_2\gamma'(t)}{\mathrm{i}\pi}\int_\Gamma \frac{b_0[\gamma(\tau)]\mu_4(\tau)}{\gamma(\tau)-\gamma(t)}\mathrm{d}\tau+$$

$$\nu_1\nu_2\gamma'(t)\int_\Gamma m[\gamma(\tau),\gamma(t)]\mu_4(\tau)\mathrm{d}\tau=0,$$

其中 $\mu_1(t)=\varphi(t)$, $\mu_2(t)=\nu_1\alpha'(t)\varphi[\alpha(t)]$, $\mu_3(t)=\nu_2\beta'(t)\varphi[\beta(t)]$, $\mu_4(t)=\nu_1\nu_2\gamma'(t)\varphi[\gamma(t)]$. 如果分离出其特征部分,并写成算子形式,将有

$$\mathscr{M}'\equiv \boldsymbol{p}'(t)\mathscr{I}-\boldsymbol{q}'(t)\mathscr{S}+\mathscr{D}^{**}, \tag{1.14}$$

其中

$$p'(t)=\begin{pmatrix} a_0(t) & a_1[\alpha(t)] & a_2[\beta(t)] & a_3[\gamma(t)] \\ a_1(t) & a_0[\alpha(t)] & a_3[\beta(t)] & a_2[\gamma(t)] \\ a_2(t) & a_3[\alpha(t)] & a_0[\beta(t)] & a_1[\gamma(t)] \\ a_3(t) & a_2[\alpha(t)] & a_1[\beta(t)] & a_0[\gamma(t)] \end{pmatrix},$$

$$q'(t)=\begin{pmatrix} b_0(t) & b_1[\alpha(t)] & b_2[\beta(t)] & b_3[\gamma(t)] \\ \nu_1 b_1(t) & \nu_1 b_0[\alpha(t)] & \nu_1 b_3[\beta(t)] & \nu_1 b_2[\gamma(t)] \\ \nu_2 b_2(t) & \nu_2 b_3[\alpha(t)] & \nu_2 b_0[\beta(t)] & \nu_2 b_1[\gamma(t)] \\ \nu_1\nu_2 b_3(t) & \nu_1\nu_2 b_2[\alpha(t)] & \nu_1\nu_2 b_1[\beta(t)] & \nu_1\nu_2 b_0[\gamma(t)] \end{pmatrix}.$$

这说明算子(1.14)是算子(1.5)的相联算子，也就是说，对应方程组(1.13)是方程组(1.4)的相联方程组.

应该指出，方程(1.10)～(1.12)的对应方程组也将是方程组(1.13)，这时候，只要分别假设

$$\mu_1(t)=\omega(t),\ \mu_2(t)=-\nu_1\alpha'(t)\omega[\alpha(t)],$$
$$\mu_3(t)=-\nu_2\beta'(t)\omega[\beta(t)],\ \mu_4(t)=\nu_1\nu_2\gamma'(t)\omega[\gamma(t)];$$
$$\mu_1(t)=\omega(t),\ \mu_2(t)=-\nu_1\alpha'(t)\omega[\alpha(t)],$$
$$\mu_3(t)=\nu_2\beta'(t)\omega[\beta(t)],\ \mu_4(t)=-\nu_1\nu_2\gamma'(t)\omega[\gamma(t)];$$

和

$$\mu_1(t)=\omega(t),\ \mu_2(t)=\nu_1\alpha'(t)\omega[\alpha(t)],$$
$$\mu_3(t)=-\nu_2\beta'(t)\omega[\beta(t)],\ \mu_4(t)=-\nu_1\nu_2\gamma'(t)\omega[\gamma(t)];$$

就可以得到相应的结论.

§ 2.　可解性问题

对应方程组(1.4)和(1.13)是带柯西核的奇异积分方程组，它们的可解性理论是已知的[5][6]，现在我们讨论奇异积分方程(1.1)(1.6)～(1.8)的可解性问题.

首先讨论齐次方程(1.1)(即 $g(t)\equiv 0$ 的情形). 我们知道，如果齐次方程(1.1)有非零解 $\varphi(t)$，那么向量

$$\boldsymbol{\rho}(t)=\{\rho_1(t),\rho_2(t),\rho_3(t),\rho_4(t)\} \tag{2.1}$$

将是齐次对应方程组(1.4)(即 $g(t)=g[\alpha(t)]=g[\beta(t)]=g[\gamma(t)]\equiv 0$ 的情形)的解. 这里

$\rho_1(t)=\varphi(t),\rho_2(t)=\varphi[\alpha(t)],\rho_3(t)=\varphi[\beta(t)],\rho_4(t)=\varphi=[\gamma(t)].$

显然，这组解满足条件

$$\rho_1(t)=\rho_1(t),\rho_2(t)=\rho_1[\alpha(t)],\rho_3(t)=\rho_3[\beta(t)],\rho_4(t)=\rho_1[\gamma(t)]. \tag{2.2}$$

同理，如果齐次方程(1.6)(1.7)或(1.8)有非零解 $x(t)$，那么向量(2.1)也将是齐次对应方程组(1.4)的解，这里应该分别规定：

$\rho_1(t)=x(t),\rho_2(t)=-x[\alpha(t)],\rho_3(t)=-x[\beta(t)],\rho_4(t)=x[\gamma(t)];$

$\rho_1(t)=x(t),\rho_2(t)=-x[\alpha(t)],\rho_3(t)=x[\beta(t)],\rho_4(t)=-x[\gamma(t)];$

或 $\rho_1(t)=x(t),\rho_2(t)=x[\alpha(t)],\rho_3(t)=-x[\beta(t)],\rho_4(t)=-x[\gamma(t)].$

这时，解将分别满足条件

$$\rho_1(t)=\rho_1(t),\rho_2(t)=-\rho_1[\alpha(t)],\rho_3(t)=-\rho_1[\beta(t)],\rho_4(t)=\rho_1[\gamma(t)]; \tag{2.3}$$

$$\rho_1(t)=\rho_1(t),\rho_2(t)=-\rho_1[\alpha(t)],\rho_3(t)=\rho_1[\beta(t)],\rho_4(t)=-\rho_1[\gamma(t)]; \tag{2.4}$$

$$\rho_1(t)=\rho_1(t),\rho_2(t)=\rho_1[\alpha(t)],\rho_3(t)=-\rho_1[\beta(t)],\rho_4(t)=-\rho_1[\gamma(t)]. \tag{2.5}$$

反过来，如果向量(2.1)是齐次对应方程组(1.4)的非零解，并且它满足条件(2.2)～(2.5)中的一个，那么其第一个分量 $\rho_1(t)$ 将相应地是齐次方程(1.1)(1.6)(1.7)或(1.8)的非零解.

为了确定，我们假设齐次对应方程组具有 l 个线性无关解，而齐次方程(1.1)(1.6)～(1.8)分别具有 l_1,l_2,l_3 和 l_4 个线性无关解，于是将有以下结果成立：

定理 1 齐次对应方程组(1.4)的线性无关解个数 l 等于齐次方程(1.1)(1.6)～(1.8)的线性无关解个数 l_1,l_2,l_3,l_4 之和，即

$$l=l_1+l_2+l_3+l_4=\sum_{j=1}^{4}l_j. \tag{2.6}$$

证 假设 $\rho_1^{(k,j)}(t)\quad j=1,2,\cdots,l_k,k=1,2,3,4$ 分别是齐次方程(1.1)(1.6)～(1.8)的线性无关解的完备系(也叫作基本解系)，我们讨论 $l_1+l_2+l_3+l_4$ 个向量

$$\boldsymbol{\rho}^{(k,j)}(t)=\{\rho_1^{(k,j)}(t),\rho_2^{(k,j)}(t),\rho_3^{(k,j)}(t),\rho_4^{(k,j)}(t)\},$$
$$j=1,2,\cdots,l_k,\ k=1,2,3,4,$$

它们分别满足条件(2.2) ～ (2.5).

我们将证明这 $l_1+l_2+l_3+l_4$ 个向量是线性无关的，这就需要证明，只要

$$\sum_{j=1}^{l_1} a_j\rho^{(1,j)}(t)+\sum_{j=1}^{l_2} b_j\rho^{(2,j)}(t)+\sum_{j=1}^{l_3} c_j\rho^{(3,j)}(t)+\sum_{j=1}^{l_4} d_j\rho^{(4,j)}(t)=0, \tag{2.7}$$

就一定保证所有的 a_j,b_j,c_j,d_j 都等于零.为此，可以把(2.7) 写成对于其分量成立的四个等式

$$\sum_{j=1}^{l_1} a_j\rho_k^{(1,j)}(t)+\sum_{j=1}^{l_2} b_j\rho_k^{(2,j)}(t)+\sum_{j=1}^{l_3} c_j\rho_k^{(3,j)}(t)+\sum_{j=1}^{l_4} d_j\rho_k^{(4,j)}(t)=0,\quad k=1,2,3,4 \tag{2.8}$$

我们把 $k=2,3,4$ 时的等式中的 t 分别换成 $\alpha(t),\beta(t),\gamma(t)$，并利用条件(2.2) ～ (2.5) 得到等式

$$\sum_{j=1}^{l_1} a_j\rho_1^{(1,j)}(t)-\sum_{j=1}^{l_2} b_j\rho_1^{(2,j)}(t)-\sum_{j=1}^{l_3} c_j\rho_1^{(3,j)}(t)+\sum_{j=1}^{l_4} d_j\rho_1^{(4,j)}(t)=0,$$

$$\sum_{j=1}^{l_1} a_j\rho_1^{(1,j)}(t)-\sum_{j=1}^{l_2} b_j\rho_1^{(2,j)}(t)+\sum_{j=1}^{l_3} c_j\rho_1^{(3,j)}(t)-\sum_{j=1}^{l_4} d_j\rho_1^{(4,j)}(t)=0,$$

$$\sum_{j=1}^{l_1} a_j\rho_1^{(1,j)}(t)+\sum_{j=1}^{l_2} b_j\rho_1^{(2,j)}(t)-\sum_{j=1}^{l_3} c_j\rho_1^{(3,j)}(t)-\sum_{j=1}^{l_4} d_j\rho_1^{(4,j)}(t)=0.$$

然后把这三个方程与(2.8) 中第一个方程($k=1$) 联立就可以解得

$$\sum_{j=1}^{l_1} a_j\rho_1^{(1,j)}(t)=0,\quad \sum_{j=1}^{l_2} b_j\rho_1^{(2,j)}(t)=0,$$

$$\sum_{j=1}^{l_3} c_j\rho_1^{(3,j)}(t)=0,\quad \sum_{j=1}^{l_4} d_j\rho_1^{(4,j)}(t)=0,$$

但 $\rho_1^{(k,j)}(t),j=1,2,\cdots,l_k,k=1,2,3,4$ 分别是线性无关的，从而得到所有的 a_j,b_j,c_j,d_j 都等于零. 从而，我们得到了 $l\geqslant l_1+l_2+l_3+l_4$ 的结论，下面将证明一定还有等式 $l=l_1+l_2+l_3+l_4$ 成立.

假设向量

$$\boldsymbol{\rho}_j^0(t)=\{\rho_1^{(0,j)}(t),\rho_2^{(0,j)}(t),\rho_3^{(0,j)}(t),\rho_4^{(0,j)}(t)\},\quad j=1,2,\cdots,l, \tag{2.9}$$

是齐次对应方程组(1.4) 的一组基本解系，通过直接验证可以知道:向量

$$\boldsymbol{\rho}_j^1(t)=\{\rho_2^{(0,j)}[\alpha(t)],\rho_1^{(0,j)}[\alpha(t)],\rho_4^{(0,j)}[\alpha(t)],\rho_3^{(0,j)}[\alpha(t)]\},\tag{2.10}$$

$$\boldsymbol{\rho}_j^2(t)=\{\beta_3^{(0,j)}[\beta(t)],\rho_4^{(0,j)}[\beta(t)],\rho_1^{(0,j)}[\beta(t)],\rho_2^{(0,j)}[\beta(t)]\},\tag{2.11}$$

$$\boldsymbol{\rho}_j^3(t)=\{\rho_4^{(0,j)}[\alpha(t)],\rho_3^{(0,j)}[\alpha(t)],\rho_2^{(0,j)}[\gamma(t)],\rho_1^{(0,j)}[\gamma(t)]\},\tag{2.12}$$

$$j=1,2,\cdots,l.$$

也都是齐次对应方程组(1.4)的基本解系.

我们考虑以下四组向量

$$\boldsymbol{\rho}^{(1,j)}(t)=\boldsymbol{\rho}_j^0(t)+\boldsymbol{\rho}_j^1(t)+\boldsymbol{\rho}_j^2(t)+\boldsymbol{\rho}_j^3(t),\tag{2.13}$$

$$\boldsymbol{\rho}^{(2,j)}(t)=-\boldsymbol{\rho}_j^0(t)+\boldsymbol{\rho}_j^1(t)+\boldsymbol{\rho}_j^2(t)-\boldsymbol{\rho}_j^3(t),\tag{2.14}$$

$$\boldsymbol{\rho}^{(3,j)}(t)=-\boldsymbol{\rho}_j^0(t)+\boldsymbol{\rho}_j^1(t)-\boldsymbol{\rho}_j^2(t)+\boldsymbol{\rho}_j^3(t),\tag{2.15}$$

$$\boldsymbol{\rho}^{(4,j)}(t)=-\boldsymbol{\rho}_j^0(t)-\boldsymbol{\rho}_j^1(t)+\boldsymbol{\rho}_j^2(t)+\boldsymbol{\rho}_j^3(t),\tag{2.16}$$

$$j=1,2,\cdots,l.$$

显然,向量组(2.13)～(2.16)仍是齐次对应方程组(1.4)的解,它们分别满足条件(2.2)～(2.5).由于向量组(2.10)～(2.12)都是齐次对应方程组(1.4)的解,于是有

$$\boldsymbol{\rho}_j^1(t)=\sum_{k=1}^{l}\lambda_{jk}\boldsymbol{\rho}_k^0(t),\ \boldsymbol{\rho}_j^2(t)=\sum_{k=1}^{l}\mu_{jk}\boldsymbol{\rho}_k^0(t),$$

$$\boldsymbol{\rho}_j^3(t)=\sum_{k=1}^{l}\nu_{jk}\boldsymbol{\rho}_k^0(t),\qquad j=1,2,\cdots,l.$$

另外,显然有

$$\boldsymbol{\rho}_j^0(t)=\sum_{k=1}^{l}\delta_{jk}\boldsymbol{\rho}_k^0(t),$$

其中 $\delta_{jk}=\begin{cases}1, & j=k,\\ 0, & j\neq k.\end{cases}$ 容易看出:向量组 $\boldsymbol{\rho}_j^0(t),\boldsymbol{\rho}_j^1(t),\boldsymbol{\rho}_j^2(t),\boldsymbol{\rho}_j^3(t)$ 之间的这种线性变换的系数矩阵是

$$\begin{pmatrix} 1 & 0 & 0 & \cdots & 0 \\ \lambda_{11} & \lambda_{12} & \lambda_{13} & \cdots & \lambda_{1l} \\ \mu_{11} & \mu_{12} & \mu_{13} & \cdots & \mu_{1l} \\ \nu_{11} & \nu_{12} & \nu_{13} & \cdots & \nu_{1l} \\ \vdots & \vdots & \vdots & & \vdots \\ 1 & 0 & 0 & & 1 \\ \lambda_{l1} & \lambda_{l2} & \lambda_{l3} & \cdots & \lambda_{ll} \\ \mu_{l1} & \mu_{l2} & \mu_{l3} & \cdots & \mu_{ll} \\ \nu_{l1} & \nu_{l2} & \nu_{l3} & \cdots & \nu_{ll} \end{pmatrix} \tag{2.17}$$

它的秩显然等于 l(它是 $4l\times l$ 阶矩阵).

另外,向量组(2.13)～(2.16)通过向量组(2.9)可以线性表示成以下形式

$$\rho^{(1,j)}(t)=\rho_j^0(t)+\sum_{k=1}^{l}\lambda_{jk}\rho_k^0(t)+\sum_{k=1}^{l}\mu_{jk}\rho_k^0(t)+\sum_{k=1}^{l}\nu_{jk}\rho_k^0(t),$$

$$\rho^{(2,j)}(t)=-\rho_j^0(t)+\sum_{k=1}^{l}\lambda_{jk}\rho_k^0(t)+\sum_{k=1}^{l}\mu_{jk}\rho_k^0(t)-\sum_{k=1}^{l}\nu_{jk}\rho_k^0(t),$$

$$\rho^{(3,j)}(t)=-\rho_j^0(t)+\sum_{k=1}^{l}\lambda_{jk}\rho_k^0(t)-\sum_{k=1}^{l}\mu_{jk}\rho_k^0(t)+\sum_{k=1}^{l}\nu_{jk}\rho_k^0(t),$$

$$\rho^{(4,j)}(t)=-\rho_j^0(t)-\sum_{k=1}^{l}\lambda_{jk}\rho_k^0(t)+\sum_{k=1}^{l}\mu_{jk}\rho_k^0(t)+\sum_{k=1}^{l}\nu_{jk}\rho_k^0(t).$$

这一线性变换的系数矩阵是

$$\begin{pmatrix} 1+\lambda_{11}+\mu_{11}+\nu_{11} & \lambda_{12}+\mu_{12}+\nu_{12} & \cdots & \lambda_{1l}+\mu_{1l}+\nu_{1l} \\ -1+\lambda_{11}+\mu_{11}-\nu_{11} & \lambda_{12}+\mu_{12}-\nu_{12} & \cdots & \lambda_{1l}+\mu_{1l}-\nu_{1l} \\ -1+\lambda_{11}-\mu_{11}+\nu_{11} & \lambda_{12}-\mu_{12}+\nu_{12} & \cdots & \lambda_{1l}-\mu_{1l}+\nu_{1l} \\ -1-\lambda_{11}+\mu_{11}+\nu_{11} & -\lambda_{12}-\mu_{12}+\nu_{12} & \cdots & -\lambda_{1l}+\mu_{1l}+\nu_{1l} \\ \lambda_{21}+\mu_{21}+\nu_{21} & 1+\lambda_{22}+\mu_{22}+\nu_{22} & \cdots & \lambda_{2l}+\mu_{2l}+\nu_{2l} \\ \lambda_{21}+\mu_{21}-\nu_{21} & -1+\lambda_{22}+\mu_{22}-\nu_{22} & \cdots & \lambda_{2l}+\mu_{2l}-\nu_{2l} \\ \lambda_{21}-\mu_{21}+\nu_{21} & -1+\lambda_{22}-\mu_{22}+\nu_{22} & \cdots & \lambda_{2l}-\mu_{2l}+\nu_{2l} \\ -\lambda_{21}+\mu_{21}+\nu_{21} & -1-\lambda_{22}+\mu_{22}+\nu_{22} & \cdots & -\lambda_{2l}+\mu_{2l}+\nu_{2l} \\ \vdots & \vdots & & \vdots \\ \lambda_{l1}+\mu_{l1}+\nu_{l1} & \lambda_{l2}+\mu_{l2}+\nu_{l2} & \cdots & 1+\lambda_{ll}+\mu_{ll}+\nu_{ll} \\ \lambda_{l1}+\mu_{l1}-\nu_{l1} & \lambda_{l2}+\mu_{l2}-\nu_{l2} & \cdots & -1+\lambda_{ll}+\mu_{ll}-\nu_{ll} \\ \lambda_{l1}-\mu_{l1}+\nu_{l1} & \lambda_{l2}-\mu_{l2}+\nu_{l2} & \cdots & -1+\lambda_{ll}-\mu_{ll}+\nu_{ll} \\ -\lambda_{l1}+\mu_{l1}+\nu_{l1} & -\lambda_{l2}+\mu_{l2}+\nu_{l2} & \cdots & -1-\lambda_{ll}+\mu_{ll}+\nu_{ll} \end{pmatrix}, \tag{2.18}$$

这个矩阵还可以写成以下对角线分块矩阵

$$\begin{pmatrix} 1 & 1 & 1 & 1 & 0 & 0 & 0 & 0 & \cdots & 0 & 0 & 0 & 0 \\ -1 & 1 & 1 & -1 & 0 & 0 & 0 & 0 & \cdots & 0 & 0 & 0 & 0 \\ -1 & 1 & -1 & 1 & 0 & 0 & 0 & 0 & \cdots & 0 & 0 & 0 & 0 \\ -1 & -1 & 1 & 1 & 0 & 0 & 0 & 0 & \cdots & 0 & 0 & 0 & 0 \\ 0 & 0 & 0 & 0 & 1 & 1 & 1 & 1 & \cdots & 0 & 0 & 0 & 0 \\ 0 & 0 & 0 & 0 & -1 & 1 & 1 & -1 & \cdots & 0 & 0 & 0 & 0 \\ 0 & 0 & 0 & 0 & -1 & 1 & -1 & 1 & \cdots & 0 & 0 & 0 & 0 \\ 0 & 0 & 0 & 0 & -1 & -1 & 1 & 1 & \cdots & 0 & 0 & 0 & 0 \\ \vdots & \vdots & \vdots & \vdots & \vdots & \vdots & \vdots & \vdots & & \vdots & \vdots & \vdots & \vdots \\ 0 & 0 & 0 & 0 & 0 & 0 & 0 & 0 & \cdots & 1 & 1 & 1 & 1 \\ 0 & 0 & 0 & 0 & 0 & 0 & 0 & 0 & \cdots & -1 & 1 & 1 & -1 \\ 0 & 0 & 0 & 0 & 0 & 0 & 0 & 0 & \cdots & -1 & 1 & -1 & 1 \\ 0 & 0 & 0 & 0 & 0 & 0 & 0 & 0 & \cdots & -1 & -1 & 1 & 1 \end{pmatrix} \tag{2.19}$$

左乘矩阵(2.17)的乘积.矩阵(2.19)是 $4l\times 4l$ 阶非退化矩阵,而矩阵(2.17)的秩等于 l,所以矩阵(2.18)的秩也等于 l(参看[4]).这说明把向量组(2.13) ～ (2.16) 合在一起其中必可以选出 l 个是线性无关的(它们可以作为齐次对应方程组(1.4) 的基本解系),假设

$$\boldsymbol{\rho}^{(1,j)}(t),\ j=1,2,\cdots,m_1,$$

$$\boldsymbol{\rho}^{(2,j)}(t),\ j=m_1+1,m_1+2,\cdots,m_1+m_2,$$

$$\boldsymbol{\rho}^{(3,j)}(t),\ j=m_1+m_2+1,\cdots,m_1+m_2+m_3,$$

$$\boldsymbol{\rho}^{(4,j)}(t),\ j=m_1+m_2+m_3+1,\cdots,l,$$

就是用某种方法选出的一组线性无关解,其中前 m_1 个选自向量组(2.13),再下面 m_2 个选自向量组(2.14),然后再下面 m_3 个选自向量组(2.15),最后 $l-(m_1+m_2+m_3)$ 个选是向量组(2.16). 我们将证明一定有 $m_1=l_1, m_2=l_2, m_3=l_3, l-(m_1+m_1+m_3)=l_4$,即 m_1,m_2,m_3 是与具体选法无关的.

事实上,如果 $\boldsymbol{\rho}(t)=\{\rho_1(t),\rho_2(t),\rho_3(t),\rho_4(t)\}$ 是齐次对应方程组(1.4) 的任意解,它满足条件(2.2),那么一定有

$$\boldsymbol{\rho}(t)=\sum_{j=1}^{m_1}\beta_j\boldsymbol{\rho}^{(1,j)}(t)+\sum_{j=m_1+1}^{m_1+m_2}\beta_j\boldsymbol{\rho}^{(2,j)}(t)+\sum_{j=m_1+m_2+1}^{m_1+m_2+m_3}\beta_j\boldsymbol{\rho}^{(3,j)}(t)+\sum_{l=m_1+m_2+m_3+1}^{l}\beta_j\boldsymbol{\rho}^{(4,j)}(t).$$

如果按其分量写成四个等式,将有

$$-\rho_k(t)+\sum_{j=1}^{m_1}\beta_j\rho_1^{(1,j)}(t)+\sum_{j=m_1+1}^{m_1+m_2}\beta_j\rho_1^{(2,j)}(t)+\sum_{j=m_1+m_2+1}^{m_1+m_2+m_3}\beta_j\rho_1^{(3,j)}(t)+\sum_{j=m_1+m_2+m_3+1}^{l}\beta_j\rho_1^{(4,j)}(t)=0,\ k=1,2,3,4.$$

在第二、三、四个等式(即 $k=2,3,4$ 情形) 中把 t 分别换成 $\alpha(t),\beta(t),\gamma(t)$,再利用条件(2.2) ～ (2.5) 可以得到

$$-\rho_1(t)+\sum_{j=1}^{m_1}\beta_j\rho_1^{(1,j)}(t)-\sum_{i=m_1+1}^{m_1+m_2}\beta_j\rho_1^{(2,j)}(t)-\sum_{j=m_1+m_2+1}^{m_1+m_2+m_3}\beta_j\rho_1^{(3,j)}(t)+\sum_{j=m_1+m_2+m_3+1}^{l}\beta_j\rho_1^{(4,j)}(t)=0,$$

$$-\rho_1(t)+\sum_{j=1}^{m_1}\beta_j\rho_1^{(1,j)}(t)-\sum_{i=m_1+1}^{m_1+m_2}\beta_j\rho_1^{(2,j)}(t)+\sum_{j=m_1+m_2+1}^{m_1+m_2+m_3}\beta_j\rho_1^{(3,j)}(t)-\sum_{j=m_1+m_2+m_3+1}^{l}\beta_j\rho_1^{(4,j)}(t)=0,$$

和 $$-\rho_1(t)+\sum_{j=1}^{m_1}\beta_j\rho_1^{(1,j)}(t)+\sum_{i=m_1+1}^{m_1+m_2}\beta_j\rho_1^{(2,j)}(t)-\sum_{j=m_1+m_2+1}^{m_1+m_2+m_3}\beta_j\rho_1^{(3,j)}(t)-\sum_{j=m_1+m_2+m_3+1}^{l}\beta_j\rho_1^{(4,j)}(t)=0.$$

易得

$$\rho_1(t)=\sum_{j=1}^{m_1}\beta_j\rho_1^{(1,j)}(t),\ \sum_{j=m_1+1}^{m_1+m_2}\beta_j\rho_1^{(2,j)}(t)=0,$$

$$\sum_{j=m_1+m_2+1}^{m_1+m_2+m_3}\beta_j\rho_1^{(3,j)}(t)=0,$$

$$\sum_{j=m_1+m_2+m_3+1}^{l}\beta_j\rho_1^{(4,j)}(t)=0,$$

从而推出 $\beta_j=0, j=m_1+1, m_1+2, \cdots, l$,而且有

$$\boldsymbol{\rho}(t)=\sum_{j=1}^{m_1}\beta_j\boldsymbol{\rho}^{(1,j)}(t),$$

这说明 $\boldsymbol{\rho}^{(1,j)}(t), j=1,2,\cdots,m_1$ 就是齐次对应方程组(1.4)满足条件(2.2)的基本解系,从而可以知道 $\rho_1^{(1,j)}(t), j=1,2,\cdots,m_1$ 将是方程(1.1)的基本解系,于是有 $l_1=m_1$.

同理可证 $l_2=m_2, l_3=m_3$ 和 $l_4=l-(m_1+m_2+m_3)$,也就是 $l=l_1+l_2+l_3+l_4$.

定理 2 齐次对应方程组(1.13)的线性无关解的个数 l^* 等于齐次相联方程(1.9)和齐次伴随相联方程(1.10)～(1.12)的线性无关解的个数 $l_1^*, l_2^*, l_3^*, l_4^*$ 之和,即 $l^*=l_1^*+l_2^*+l_3^*+l_4^*$.

证明方法与定理 1 完全类似,这时候,代替条件(2.2)～(2.5)应该利用以下条件:

$$\mu_1(t)=\mu_1(t),\ \mu_2(t)=\nu_1\alpha'(t)\mu_1[\alpha(t)],$$
$$\mu_3(t)=\nu_2\beta'(t)\mu_1[\beta(t)],\ \mu_4(t)=\nu_1\nu_2\gamma'(t)\mu_1[\gamma(t)]; \tag{2.2$'$}$$

$$\mu_1(t)=\mu_1(t),\ \mu_2(t)=-\nu_1\alpha'(t)\mu_1[\alpha(t)],$$
$$\mu_3(t)=-\nu_2\beta'(t)\mu_1[\beta(t)],\ \mu_4(t)=\nu_1\nu_2\gamma'(t)\mu_1[\gamma(t)]; \tag{2.3$'$}$$

$$\mu_1(t)=\mu_1(t),\ \mu_2(t)=-\nu_1\alpha'(t)\mu_1[\alpha(t)],$$
$$\mu_3(t)=\nu_2\beta'(t)\mu_1[\beta(t)],\ \mu_4(t)=-\nu_1\nu_2\gamma'(t)\mu_1[\gamma(t)]; \tag{2.4$'$}$$

$$\mu_1(t)=\mu_1(t),\ \mu_2(t)=\nu_1\alpha'(t)\mu_1[\alpha(t)],$$
$$\mu_3(t)=-\nu_2\beta'(t)\mu_1[\beta(t)],\ \mu_4(t)=-\nu_1\nu_2\gamma'(t)\mu_1[\gamma(t)]. \tag{2.5$'$}$$

这样很容易就可以得到所需要的结论.

定理 3　非齐次奇异积分方程(1.1)可解的完分必要条件是非齐次对应方程组(1.4)可解.

如果非齐次方程(1.1)有解 $\varphi(t)$,那么函数组

$$\rho_1(t)=\varphi(t),\rho_2(t)=\varphi[\alpha(t)],\rho_3(t)=\varphi[\beta(t)],\rho_4(t)=\varphi[\gamma(t)]$$

将是非齐次对应方程组(1.4)的解.

反过来,如果非齐次对应方程组(1.4)有解 $\rho(t)=\{\rho_1(t),\rho_2(t),\rho_3(t),\rho_4(t)\}$,那么,可以直接验证向量

$$\{\rho_2[\alpha(t)],\rho_1[\alpha(t)],\rho_4[\alpha(t)],\rho_3[\alpha(t)]\},$$
$$\{\rho_3[\beta(t)],\rho_4[\beta(t)],\rho_1[\beta(t)],\rho_2[\beta(t)]\},$$
$$\{\rho_4[\gamma(t)],\rho_3[\gamma(t)],\rho_2[\gamma(t)],\rho_1[\gamma(t)]\}$$

也都是非齐次对应方程组(1.4)的解. 我们考虑向量

$$\boldsymbol{\rho}^*(t)=\{\boldsymbol{\rho}_1^*(t),\boldsymbol{\rho}_2^*(t),\boldsymbol{\rho}_3^*(t),\boldsymbol{\rho}_4^*(t)\},$$

这里

$$\boldsymbol{\rho}_1^*(t)=\frac{1}{4}\{\rho_1(t)+\rho_2[\alpha(t)]+\rho_3[\beta(t)]+\rho_4[\gamma(t)]\},$$

$$\boldsymbol{\rho}_2^*(t)=\frac{1}{4}\{\rho_2(t)+\rho_1[\alpha(t)]+\rho_4[\beta(t)]+\rho_3[\gamma(t)]\},$$

$$\boldsymbol{\rho}_3^*(t)=\frac{1}{4}\{\rho_3(t)+\rho_4[\alpha(t)]+\rho_1[\beta(t)]+\rho_2[\gamma(t)]\},$$

$$\boldsymbol{\rho}_4^*(t)=\frac{1}{4}\{\rho_4(t)+\rho_3[\alpha(t)]+\rho_2[\beta(t)]+\rho_1[\gamma(t)]\}.$$

显然,向量 $\boldsymbol{\rho}^*(t)$ 仍应该是非齐次对应方程组(1.4)的解,而这个解满足条件(2.2),从而 $\varphi(t)=\rho_1^*(t)$ 就是奇异积分方程(1.1)的解. 定理 3 得证.

§3.　算子 $\mathscr{K}$ 与伴随算子 $\mathscr{T}_j$ 之间的关系

定理 4　如果 $\alpha(t)$ 和 $\beta(t)$ 都是 Carleman 正位移,对于算子 $\mathscr{K}$ 和 $\mathscr{T}_j$,$j=1,2,3$,将有关系式

$$u_j\mathscr{K}-\mathscr{T}_ju_j=\mathscr{D}_j \tag{3.1}$$

成立,其中 $u_1(t)=\alpha(t)+\beta(t)-\gamma(t)-t$,$u_2(t)=\alpha(t)-\beta(t)+\gamma(t)-t$,$u_3(t)=-\alpha(t)+\beta(t)+\gamma(t)-t$,而 $\mathscr{D}_j$ 是完全连续算子.

根据条件 $\alpha(t),\beta(t)$ 以及 $\gamma(t)$ 都是正位移，所以它们都没有不动点[1]，即 $\alpha(t)\neq t,\beta(t)\neq t$，而且 $\gamma(t)\neq t,\gamma(t)\neq\alpha(t),\gamma(t)\neq\beta(t)$. 显然，有 $u_j(t)\neq 0$，而且还有

$$u_1[\alpha(t)]=-u_1(t),u_1[\beta(t)]=-u_1(t),u_1[\gamma(t)]=u_1(t),\quad(3.2)$$

$$u_2[\alpha(t)]=-u_2(t),u_2[\beta(t)]=u_2(t),\quad u_2[\gamma(t)]=-u_2(t),\quad(3.3)$$

$$u_3[\alpha(t)]=u_3(t),\quad u_3[\beta(t)]=-u_3(t),u_3[\gamma(t)]=-u_3(t).\quad(3.4)$$

通过直接计算 $u_1\mathscr{K}$ 和 $\mathscr{T}_1u_1$，并分离出特征部分，可以得到

$$(u_1\mathscr{K}\varphi)(t)\equiv u_1(t)a_0(t)\varphi(t)+u_1(t)a_1(t)\varphi[\alpha(t)]+$$

$$u_1(t)a_2(t)\varphi[\beta(t)]+u_1(t)a_3(t)\varphi[\gamma(t)]+\frac{u_1(t)b_0(t)}{\mathrm{i}\pi}\int_\Gamma\frac{\varphi(\tau)}{\tau-t}\mathrm{d}\tau+$$

$$\frac{u_1(t)b_1(t)}{\mathrm{i}\pi}\int_\Gamma\frac{\varphi[\alpha(\tau)]}{\tau-t}\mathrm{d}\tau+\frac{u_1(t)b_2(t)}{\mathrm{i}\pi}\int_\Gamma\frac{\varphi[\beta(\tau)]}{\tau-t}\mathrm{d}\tau+$$

$$\frac{u_1(t)b_3(t)}{\mathrm{i}\pi}\int_\Gamma\frac{\varphi[\gamma(\tau)]}{\tau-t}\mathrm{d}\tau+(\mathscr{D}_1^*\varphi)(t),$$

$$(\mathscr{T}_1u_1\varphi)(t)\equiv a_0(t)u_1(t)\varphi(t)-u_1[\alpha(t)]a_1(t)\varphi[\alpha(t)]-$$

$$a_2(t)u_1[\beta(t)]\varphi[\beta(t)]+a_3(t)u_1[\gamma(t)]\varphi[\gamma(t)]+\frac{b_0(t)u_1(t)}{\mathrm{i}\pi}\int_\Gamma\frac{\varphi(\tau)}{\tau-t}\mathrm{d}\tau-$$

$$\frac{b_1(t)u_1[\alpha(t)]}{\mathrm{i}\pi}\int_\Gamma\frac{\varphi[\alpha(t)]}{\tau-t}\mathrm{d}\tau-\frac{b_2(t)u_1[\beta(t)]}{\mathrm{i}\pi}\int_\Gamma\frac{\varphi[\beta(t)]}{\tau-t}\mathrm{d}\tau+$$

$$\frac{b_3(t)u_1[\gamma(t)]}{\mathrm{i}\pi}\int_\Gamma\frac{\varphi[\gamma(t)]}{\tau-t}\mathrm{d}\tau+(\mathscr{D}_2^*\varphi)(t),$$

这里 $\mathscr{D}_1^*,\mathscr{D}_2^*$ 都是完全连续算子. 只要利用等式(3.2) 马上可以得到：$u_1\mathscr{K}-\mathscr{T}_1u_1=\mathscr{D}_1$ 是完全连续算子. 同理，只需利用等式(3.3) (3.4) 就可以证明定理 4 的另外两条结论($j=2,3$).

定理 5 如果 $\alpha(t),\beta(t)$ 都是反位移，对于算子 $\mathscr{K}$ 和 $\mathscr{T}_j,j=1,2,3$，将有关系式

$$\mathscr{S}v_j\mathscr{K}-\mathscr{T}_jv_j\mathscr{S}=\mathscr{D}_j,j=1,2,3\quad(3.5)$$

成立，其中 $v_1(t)\equiv 1,v_2(t)=-\alpha(t)+\beta(t)+\gamma(t)-t=u_3(t),v_3(t)=\alpha(t)-\beta(t)+\gamma(t)-t=u_2(t)$，$\mathscr{T}$ 是奇异积分算子，$\mathscr{D}_j,j=1,2,3$，是完全连续算子.

根据条件 $\alpha(t),\beta(t)$ 都是反位移，从而 $\gamma(t)$ 是正位移，所以 $\gamma(t)\neq t$，又 $\alpha(t)\neq\beta(t)$，于是有 $v_j(t)\neq 0,j=2,3$，而且有

$$v_2[\alpha(t)]=v_2(t),v_2[\beta(t)]=-v_2(t),v_2[\gamma(t)]=-v_2(t);\quad(3.4)'$$

$$v_3[\alpha(t)]=-v_3(t),v_3[\beta(t)]=v_3(t),v_3[\gamma(t)]=-v_3(t). \quad (3.3)'$$

通过直接计算算子$\mathscr{T}v_j\mathscr{K}$和$\mathscr{J}_jv_j\mathscr{S}$，考虑到$\alpha(t)$，$\beta(t)$是反位移，$\gamma(t)$是正位移，再利用 Poincarè-Bertrand 公式，并分离出特征部分，我们将得到

$$(\mathscr{S}v_i\mathscr{K}_\varphi)(t)\equiv b_0(t)v_j(t)\varphi(t)-b_1(t)v_j(t)\varphi[\alpha(t)]-$$

$$b_2(t)v_j(t)\varphi[\beta(t)]+b_3(t)v_j(t)\varphi[\gamma(t)]+\frac{a_0(t)v_j(t)}{\mathrm{i}\pi}\int_\Gamma\frac{\varphi(\tau)}{\tau-t}\mathrm{d}\tau+$$

$$\frac{a_1(t)v_j(t)}{\mathrm{i}\pi}\int_\Gamma\frac{\varphi[\alpha(\tau)]}{\tau-t}\mathrm{d}\tau+\frac{a_2(t)v_j(t)}{\mathrm{i}\pi}\int_\Gamma\frac{\varphi[\beta(\tau)]}{\tau-t}\mathrm{d}\tau+$$

$$\frac{a_3(t)v_j(t)}{\mathrm{i}\pi}\int_\Gamma\frac{\varphi[\gamma(\tau)]}{\tau-t}\mathrm{d}\tau+(\mathscr{D}_j^*\varphi)(t),\ j=1,2,3,$$

$$(\mathscr{T}_1\mathscr{S}\varphi)(t)\equiv b_0(t)\varphi(t)-b_1(t)\varphi[\alpha(t)]-b_2(t)\varphi[\beta(t)]+$$

$$b_3(t)\varphi[\gamma(t)]+\frac{a_0(t)}{\mathrm{i}\pi}\int_\Gamma\frac{\varphi(\tau)}{\tau-t}\mathrm{d}\tau+\frac{a_1(t)}{\mathrm{i}\pi}\int_\Gamma\frac{\varphi[\alpha(\tau)]}{\tau-t}\mathrm{d}\tau+\frac{a_2(t)}{\mathrm{i}\pi}\int_\Gamma\frac{\varphi[\beta(\tau)]}{\tau-t}\mathrm{d}\tau+$$

$$\frac{a_3(t)}{\mathrm{i}\pi}\int_\Gamma\frac{\varphi[\gamma(\tau)]}{\tau-t}\mathrm{d}\tau+(\mathscr{D}_1^{**}\varphi)(t),$$

$$(\mathscr{T}_2v_2\mathscr{S}\varphi)(t)\equiv b_0(t)v_2(t)\varphi(t)-b_1(t)v_2[\alpha(t)]\varphi[\alpha(t)]+$$

$$b_2(t)v_2[\beta(t)]\varphi[\beta(t)]-b_3(t)v_2[\gamma(t)]\varphi[\gamma(t)]+\frac{a_0(t)v_2(t)}{\mathrm{i}\pi}\int_\Gamma\frac{\varphi(\tau)}{\tau-t}\mathrm{d}\tau+$$

$$\frac{a_1(t)v_2[\alpha(t)]}{\mathrm{i}\pi}\int_\Gamma\frac{\varphi[\alpha(\tau)]}{\tau-t}\mathrm{d}\tau-\frac{a_2(t)v_2[\beta(t)]}{\mathrm{i}\pi}\int_\Gamma\frac{\varphi[\beta(\tau)]}{\tau-t}\mathrm{d}\tau-$$

$$\frac{a_3(t)v_2[\gamma(t)]}{\mathrm{i}\pi}\int_\Gamma\frac{\varphi[\gamma(\tau)]}{\tau-t}\mathrm{d}\tau+(\mathscr{D}_2^{**}\varphi)(t),$$

$$(\mathscr{T}_3v_3\mathscr{S}\varphi)(t)\equiv b_0(t)v_3(t)\varphi(t)+b_1(t)v_3[\alpha(t)]\varphi[\alpha(t)]-$$

$$b_2(t)v_3[\beta(t)]\varphi[\beta(t)]-b_3(t)v_3[\gamma(t)]\varphi[\gamma(t)]+\frac{a_0(t)v_3(t)}{\mathrm{i}\pi}\int_\Gamma\frac{\varphi(\tau)}{\tau-t}\mathrm{d}\tau-$$

$$\frac{a_1(t)v_3[\alpha(t)]}{\mathrm{i}\pi}\int_\Gamma\frac{\varphi[\alpha(\tau)]}{\tau-t}\mathrm{d}\tau+\frac{a_2(t)v_3[\beta(t)]}{\mathrm{i}\pi}\int_\Gamma\frac{\varphi[\beta(\tau)]}{\tau-t}\mathrm{d}\tau-$$

$$\frac{a_3(t)v_3[\gamma(t)]}{\mathrm{i}\pi}\int_\Gamma\frac{\varphi[\gamma(\tau)]}{\tau-t}\mathrm{d}\tau+(\mathscr{D}_3^{**}\varphi)(t),$$

这里$\mathscr{D}_j^*$，$\mathscr{D}_j^{**}$，$j=1,2,3$都是完全连续算子，利用(3.3)′(3.4)′不难得到定理 5 的结论.

定理6 如果$\alpha(t)$，$\beta(t)$中一个是正位移，另一个是反位移，无妨假设$\alpha(t)$是正位移，$\beta(t)$是反位移，那么对于算子$\mathscr{K}$和$\mathscr{T}_j$，$j=1,2,3$，将有关系式

$$\mathscr{S}W_j\mathscr{K}-\mathscr{T}_jW_j\mathscr{S}=\mathscr{D}_j,\ j=1,2,3 \quad (3.6)$$

成立，其中$w_1(t)=\alpha(t)-\beta(t)+\gamma(t)-t=u_2(t)$，$w_2(t)=\alpha(t)+\beta(t)-\gamma(t)-t=u_1(t)$，$w_3\equiv 1$，$\mathscr{S}$是奇异积分算子，$\mathscr{D}_j$，$j=1,2,3$是完全连续算子.

根据条件$\alpha(t)$是正位移，$\beta(t)$是反位移，所以$\gamma(t)$是反位移，从而知道$\alpha(t)\neq t$，$\gamma(t)\neq\beta(t)$，于是显然有$w_1(t)\neq 0$，$w_2(t)\neq 0$，而且

$$w_1[\alpha(t)]=-w_1(t),w_1[\beta(t)]=w_1(t),\quad w_1[\gamma(t)]=-w_1(t),\tag{3.3}''$$

$$w_2[\alpha(t)]=-w_2(t),w_2[\beta(t)]=-w_2(t),w_2[\gamma(t)]=w_2(t).\tag{3.2}'$$

通过直接计算算子$\mathscr{S}w_j\mathscr{K}$和$\mathscr{T}_jw_j\mathscr{S}$，考虑到$\alpha(t)$是正位移，$\beta(t)$，$\gamma(t)$是反位移的条件，再利用Poincarè-Bertrand公式，并分离出特征部分，我们将得到

$$\begin{aligned}(\mathscr{S}w_j\mathscr{K}\varphi)(t)&\equiv b_0(t)w_j(t)\varphi(t)+b_1(t)w_j(t)\varphi[\alpha(t)]-\\&b_2(t)w_j(t)\varphi[\beta(t)]-b_3(t)w_j(t)\varphi[\gamma(t)]+\frac{a_0(t)w_j(t)}{\mathrm{i}\pi}\int_\Gamma\frac{\varphi(\tau)}{\tau-t}\mathrm{d}\tau+\\&\frac{a_1(t)w_j(t)}{\mathrm{i}\pi}\int_\Gamma\frac{\varphi[\alpha(\tau)]}{\tau-t}\mathrm{d}\tau+\frac{a_2(t)w_j(t)}{\mathrm{i}\pi}\int_\Gamma\frac{\varphi[\beta(\tau)]}{\tau-t}\mathrm{d}\tau+\\&\frac{a_3(t)w_j(t)}{\mathrm{i}\pi}\int_\Gamma\frac{\varphi[\gamma(\tau)]}{\tau-t}\mathrm{d}\tau+(\mathscr{D}_j^*\varphi)(t),j=1,2,3,\end{aligned}$$

$$\begin{aligned}(\mathscr{T}_1w_1\mathscr{S}\varphi)(t)&\equiv b_0(t)w_1(t)\varphi(t)-b_1(t)w_1(t)\varphi[\alpha(t)]-\\&b_2(t)w_1[\beta(t)]\varphi[\beta(t)]+b_3(t)w_1[\gamma(t)]\varphi[\gamma(t)]+\frac{a_0(t)w_1(t)}{\mathrm{i}\pi}\int_\Gamma\frac{\varphi(\tau)}{\tau-t}\mathrm{d}\tau-\\&\frac{a_1(t)w_1[\alpha(t)]}{\mathrm{i}\pi}\int_\Gamma\frac{\varphi[\alpha(\tau)]}{\tau-t}\mathrm{d}\tau+\frac{a_2(t)w_1[\beta(t)]}{\mathrm{i}\pi}\int_\Gamma\frac{\varphi[\beta(\tau)]}{\tau-t}\mathrm{d}\tau-\\&\frac{a_3(t)w_1[\gamma(t)]}{\mathrm{i}\pi}\int_\Gamma\frac{\varphi[\gamma(\tau)]}{\tau-t}\mathrm{d}\tau+(\mathscr{D}_1^{**}\varphi)(t),\end{aligned}$$

$$\begin{aligned}(\mathscr{T}_2w_2\mathscr{S}\varphi)(t)&\equiv b_0(t)w_2(t)\varphi(t)-b_1(t)w_2[\alpha(t)]\varphi[\alpha(t)]+\\&b_2(t)w_2[\beta(t)]\varphi[\beta(t)]-b_3(t)w_2[\gamma(t)]\varphi[\gamma(t)]+\frac{a_0(t)w_2(t)}{\mathrm{i}\pi}\int_\Gamma\frac{\varphi(\tau)}{\tau-t}\mathrm{d}\tau-\\&\frac{a_1(t)w_2[\alpha(t)]}{\mathrm{i}\pi}\int_\Gamma\frac{\varphi[\alpha(\tau)]}{\tau-t}\mathrm{d}\tau-\frac{a_2(t)w_2[\beta(t)]}{\mathrm{i}\pi}\int_\Gamma\frac{\varphi[\beta(\tau)]}{\tau-t}\mathrm{d}\tau+\\&\frac{a_3(t)w_2[\gamma(t)]}{\mathrm{i}\pi}\int_\Gamma\frac{\varphi[\gamma(\tau)]}{\tau-t}\mathrm{d}\tau+(\mathscr{D}_2^{**}\varphi)(t),\end{aligned}$$

$$\begin{aligned}(\mathscr{T}_3\mathscr{S}\varphi)(t)&\equiv b_0(t)\varphi(t)+b_1(t)\varphi[\alpha(t)]-b_2(t)\varphi[\beta(t)]-b_3(t)\varphi[\gamma(t)]+\\&\frac{a_0(t)}{\mathrm{i}\pi}\int_\Gamma\frac{\varphi(\tau)}{\tau-t}\mathrm{d}\tau+\frac{a_1(t)}{\mathrm{i}\pi}\int_\Gamma\frac{\varphi[\alpha(\tau)]}{\tau-t}\mathrm{d}\tau+\frac{a_2(t)}{\mathrm{i}\pi}\int_\Gamma\frac{\varphi[\beta(\tau)]}{\tau-t}\mathrm{d}\tau+\end{aligned}$$

$$\frac{a_3(t)}{\mathrm{i}\pi}\int_\Gamma \frac{\varphi[\gamma(\tau)]}{\tau - t}\mathrm{d}\tau + (\mathscr{D}_3^{**}\varphi)(t),$$

这里 $\mathscr{D}_j^*$，$\mathscr{D}_j^{**}$，$j=1,2,3$ 都是完全连续算子，只要利用(3.3)″(3.2)′就不难得到定理 6 的结论.

定理 7　如果 $\mathscr{K}$ 或者 $\mathscr{T}_j(j=1,2,3)$ 中有一个是 Noether 算子，那么其余三个也必定是 Noether 算子，并且有

$$\mathrm{Ind}\mathscr{K} = \mathrm{Ind}\mathscr{T}_j, j=1,2,3. \tag{3.7}$$

(1) 如果 $\alpha(t)$，$\beta(t)$ 都是正位移，那么利用(3.1)很容易得到定理 7 的第一部分结论，另外

$$\mathrm{Ind}\mathscr{K} = \mathrm{Ind}u_j\mathscr{K} = \mathrm{Ind}(\mathscr{T}_i u_i + \mathscr{D}_j) = \mathrm{Ind}\mathscr{T}_j u_j = \mathrm{Ind}\mathscr{T}_j, j=1,2,3.$$

(2) 如果 $\alpha(t)$，$\beta(t)$ 都是反位移，那么利用(3.5)很容易得到定理 7 的第一部分结论，再根据 $\mathscr{S}$ 是有界线性算子将有

$$\mathrm{Ind}\mathscr{S}v_j\mathscr{K} = \mathrm{Ind}\mathscr{T}_j v_j \mathscr{S} \quad j=1,2,3，从而有$$

$$\mathrm{Ind}\mathscr{S} + \mathrm{Ind}v_j\mathscr{K} = \mathrm{Ind}\mathscr{T}_j v_j + \mathrm{Ind}\mathscr{S}$$

或

$$\mathrm{Ind}v_j\mathscr{K} = \mathrm{Ind}\mathscr{T}_j v_j,$$

最后有

$$\mathrm{Ind}\mathscr{K} = \mathrm{Ind}\mathscr{T}_j, j=1,2,3.$$

(3) 如果 $\alpha(t)$ 是正位移，$\beta(t)$ 是反位移时，那么利用(3.6)，仿照情形(2)不难得到定理 7 的结论.

§4.　Noether 条件和指数公式

现在我们讨论带两个 Carleman 位移的奇异积分方程(1.1)的 Noether 理论，上面已经知道奇异积分方程(1.1)与对应方程组(1.4)在一定意义下是等价的. 而方程组(1.4)的 Noether 理论是已知的[6]，也就是说，我们知道对应方程组(1.4)的 Noether 条件是

$$\det(\boldsymbol{p}(t) - \boldsymbol{q}(t)) = \det M(t, -1) \neq 0,$$
$$\det(\boldsymbol{p}(t) + \boldsymbol{q}(t)) = \det M(t, 1) \neq 0, \tag{4.1}$$

其中 $p(t)$，$q(t)$ 是(1.5)中所规定的矩阵，而方程组(1.4)的指数公式是

$$\mathrm{Ind}\mathscr{M} = \frac{1}{2\pi}\left\{\arg\frac{\det(\boldsymbol{p}(t) - \boldsymbol{q}(t))}{\det(\boldsymbol{p}(t) + \boldsymbol{q}(t))}\right\}_\Gamma. \tag{4.2}$$

定理 8　为了使奇异积分方程(1.1)是 Noether 方程的充分条件是满足条件(4.1).

事实上，当满足条件(4.1)时，方程组(1.4)显然是 Noether 方程组，

从而方程组(1.14)和它的相联方程组(1.13)的线性无关解的个数 l 和 l^* 都是有限的．于是，根据定理1、定理2可知 l_1 和 l_1^* 都是有限的．除此以外，还可以由方程组(1.4)的正规可解性得出方程(1.1)是正规可解的.

我们知道，方程组(1.4)可解的充分必要条件是满足

$$\int_\Gamma \boldsymbol{G}(t)\boldsymbol{W}(t)\mathrm{d}t = 0, \tag{4.3}$$

其中向量 $\boldsymbol{G}(t) = \{g(t), g[\alpha(t)], g[\beta(t)], g[\gamma(t)]\}$，向量 $\boldsymbol{W}(t) = \{\mu_1(t), \mu_2(t), \mu_3(t), \mu_4(t)\}$ 是相联齐次方程组的任意解，条件(4.3)还可以写成

$$\int_\Gamma \{g(t)\mu_1(t) + g[\alpha(t)]\mu_2(t) + g[\beta(t)]\mu_3(t) + g[\gamma(t)]\mu_4(t)\}\mathrm{d}t = 0, \tag{4.4}$$

根据定理3知道，这也是方程(1.1)可解的充分必要条件，对手满足条件(2.2)′的解 $\boldsymbol{W}(t)$ 来说，条件(4.4)还可以改写成

$$\int_\Gamma g(t)\mu_1(t)\mathrm{d}t + \nu_1\int_\Gamma g[\alpha(t)]\alpha'(t)\mu_1[\alpha(t)]\mathrm{d}t + \nu_2\int_\Gamma g[\beta(t)]\beta'(t)\mu_1[\beta(t)]\mathrm{d}t +$$
$$\nu_1\nu_2\int_\Gamma g[\gamma(t)]\gamma'(t)\mu_1[\gamma(t)]\mathrm{d}t = 0.$$

如果把上式中第二、三、四个积分中的 t 分别换成 $\alpha(t)$，$\beta(t)$ 和 $\gamma(t)$，我们得到

$$4\int_\Gamma g(t)\mu_1(t)\mathrm{d}t = 0,$$

这里 $\mu_1(t) = \varphi(t)$ 是相联方程(1.9)的任意解．

对于满足条件(2.3)′(2.4)′或(2.5)′的解 $\boldsymbol{W}(t)$ 来说，条件(4.4)还可以分别改写成

$$\int_\Gamma g(t)\mu_1(t)\mathrm{d}t - \nu_1\int_\Gamma g[\alpha(t)]\alpha'(t)\mu_1[\alpha(t)]\mathrm{d}t -$$
$$\nu_2\int_\Gamma g[\beta(t)]\beta'(t)\mu_1[\beta(t)]\mathrm{d}t + \nu_1\nu_2\int_\Gamma g[\gamma(t)]\gamma'(t)\mu_1[\gamma(t)]\mathrm{d}t = 0,$$

$$\int_\Gamma g(t)\mu_1(t)\mathrm{d}t - \nu_1\int_\Gamma g[\alpha(t)]\alpha'(t)\mu_1[\alpha(t)]\mathrm{d}t +$$
$$\nu_2\int_\Gamma g[\beta(t)]\beta'(t)\mu_1[\beta(t)]\mathrm{d}t - \nu_1\nu_2\int_\Gamma g[\gamma(t)]\gamma'(t)\mu_1[\gamma(t)]\mathrm{d}t = 0,$$

或

$$\int_\Gamma g(t)\mu_1(t)\mathrm{d}t + \nu_1\int_\Gamma g[\alpha(t)]\alpha'(t)\mu_1[\alpha(t)]\mathrm{d}t -$$

$$\nu_2\int_\Gamma g[\beta(t)]\beta'(t)\mu_1[\beta(t)]\mathrm{d}t-\nu_1\nu_2\int_\Gamma g[\gamma(t)]\gamma'(t)\mu_1[\gamma(t)]\mathrm{d}t=0.$$

把上面三式中的第二、三、四个积分中的 t 分别用 $\alpha(t)$, $\beta(t)$ 和 $\gamma(t)$ 来代替，马上看出，这时候，条件(4.4)将自动满足．于是，方程(1.1)可解的充分必要条件就可以写成

$$\int_\Gamma g(t)\varphi_k(t)\mathrm{d}t=0,\quad k=1,2,\cdots,l_1^*,\tag{4.5}$$

其中 $\varphi_k(t)$, $k=1,2,\cdots,l_1^*$ 是相联方程(1.9)的基本解系．这刚好说明方程(1.1)是 Noether 方程．

定理 9　满足 Noether 条件时，奇异积分方程(1.1)的指数可以按以下公式计算：

$$\mathrm{Ind}\mathscr{K}=\frac{1}{4}\mathrm{Ind}\mathscr{M}=\frac{1}{8\pi}\left\{\arg\frac{\det(\boldsymbol{p}(t)-\boldsymbol{q}(t))}{\det(\boldsymbol{p}(t)+\boldsymbol{q}(t))}\right\}_\Gamma\tag{4.6}$$

证　只要能够证明

$$\mathrm{Ind}\mathscr{L}=\mathrm{Ind}\mathscr{K}+\mathrm{Ind}\mathscr{T}_1+\mathrm{Ind}\mathscr{T}_2+\mathrm{Ind}\mathscr{T}_3$$

即可．事实上，根据 $\mathrm{Ind}\mathscr{M}=l-l^*$, $\mathrm{Ind}\mathscr{K}=l_1-l_1^*$, $\mathrm{Ind}\mathscr{T}_j=l_{j+1}-l_{j+1}^*$, $j=1,2,3$，和定理 1、定理 2 的结论 $l=l_1+l_2+l_3+l_4$ 和 $l^*=l_1^*+l_2^*+l_3^*+l_4^*$ 很容易得到 $l-l^*=(l_1-l_1^*)+(l_2-l_2^*)+(l_3-l_3^*)+(l_4-l_4^*)$．再根据 $\mathrm{Ind}\mathscr{K}=\mathrm{Ind}\mathscr{T}_j$, $j=1,2,3$，马上就可以得到公式(4.6)．

参考文献

[1] Литвинчук Г С. Краевые задачи и сингулярные интеяральные уравнения со сдвигом . 1977.

[2] Литвинчук Г С. Оъ индексе и нормадъной разрешимости одного класса функдионадъных уравнений. ДАН СССР, 1963, 149(5): 1 029-1 032.

[3] Литвинчук Г С. Теоремы нетера для одного кдасса сингулярных интегралъных уравнений со сдвигом и сопряжением. ДАН СССР, 1965, 162(1): 26-29.

[4] Гантмахер Ф Р. Теория матриц. 1967.

[5] Мусхелишвили Н И. Сингулярные интеяральные уравнения. 1968.（1962 年版有中译本，上海科技出版社）

[6] Векуа И Н. Системы сингулярных интеяралъных уравнений и некоторые граничные задачи. 1970.（1950 年版有中译本，上海科技出版社）

数学年刊，
1981，2A(1)：91-100.

带两个 Carleman 位移的奇异积分方程 Noether 可解的充分必要条件①

The Sufficient and Necessary Conditions for Noether's Solvability of Singular Integral Equations with Two Carleman's Shifts

Abstract In this paper we consider the problem of solvability of singular integral equtions with two Carleman's shifts

$$(\mathscr{K}\varphi)(t) \equiv a_0(t)\varphi(t) + a_1(t)\varphi[\alpha(t)] + a_2(t)\varphi[\beta(t)] + a_3(t)\varphi[\gamma(t)] + \frac{b_0(t)}{i\pi}\int_\Gamma \frac{\varphi(t)}{\tau - t}d\tau + \frac{b_1(t)}{i\pi}\int_\Gamma \frac{\varphi(t)}{\tau - \alpha(t)}d\tau + \frac{b_2(t)}{i\pi}\int_\Gamma \frac{\varphi(t)}{\tau - \beta(t)}d\tau + \frac{b_3(t)}{i\pi}\int_\Gamma \frac{\varphi(t)}{\tau - \gamma(t)}d\tau + \int_\Gamma K(t,\tau)\varphi(\tau)d\tau = g(t). \quad (0.1)$$

Suppose that Γ is a closed simple Lyapunoff's curve and $\alpha(t)$, $\beta(t)$ which satisfy Carleman's conditions and $\alpha[\beta(t)]=\beta[\alpha(t)]$ are two different homeomorphisms of Γ onto itself, and that $a_k(t)$, $b_k(t)$, $k=0,1,2,3$, belong to the space $H_\mu(\Gamma)$, $g(t)$ belongs to the space $L_p(\Gamma)$, $p>1$ and $K(t,\tau)$ has only weak singularity.

The following main results are obtained:

(1) Singular integral equation (0.1) is solvable if and only if the Noether's conditions

---

① 收稿日期：1979-12-28.

$$\det(\boldsymbol{p}(t) \pm \boldsymbol{q}(t)) \neq 0$$

are satisfied.

(2) Index of singular integral equation (0.1) is calculated by the formula

$$\operatorname{Ind}\mathscr{K} = \frac{1}{8\pi}\left\{\arg\frac{\det(\boldsymbol{p}(t) - \boldsymbol{q}(t))}{\det(\boldsymbol{p}(t) + \boldsymbol{q}(t))}\right\}_\Gamma,$$

where $\boldsymbol{p}(t)$ and $\boldsymbol{q}(t)$ are matrices of coefficients of so-called corresponding system of equations.

All these results have been generalized for systems of singular integral equtions with two Carleman's shifts and complex conjugate of unknown functions.

带 Carleman 位移的奇异积分方程理论，近年来得到了很大发展. 在[1]中建立了这种奇异积分方程的 Noether 理论，所用的基本方法是建立所谓的对应方程组(是不带位移的奇异积分方程组，它的理论是已知的，参看[2][3]). 在[4]中讨论了带两个 Carleman 位移的奇异积分方程 Noether 可解的充分条件，并给出了计算指数的公式. 本文目的是在文[4]的基础上，利用不同的方法解决带两个 Carleman 位移的奇异积分方程 Noether 可解的充分必要条件问题，并把所得结果对带两个 Carleman 位移及未知函数复共轭值的奇异积分方程进行推广.

§1.

假设 $\Gamma = \Gamma_0 + \Gamma_1 + \Gamma_2 + \cdots + \Gamma_m$ 是由 $m+1$ 条简单闭 Ляпунов 曲线组成，它围出一个连通区域 D^+. 假设 $\alpha(t), \beta(t)$ 是把 Γ 仍映射成它自身的两个不同的同胚，它们可以是正位移或反位移. 在 Γ 上它们都满足 Carleman 条件 K_2，亦即满足条件 $\alpha[\alpha(t)] = t, \beta[\beta(t)] = t$. 此外，还假设 $\alpha[\beta(t)] = \beta[\alpha(t)]$. 为了方便，我们记 $\gamma(t) = \alpha[\beta(t)] = \beta[\alpha(t)]$，显然，$\gamma(t)$ 也是满足 Carleman 条件 K_2 的位移. 关于位移 $\alpha(t), \beta(t)$，我们还要求 $\alpha'(t), \beta'(t) \in H_\mu(\Gamma)$，而且 $\alpha'(t) \cdot \beta'(t) \neq 0$.

我们将讨论带两个 Carleman 位移的奇异积分方程

$$(\mathscr{K}\varphi)(t) \equiv (\mathscr{K}^0\varphi)(t) + \int_\Gamma K(t,\tau)\varphi(\tau)\mathrm{d}\tau = g(t), \qquad (1.1)$$

这里

$$(\mathscr{K}^0\varphi)(t)\equiv a_0(t)\varphi(t)+a_1(t)\varphi[\alpha(t)]+a_2(t)\varphi[\beta(t)]+$$

$$a_3(t)\varphi[\gamma(t)]+\frac{b_0(t)}{\mathrm{i}\pi}\int_\Gamma\frac{\varphi(\tau)}{\tau-t}\mathrm{d}\tau+\frac{b_1(t)}{\mathrm{i}\pi}\int_\Gamma\frac{\varphi(\tau)}{\tau-\alpha(t)}\mathrm{d}\tau+$$

$$\frac{b_2(t)}{\mathrm{i}\pi}\int_\Gamma\frac{\varphi(\tau)}{\tau-\beta(t)}\mathrm{d}\tau+\frac{b_3(t)}{\mathrm{i}\pi}\int_\Gamma\frac{\varphi(\tau)}{\tau-\gamma(t)}\mathrm{d}\tau,\tag{1.2}$$

其中 $a_k(t),b_k(t),k=0,1,2,3$，属于空间 $H_\mu(\Gamma)$，$g(t)$ 属于空间 $L_p(\Gamma)$，$p>1$，而 $K(t,\tau)$ 最多只具有弱奇异性．如果 $K(t,\tau)\equiv 0$，我们将得到特征方程 $(\mathscr{K}^0\varphi)(t)=g(t)$．如果引入算子表示法，还可以把方程(1.1)写成如下形式

$$\mathscr{K}\equiv a_0(t)\mathscr{I}+a_1(t)\mathscr{W}_1+a_2(t)\mathscr{W}_2+a_3(t)\mathscr{W}_3+b_0(t)\mathscr{I}+$$

$$b_1(t)\mathscr{W}_1\mathscr{I}+b_2(t)\mathscr{W}_2\mathscr{I}+b_3(t)\mathscr{W}_3\mathscr{I}+\mathscr{D},\tag{1.3}$$

其中 $\mathscr{I}$ 是恒等算子，$\mathscr{W}_k(k=1,2,3)$ 是位移算子，$\mathscr{S}$ 是奇异积分算子，即

$$(\mathscr{W}_1\varphi)(t)\equiv\varphi[\alpha(t)],\ (\mathscr{W}_2\varphi)(t)\equiv\varphi[\beta(t)],\ (\mathscr{W}_3\varphi)(t)\equiv\varphi[\gamma(t)],$$

$$(\mathscr{I}\varphi)(t)\equiv\frac{1}{\mathrm{i}\pi}\int_\Gamma\frac{\varphi(t)}{\tau-t}\mathrm{d}\tau,\ (\mathscr{W}_1\mathscr{I}\varphi)(t)\equiv\frac{1}{\mathrm{i}\pi}\int_\Gamma\frac{\varphi(t)}{\tau-\alpha(t)}\mathrm{d}\tau,$$

$$(\mathscr{W}_2\mathscr{I}\varphi)(t)\equiv\frac{1}{\mathrm{i}\pi}\int_\Gamma\frac{\varphi(t)}{\tau-\beta(t)}\mathrm{d}\tau,\ (\mathscr{W}_3\mathscr{I}\varphi)(t)\equiv\frac{1}{\mathrm{i}\pi}\int_\Gamma\frac{\varphi(t)}{\tau-\gamma(t)}\mathrm{d}\tau,$$

而 $\mathscr{D}$ 是完全连续算子．

如果把奇异积分方程(1.1)中的 t 分别换成 $\alpha(t)$，$\beta(t)$ 和 $\gamma(t)$，并把这样得到的三个方程与方程(1.1)联立，再假设

$$\varphi(t)=\rho_1(t),\varphi[\alpha(t)]=\rho_2(t),\varphi[\beta(t)]=\rho_3(t),\varphi[\gamma(t)]=\rho_4(t),$$

最后得到方程组

$$a_0(t)\rho_1(t)+a_1(t)\rho_2(t)+a_2(t)\rho_3(t)+a_3(t)\rho_4(t)+$$

$$\frac{b_0(t)}{\mathrm{i}\pi}\int_\Gamma\frac{\rho_1(\tau)}{\tau-t}\mathrm{d}\tau+\frac{\nu_1 b_1(t)}{\mathrm{i}\pi}\int_\Gamma\frac{\alpha'(\tau)}{\alpha(\tau)-\alpha(t)}\rho_2(\tau)\mathrm{d}\tau+$$

$$\frac{\nu_2 b_2(t)}{\mathrm{i}\pi}\int_\Gamma\frac{\beta'(\tau)}{\beta(\tau)-\beta(t)}\rho_3(\tau)\mathrm{d}\tau+\frac{\nu_1\nu_2 b_3(t)}{\mathrm{i}\pi}\int_\Gamma\frac{\gamma'(\tau)}{\gamma(\tau)-\gamma(t)}\rho_4(\tau)\mathrm{d}\tau+$$

$$\int_\Gamma K(t,\tau)\rho_1(\tau)\mathrm{d}\tau=g(t),$$

$$a_1[\alpha(t)]\rho_1(t)+a_0[\alpha(t)]\rho_2(t)+a_3[\alpha(t)]\rho_3(t)+a_2[\alpha(t)]\rho_4(t)+$$

$$\frac{b_1[\alpha(t)]}{\mathrm{i}\pi}\int_\Gamma\frac{\rho_1(\tau)}{\tau-t}\mathrm{d}\tau+\frac{\nu_1 b_0(t)}{\mathrm{i}\pi}\int_\Gamma\frac{\alpha'(\tau)}{\alpha(\tau)-\alpha(t)}\rho_2(\tau)\mathrm{d}\tau+$$

$$\frac{\nu_2 b_3[\alpha(t)]}{\mathrm{i}\pi}\int_\Gamma \frac{\beta'(\tau)}{\beta(\tau)-\beta(t)}\rho_3(\tau)\mathrm{d}\tau+\frac{\nu_1\nu_2 b_2[\alpha(t)]}{\mathrm{i}\pi}\int_\Gamma \frac{\gamma'(\tau)}{\gamma(\tau)-\gamma(t)}\rho_4(\tau)\mathrm{d}\tau+$$

$$\int_\Gamma K[\alpha(t),\tau]\rho_1(\tau)\mathrm{d}\tau=g[\alpha(t)], \tag{1.4}$$

$$a_2[\beta(t)]\rho_1(t)+a_3[\beta(t)]\rho_2(t)+a_0[\beta(t)]\rho_3(t)+a_1[\beta(t)]\rho_4(t)+$$

$$\frac{b_2[\beta(t)]}{\mathrm{i}\pi}\int_\Gamma \frac{\rho_1(\tau)}{\tau-t}\mathrm{d}\tau+\frac{\nu_1 b_3[\beta(t)]}{\mathrm{i}\pi}\int_\Gamma \frac{\alpha'(\tau)}{\alpha(\tau)-\alpha(t)}\rho_2(\tau)\mathrm{d}\tau+$$

$$\frac{\nu_2 b_0[\beta(t)]}{\mathrm{i}\pi}\int_\Gamma \frac{\beta'(\tau)}{\beta(\tau)-\beta(t)}\rho_3(\tau)\mathrm{d}\tau+\frac{\nu_1\nu_2 b_1[\beta(t)]}{\mathrm{i}\pi}\int_\Gamma \frac{\gamma'(\tau)}{\gamma(\tau)-\gamma(t)}\rho_4(\tau)\mathrm{d}\tau+$$

$$\int_\Gamma K[\beta(t),\tau]\rho_1(\tau)\mathrm{d}\tau=g[\beta(t)],$$

$$a_3[\gamma(t)]\rho_1(t)+a_2[\gamma(t)]\rho_2(t)+a_1[\gamma(t)]\rho_3(t)+a_0[\gamma(t)]\rho_4(t)+$$

$$\frac{b_3[\gamma(t)]}{\mathrm{i}\pi}\int_\Gamma \frac{\rho_1(\tau)}{\tau-t}\mathrm{d}\tau+\frac{\nu_1 b_2[\gamma(t)]}{\mathrm{i}\pi}\int_\Gamma \frac{\alpha'(\tau)}{\alpha(\tau)-\alpha(t)}\rho_2(\tau)\mathrm{d}\tau+$$

$$\frac{\nu_2 b_1[\gamma(t)]}{\mathrm{i}\pi}\int_\Gamma \frac{\beta'(\tau)}{\beta(\tau)-\beta(t)}\rho_3(\tau)\mathrm{d}\tau+\frac{\nu_1\nu_2 b_0[\gamma(t)]}{\mathrm{i}\pi}\int_\Gamma \frac{\gamma'(\tau)}{\gamma(\tau)-\gamma(t)}\rho_4(\tau)\mathrm{d}\tau+$$

$$\int_\Gamma K[\gamma(t),\tau]\rho_1(\tau)\mathrm{d}\tau=g[\gamma(t)],$$

其中 $\nu_1=+1$ 或 -1，这依赖于 $\alpha(t)$ 是正位移或反位移，而 $\nu_2=+1$ 或 -1 这依赖于 $\beta(t)$ 是正位移或反位移.

方程组(1.4) 实际上是不带位移的奇异积分方程组，如果写成算子形式，将有

$$\mathscr{L}\equiv \boldsymbol{p}(t)\mathscr{I}+\boldsymbol{q}(t)\mathscr{S}+\mathscr{D}^*, \tag{1.5}$$

其中

$$\boldsymbol{p}(t)=\begin{pmatrix} a_0(t) & a_1(t) & a_2(t) & a_3(t)\\ a_1[\alpha(t)] & a_0[\alpha(t)] & a_3[\alpha(t)] & a_2[\alpha(t)]\\ a_2[\beta(t)] & a_3[\beta(t)] & a_0[\beta(t)] & a_1[\beta(t)]\\ a_3[\gamma(t)] & a_2[\gamma(t)] & a_1[\gamma(t)] & a_0[\gamma(t)]\end{pmatrix}, \tag{1.6}$$

$$\boldsymbol{q}(t)=\begin{pmatrix} b_0(t) & \nu_1 b_1(t) & \nu_2 b_2(t) & \nu_1\nu_2 b_3(t)\\ b_1[\alpha(t)] & \nu_1 b_0[\alpha(t)] & \nu_2 b_3[\alpha(t)] & \nu_1\nu_2 b_2[\alpha(t)]\\ b_2[\beta(t)] & \nu_1 b_3[\beta(t)] & \nu_2 b_0[\beta(t)] & \nu_1\nu_2 b_1[\beta(t)]\\ b_3[\gamma(t)] & \nu_1 b_2[\gamma(t)] & \nu_2 b_1[\gamma(t)] & \nu_1\nu_2 b_0[\gamma(t)]\end{pmatrix}, \tag{1.7}$$

$\mathscr{D}^*$ 是完全连续算子，或者

$$\mathscr{M}\equiv M(t,1)\mathscr{P}+M(t,-1)\mathscr{Q}+\mathscr{D}^*,$$

其中 $M(t,j)=\boldsymbol{p}(t)+j\boldsymbol{q}(t), j=\pm1$，而

$$\mathscr{P}=\frac{1}{2}(\mathscr{I}+\mathscr{S}),\quad \mathscr{Q}=\frac{1}{2}(\mathscr{I}-\mathscr{S}).$$

以后我们把方程组(1.4) 叫作方程(1.1) 的对应方程组，$\mathscr{L}$ 叫作对应算子，而 $M(t,j)$ 叫作算子 $\mathscr{L}$ 的标符，也叫作算子 $\mathscr{K}$ 的标符.

为了方便，我们还引入方程(1.1) 的三个伴随方程

$$(\mathscr{T}_1x)(t)\equiv a_0(t)x(t)-a_1(t)x[\alpha(t)]-a_2(t)x[\beta(t)]+a_3(t)x[\gamma(t)]+$$
$$\frac{b_0(t)}{\mathrm{i}\pi}\int_\Gamma\frac{x(\tau)}{\tau-t}\mathrm{d}\tau-\frac{b_1(t)}{\mathrm{i}\pi}\int_\Gamma\frac{x(\tau)}{\tau-\alpha(t)}\mathrm{d}\tau-\frac{b_2(t)}{\mathrm{i}\pi}\int_\Gamma\frac{x(\tau)}{\tau-\beta(t)}\mathrm{d}\tau+$$
$$\frac{b_3(t)}{\mathrm{i}\pi}\int_\Gamma\frac{x(\tau)}{\tau-\gamma(t)}\mathrm{d}\tau+\int_\Gamma K(t,\tau)x(\tau)\mathrm{d}\tau=0,\tag{1.8}$$

$$(\mathscr{T}_2x)(t)\equiv a_0(t)x(t)-a_1(t)x[\alpha(t)]+a_2(t)x[\beta(t)]-a_3(t)x[\gamma(t)]+$$
$$\frac{b_0(t)}{\mathrm{i}\pi}\int_\Gamma\frac{x(\tau)}{\tau-t}\mathrm{d}\tau-\frac{b_1(t)}{\mathrm{i}\pi}\int_\Gamma\frac{x(\tau)}{\tau-\alpha(t)}\mathrm{d}\tau+\frac{b_2(t)}{\mathrm{i}\pi}\int_\Gamma\frac{x(\tau)}{\tau-\beta(t)}\mathrm{d}\tau-$$
$$\frac{b_3(t)}{\mathrm{i}\pi}\int_\Gamma\frac{x(\tau)}{\tau-\gamma(t)}\mathrm{d}\tau+\int_\Gamma K(t,\tau)x(\tau)\mathrm{d}\tau=0,\tag{1.9}$$

$$(\mathscr{T}_3x)(t)\equiv a_0(t)x(t)+a_1(t)x[\alpha(t)]-a_2(t)x[\beta(t)]-a_3(t)x[\gamma(t)]+$$
$$\frac{b_0(t)}{\mathrm{i}\pi}\int_\Gamma\frac{x(\tau)}{\tau-t}\mathrm{d}\tau+\frac{b_1(t)}{\mathrm{i}\pi}\int_\Gamma\frac{x(\tau)}{\tau-\alpha(t)}\mathrm{d}\tau-\frac{b_2(t)}{\mathrm{i}\pi}\int_\Gamma\frac{x(\tau)}{\tau-\beta(t)}\mathrm{d}\tau-$$
$$\frac{b_3(t)}{\mathrm{i}\pi}\int_\Gamma\frac{x(\tau)}{\tau-\gamma(t)}\mathrm{d}\tau+\int_\Gamma K(t,\tau)x(\tau)\mathrm{d}\tau=0.\tag{1.10}$$

如果对(1.8) ~ (1.10) 分别令

$$\rho_1(t)=x(t),\rho_2(t)=-x[\alpha(t)],\rho_3(t)=-x[\beta(t)],\rho_4(t)=x[\gamma(t)];$$
$$\rho_1(t)=x(t),\rho_2(t)=-x[\alpha(t)],\rho_3(t)=x[\beta(t)],\rho_4(t)=-x[\gamma(t)];$$
$$\rho_1(t)=x(t),\rho_2(t)=x[\alpha(t)],\rho_3(t)=-x[\beta(t)],\rho_4(t)=-x[\gamma(t)],$$

可以得到以下结论：它们的对应方程组也是方程组(1.4).

§ 2. Noether 可解的充分必要条件

由于完全连续算子并不影响奇异积分算子的 Noether 性质，[1] 不失一般性，可以认为算子 $\mathscr{K}$ 与 $\mathscr{T}_j(j=1,2,3)$ 中的 $K(t,\tau)\equiv0$. 这时候，对应奇异积分算子可以写成

$$\mathscr{M}\equiv\boldsymbol{p}(t)\mathscr{I}+\boldsymbol{q}(t)\mathscr{S}+\mathscr{D},\tag{2.1}$$

其中 $\boldsymbol{p}(t),\boldsymbol{q}(t)$ 分别由(1.6)(1.7) 规定，且

$$\mathscr{I}(\rho_1,\rho_2,\rho_3,\rho_4)=(\mathscr{I}\rho_1,\mathscr{I}\rho_2,\mathscr{I}\rho_3,\mathscr{I}\rho_4),$$
$$\mathscr{S}(\rho_1,\rho_2,\rho_3,\rho_4)=(\mathscr{S}\rho_1,\mathscr{S}\rho_2,\mathscr{S}\rho_3,\mathscr{S}\rho_4),$$
$$\mathscr{D}(\rho_1,\rho_2,\rho_3,\rho_4)=(\mu_1,\mu_2,\mu_3,\mu_4),$$

这里

$$\mu_1=b_1(t)[\mathscr{W}_1\mathscr{S}\mathscr{W}_1-\nu_1\mathscr{S}]\rho_2+b_2(t)[\mathscr{W}_2\mathscr{S}\mathscr{W}_2-\nu_2\mathscr{S}]\rho_3+ b_3(t)[\mathscr{W}_3\mathscr{S}\mathscr{W}_3-\nu_1\nu_2\mathscr{S}]\rho_4 \tag{2.2}$$

$$\mu_2=b_0[\alpha(t)][\mathscr{W}_1\mathscr{S}\mathscr{W}_1-\nu_1\mathscr{S}]\rho_2+b_3(t)[\alpha(t)][\mathscr{W}_2\mathscr{S}\mathscr{W}_2-\nu_2\mathscr{S}]\rho_3+ b_2[\alpha(t)][\mathscr{W}_3\mathscr{S}\mathscr{W}_3-\nu_1\nu_2\mathscr{S}]\rho_4 \tag{2.3}$$

$$\mu_3=b_3[\beta(t)][\mathscr{W}_1\mathscr{S}\mathscr{W}_1-\nu_1\mathscr{S}]\rho_2+b_0(t)[\beta(t)][\mathscr{W}_2\mathscr{S}\mathscr{W}_2-\nu_2\mathscr{S}]\rho_3+ b_1[\beta(t)][\mathscr{W}_3\mathscr{S}\mathscr{W}_3-\nu_1\nu_2\mathscr{S}]\rho_4 \tag{2.4}$$

$$\mu_4=b_2[\gamma(t)][\mathscr{W}_1\mathscr{S}\mathscr{W}_1-\nu_1\mathscr{S}]\rho_2+b_1(t)[\gamma(t)][\mathscr{W}_2\mathscr{S}\mathscr{W}_2-\nu_2\mathscr{S}]\rho_3+ b_0[\gamma(t)][\mathscr{W}_3\mathscr{S}\mathscr{W}_3-\nu_1\nu_2\mathscr{S}]\rho_4 \tag{2.5}$$

我们将在四维向量空间 $L_p^4(\Gamma),p>1$ 中来讨论对应奇异积分算子，我们把分别由四维向量

$$\{\rho(t),\quad \rho[\alpha(t)],\quad \rho[\beta(t)],\quad \rho[\gamma(t)]\},$$
$$\{\rho(t),\quad -\rho[\alpha(t)],\quad -\rho[\beta(t)],\quad \rho[\gamma(t)]\},$$
$$\{\rho(t),\quad -\rho[\alpha(t)],\quad \rho[\beta(t)],\quad -\rho[\gamma(t)]\},$$
$$\{\rho(t),\quad \rho[\alpha(t)],\quad -\rho[\beta(t)],\quad -\rho[\gamma(t)]\},$$

产生的子空间记作 $L_p^{4,0}(\Gamma),L_p^{4,1}(\Gamma),L_p^{4,2}(\Gamma),L_p^{4,3}(\Gamma)$，这里 $\rho(t)\in L_p(\Gamma),p>1$.

引理 1　四维向量空间 $L_p^4(\Gamma)$ 可以分解成子空间 $L_p^{4,0}(\Gamma),L_p^{4,1}(\Gamma)$，$L_p^{4,2}(\Gamma)$ 和 $L_p^{4,3}(\Gamma)$ 的直接和.

证　只要证明 $L_p^{4,0}(\Gamma)\cap L_p^{4,1}(\Gamma)\cap L_p^{4,2}(\Gamma)\cap L_p^{4,3}(\Gamma)=\{0\}$，而且任意四维向量 $\boldsymbol{\varphi}(t)=\{\varphi_1(t),\varphi_2(t),\varphi_3(t),\varphi_4(t)\}\in L_p^4(\Gamma)$ 都可以表示成 $\boldsymbol{\varphi}(t)=\boldsymbol{\psi}_0(t)+\boldsymbol{\psi}_1(t)+\boldsymbol{\psi}_2(t)+\boldsymbol{\psi}_3(t)$，其中 $\boldsymbol{\psi}_j(t)\in L_p^{4,j}(\Gamma),j=0,1,2,3$ 就可以了.

如果假设

$$\boldsymbol{\psi}_0(t)=\Big\{\frac{\varphi_1(t)+\varphi_2[\alpha(t)]+\varphi_3[\beta(t)]+\varphi_4[\gamma(t)]}{4},$$
$$\frac{\varphi_2(t)+\varphi_1[\alpha(t)]+\varphi_4[\beta(t)]+\varphi_3[\gamma(t)]}{4},$$

$$\frac{\varphi_3(t)+\varphi_4[\alpha(t)]+\varphi_1[\beta(t)]+\varphi_2[\gamma(t)]}{4},$$

$$\left.\frac{\varphi_4(t)+\varphi_3[\alpha(t)]+\varphi_2[\beta(t)]+\varphi_1[\gamma(t)]}{4}\right\},$$

$$\boldsymbol{\psi}_1(t)=\left\{\frac{\varphi_1(t)-\varphi_2[\alpha(t)]-\varphi_3[\beta(t)]+\varphi_4[\gamma(t)]}{4},\right.$$

$$\frac{\varphi_2(t)-\varphi_1[\alpha(t)]-\varphi_4[\beta(t)]+\varphi_3[\gamma(t)]}{4},$$

$$\frac{\varphi_3(t)-\varphi_4[\alpha(t)]-\varphi_1[\beta(t)]+\varphi_2[\gamma(t)]}{4},$$

$$\left.\frac{\varphi_4(t)-\varphi_3[\alpha(t)]-\varphi_2[\beta(t)]+\varphi_1[\gamma(t)]}{4}\right\},$$

$$\boldsymbol{\psi}_2(t)=\left\{\frac{\varphi_1(t)-\varphi_2[\alpha(t)]+\varphi_3[\beta(t)]-\varphi_4[\gamma(t)]}{4},\right.$$

$$\frac{\varphi_2(t)-\varphi_1[\alpha(t)]+\varphi_4[\beta(t)]-\varphi_3[\gamma(t)]}{4},$$

$$\frac{\varphi_3(t)-\varphi_4[\alpha(t)]+\varphi_1[\beta(t)]-\varphi_2[\gamma(t)]}{4},$$

$$\left.\frac{\varphi_4(t)-\varphi_3[\alpha(t)]+\varphi_2[\beta(t)]-\varphi_1[\gamma(t)]}{4}\right\},$$

$$\boldsymbol{\psi}_3(t)=\left\{\frac{\varphi_1(t)+\varphi_2[\alpha(t)]-\varphi_3[\beta(t)]-\varphi_4[\gamma(t)]}{4},\right.$$

$$\frac{\varphi_2(t)+\varphi_1[\alpha(t)]-\varphi_4[\beta(t)]-\varphi_3[\gamma(t)]}{4},$$

$$\frac{\varphi_3(t)+\varphi_4[\alpha(t)]-\varphi_1[\beta(t)]-\varphi_2[\gamma(t)]}{4},$$

$$\left.\frac{\varphi_4(t)+\varphi_3[\alpha(t)]-\varphi_2[\beta(t)]-\varphi_1[\gamma(t)]}{4}\right\},$$

不难验证，向量 $\boldsymbol{\varphi}_j(t)\in L_p^{4,j}(\Gamma)(j=0,1,2,3)$，且 $\boldsymbol{\varphi}(t)=\boldsymbol{\psi}_0(t)+\boldsymbol{\psi}_1(t)+\boldsymbol{\psi}_2(t)+\boldsymbol{\psi}_3(t)$. 此外，如果 $\boldsymbol{\rho}(t)=\{\rho_1(t),\rho_2(t),\rho_3(t),\rho_4(t)\}\in L_p^{4,0}(\Gamma)\cap L_p^{4,1}(\Gamma)\cap L_p^{4,2}(\Gamma)\cap L_p^{4,3}(\Gamma)$，则有

$$\rho_2(t)=\rho_1[\alpha(t)] \text{ 和 } \rho_2(t)=-\rho_1[\alpha(t)],$$

$$\rho_3(t)=\rho_1[\beta(t)] \text{ 和 } \rho_3(t)=-\rho_1[\beta(t)],$$

$$\rho_4(t)=\rho_1[\gamma(t)] \text{ 和 } \rho_4(t)=-\rho_1[\gamma(t)],$$

从而有 $\rho_2(t)=\rho_3(t)=\rho_4(t)\equiv 0$. 又 $\rho_1(t)=\rho_2[\alpha(t)]\equiv 0$，于是 $\rho(t)\equiv$

0. 引理 1 得证.

引理 2 子空间 $L_p^{4,j}(\Gamma)(j=0,1,2,3)$ 关于算子 $\mathscr{M}$ 是不变的, 算子 $\mathscr{M}$ 到子空间 $L_p^{4,0}(\Gamma)$ 的收缩 $\mathscr{M}_0=\mathscr{M}|_{L_p^{4,0}(\Gamma)}$ 在 Noether 意义下与算子 $\mathscr{K}$ 等价, 而算子 $\mathscr{M}$ 到子空间 $L_p^{4,j}(\Gamma)$ 的收缩 $\mathscr{M}_j=\mathscr{M}|_{L_p^{4,j}(\Gamma)}$ 在 Noether 意义下与伴随算子 $\mathscr{T}_j$ 等价, $j=1,2,3$.

证 只要证明

$$\begin{aligned}
&\mathscr{M}_0(\rho(t),\rho[\alpha(t)],\rho[\beta(t)],\rho[\gamma(t)])\\
&=\mathscr{M}(\rho(t),\rho[\alpha(t)],\rho[\beta(t)],\rho[\gamma(t)])=(e_1^{(0)},e_2^{(0)},e_3^{(0)},e_4^{(0)})\\
&=(\mathscr{K}\rho,\mathscr{W}_1\mathscr{K}\rho,\mathscr{W}_2\mathscr{K}\rho,\mathscr{W}_3\mathscr{K}\rho),
\end{aligned}\tag{2.6}$$

$$\begin{aligned}
&\mathscr{M}_1(\rho(t),-\rho[\alpha(t)],-\rho[\beta(t)],\rho[\gamma(t)])\\
&=\mathscr{M}(\rho(t),-\rho[\alpha(t)],-\rho[\beta(t)],\rho[\gamma(t)])=(e_1^{(1)},e_2^{(1)},e_3^{(1)},e_4^{(1)})\\
&=(\mathscr{T}_1\rho,-\mathscr{W}_1\mathscr{T}_1\rho,-\mathscr{W}_2\mathscr{T}_1\rho,\mathscr{W}_3\mathscr{T}_1\rho),
\end{aligned}\tag{2.7}$$

$$\begin{aligned}
&\mathscr{M}_2(\rho(t),-\rho[\alpha(t)],\rho[\beta(t)],-\rho[\gamma(t)])\\
&=\mathscr{M}(\rho(t),-\rho[\alpha(t)],-\rho[\beta(t)],-\rho[\gamma(t)])=(e_1^{(2)},e_2^{(2)},e_3^{(2)},e_4^{(2)})\\
&=(\mathscr{T}_2\rho,-\mathscr{W}_1\mathscr{T}_2\rho,-\mathscr{W}_2\mathscr{T}_2\rho,-\mathscr{W}_3\mathscr{T}_2\rho),
\end{aligned}\tag{2.8}$$

$$\begin{aligned}
&\mathscr{M}_3(\rho(t),\rho[\alpha(t)],-\rho[\beta(t)],-\rho[\gamma(t)])\\
&=\mathscr{M}(\rho(t),\rho[\alpha(t)],-\rho[\beta(t)],-\rho[\gamma(t)])=(e_1^{(3)},e_2^{(3)},e_3^{(3)},e_4^{(3)})\\
&=(\mathscr{T}_3\rho,\mathscr{W}_1\mathscr{T}_3\rho,-\mathscr{W}_2\mathscr{T}_3\rho,-\mathscr{W}_3\mathscr{T}_3\rho),
\end{aligned}\tag{2.9}$$

就可以了. 首先证明等式(2.6).

$$\begin{aligned}
e_1^{(0)}&=a_0(t)\rho(t)+a_1(t)\rho[\alpha(t)]+a_2(t)\rho[\beta(t)]+a_3(t)\rho[\gamma(t)]+\\
&\quad b_0(t)\mathscr{S}\rho(t)+\nu_1b_1(t)\mathscr{S}\rho[\alpha(t)]+\nu_2b_2(t)\mathscr{S}\rho[\beta(t)]+\\
&\quad \nu_1\nu_2b_3(t)\mathscr{S}\rho[\gamma(t)]+\mu_1{}^{①}\\
&=a_0(t)\rho(t)+a_1(t)\rho[\alpha(t)]+a_2(t)\rho[\beta(t)]+a_3(t)\rho[\gamma(t)]+\\
&\quad b_0(t)\mathscr{S}\rho(t)+b_1(t)\mathscr{W}_1\mathscr{S}\rho(t)+b_2(t)\mathscr{W}_2\mathscr{S}\rho(t)+b_3(t)\mathscr{W}_3\mathscr{S}\rho(t)\\
&=\mathscr{K}\rho,
\end{aligned}$$

$$\begin{aligned}
e_2^{(0)}&=a_1[\alpha(t)]\rho(t)+a_0[\alpha(t)]\rho[\alpha(t)]+a_3[\alpha(t)]\rho[\beta(t)]+\\
&\quad a_2[\alpha(t)]\rho[\gamma(t)]+b_1[\alpha(t)]\mathscr{S}\rho(t)+\nu_1b_0[\alpha(t)]\mathscr{S}\rho[\alpha(t)]+\\
&\quad \nu_2b_3[\alpha(t)]\mathscr{S}\rho[\beta(t)]+\nu_1\nu_2b_3[\alpha(t)]\mathscr{S}\rho[\gamma(t)]+\mu_2
\end{aligned}$$

---

① 这里 μ_1 由(2.2)规定, 只是取 $\rho_2=\rho[\alpha(t)]$, $\rho_3=\rho[\beta(t)]$, $\rho_4=\rho[\gamma(t)]$. 下面用到的 μ_2, μ_3, μ_4 也仿此利用(2.3)~(2.5)规定.

$$= \mathscr{W}_1[a_0(t)\rho(t) + a_1(t)\rho[\alpha(t)] + a_2(t)\rho[\beta(t)] + a_3(t)\rho[\gamma(t)] + b_0(t)\mathscr{S}\rho(t) + b_1(t)\mathscr{W}_1\mathscr{S}\rho(t) + b_2(t)\mathscr{W}_2\mathscr{S}\rho(t) + b_3(t)\mathscr{W}_3\mathscr{S}\rho(t)]$$

$$= \mathscr{W}_1\mathscr{K}\rho,$$

$$\begin{aligned} e_3^{(0)} &= a_2[\beta(t)]\rho(t) + a_3[\beta(t)]\rho[\alpha(t)] + a_0[\beta(t)]\rho[\beta(t)] + a_1[\beta(t)](t)\rho[\gamma(t)] + b_2[\beta(t)]\mathscr{S}\rho(t) + \nu_1 b_3[\beta(t)]\mathscr{S}\rho[\alpha(t)] + \\ &\quad \nu_2 b_0[\beta(t)]\mathscr{S}\rho[\beta(t)] + \nu_1\nu_2 b_1[\beta(t)]\mathscr{S}\rho[\gamma(t)] + \mu_3 \\ &= a_2[\beta(t)]\rho(t) + a_3[\beta(t)]\rho[\alpha(t)] + a_0[\beta(t)]\rho[\beta(t)] + a_1[\beta(t)]\rho[\gamma(t)] + b_2[\beta(t)]\mathscr{S}\rho(t) + b_3[\beta(t)]\mathscr{W}_1\mathscr{S}\rho(t) + \\ &\quad b_0[\beta(t)]\mathscr{W}_2\mathscr{S}\rho(t) + b_1[\beta(t)]\mathscr{W}_3\mathscr{S}\rho(t) \\ &= \mathscr{W}_2\mathscr{K}\rho, \end{aligned}$$

$$\begin{aligned} e_4^{(0)} &= a_3[\gamma(t)]\rho(t) + a_2[\gamma(t)]\rho[\alpha(t)] + a_1[\gamma(t)]\rho[\beta(t)] + a_0[\gamma(t)](t)\rho[\gamma(t)] + b_3[\gamma(t)]\mathscr{S}\rho(t) + \nu_1 b_2[\gamma(t)]\mathscr{S}\rho[\alpha(t)] + \\ &\quad \nu_2 b_1[\gamma(t)]\mathscr{S}\rho[\beta(t)] + \nu_1\nu_2 b_0[\gamma(t)]\mathscr{S}\rho[\gamma(t)] + \mu_4 \\ &= a_3[\gamma(t)]\rho(t) + a_2[\gamma(t)]\rho[\alpha(t)] + a_1[\gamma(t)]\rho[\beta(t)] + a_0[\gamma(t)]\rho[\gamma(t)] + b_3[\gamma(t)]\mathscr{S}\rho(t) + b_2[\gamma(t)]\mathscr{W}_1\mathscr{S}\rho(t) + \\ &\quad b_1[\gamma(t)]\mathscr{W}_2\mathscr{S}\rho(t) + b_0[\gamma(t)]\mathscr{W}_3\mathscr{S}\rho(t) \\ &= \mathscr{W}_3\mathscr{K}\rho. \end{aligned}$$

类似地可以证明等式(2.7)～(2.9)成立.例如

$$\begin{aligned} e_1^{(1)} &= a_0(t)\rho(t) - a_1(t)\rho[\alpha(t)] - a_2(t)\rho[\beta(t)] + a_3(t)\rho[\gamma(t)] + b_0(t)\mathscr{S}\rho(t) - \nu_1 b_1(t)\mathscr{S}\rho[\alpha(t)] - \nu_2 b_2(t)\mathscr{S}\rho[\beta(t)] + \\ &\quad \nu_1\nu_2 b_3(t)\mathscr{S}\rho[\gamma(t)] + \tilde{\mu}_1 \text{①} \\ &= a_0(t)\rho(t) - a_1(t)\rho[\alpha(t)] - a_2(t)\rho[\beta(t)] + a_3(t)\rho[\gamma(t)] + b_0(t)\mathscr{S}\rho(t) - b_1(t)\mathscr{W}_1\mathscr{S}\rho(t) - b_2(t)\mathscr{W}_2\mathscr{S}\rho(t) + b_3(t)\mathscr{W}_3\mathscr{S}\rho(t) \\ &= \mathscr{T}_1\rho, \end{aligned}$$

$$\begin{aligned} e_2^{(1)} &= a_1[\alpha(t)]\rho(t) - a_0[\alpha(t)]\rho[\alpha(t)] - a_3[\alpha(t)]\rho[\beta(t)] + a_2[\alpha(t)]\rho[\gamma(t)] + b_1[\alpha(t)]\mathscr{S}\rho(t) - \nu_1 b_0[\alpha(t)]\mathscr{S}\rho[\alpha(t)] - \\ &\quad \nu_2 b_3[\alpha(t)]\mathscr{S}\rho[\beta(t)] + \nu_1\nu_2 b_2[\alpha(t)]\mathscr{S}\rho[\gamma(t)] + \tilde{\mu}_2 \\ &= a_1[\alpha(t)]\rho(t) - a_0[\alpha(t)]\rho[\alpha(t)] - a_3[\alpha(t)]\rho[\beta(t)] + \end{aligned}$$

---

① 这里 $\tilde{\mu}_1$ 由(2.2)规定,只是取 $\rho_2 = -\rho[\alpha(t)]$, $\rho_3 = -\rho[\beta(t)]$, $\rho_4 = \rho[\gamma(t)]$. 下面用到的 $\tilde{\mu}_2$, $\tilde{\mu}_3$, $\tilde{\mu}_4$ 也仿此利用(2.3)～(2.5)规定.

$$a_2[\alpha(t)]\rho[\gamma(t)]+b_1[\alpha(t)]\mathscr{S}\rho(t)+b_0[\alpha(t)]\mathscr{W}_1\mathscr{S}\rho(t)-$$
$$b_3[\alpha(t)]\mathscr{W}_2\mathscr{S}\rho(t)+b_2[\alpha(t)]\mathscr{W}_3\mathscr{S}\rho(t)$$
$$=-\mathscr{W}_1\mathscr{J}_1\rho,$$

$$e_3^{(1)}=a_2[\beta(t)]\rho(t)-a_3[\beta(t)]\rho[\alpha(t)]-a_0[\beta(t)]\rho[\beta(t)]+$$
$$a_1[\beta(t)]\rho[\gamma(t)]+b_2[\beta(t)]\mathscr{S}\rho(t)-\nu_1 b_3[\beta(t)]\mathscr{S}\rho[\alpha(t)]-$$
$$\nu_2 b_0[\beta(t)]\mathscr{S}\rho[\beta(t)]+\nu_1\nu_2 b_1[\beta(t)]\mathscr{S}\rho[\gamma(t)]+\tilde{\mu}_3$$
$$=a_2[\beta(t)]\rho(t)-a_3[\beta(t)]\rho[\alpha(t)]-a_0[\beta(t)]\rho[\beta(t)]+$$
$$a_1[\beta(t)]\rho[\gamma(t)]+b_2[\beta(t)]\mathscr{S}\rho(t)-b_3[\beta(t)]\mathscr{W}_1\mathscr{S}\rho(t)-$$
$$b_0[\beta(t)]\mathscr{W}_2\mathscr{S}\rho(t)+b_3[\beta(t)]\mathscr{W}_3\mathscr{S}\rho(t)$$
$$=-\mathscr{W}_2\mathscr{J}_1\rho,$$

$$e_4^{(1)}=a_3[\gamma(t)]\rho(t)-a_2[\gamma(t)]\rho[\alpha(t)]-a_1[\gamma(t)]\rho[\beta(t)]+$$
$$a_0[\gamma(t)]\rho[\gamma(t)]+b_3[\gamma(t)]\mathscr{S}\rho(t)-\nu_1 b_2[\gamma(t)]\mathscr{S}\rho[\alpha(t)]-$$
$$\nu_2 b_1[\gamma(t)]\mathscr{S}\rho[\beta(t)]+\nu_1\nu_2 b_0[\gamma(t)]\mathscr{S}\rho[\gamma(t)]+\tilde{\mu}_4$$
$$=a_3[\gamma(t)]\rho(t)-a_2[\gamma(t)]\rho[\alpha(t)]-a_1[\gamma(t)]\rho[\beta(t)]+$$
$$a_0[\gamma(t)]\rho[\gamma(t)]+b_3[\gamma(t)]\mathscr{S}\rho(t)-b_2[\gamma(t)]\mathscr{W}_1\mathscr{S}\rho(t)-$$
$$b_1[\gamma(t)]\mathscr{W}_2\mathscr{S}\rho(t)+b_0[\gamma(t)]\mathscr{W}_3\mathscr{S}\rho(t)$$
$$=\mathscr{W}_3\mathscr{J}_1\rho.$$

从而引理 2 得证.

定理 1　为了使算子

$$\mathscr{K}\equiv a_0(t)\mathscr{I}+a_1(t)\mathscr{W}_1+a_2(t)\mathscr{W}_2+a_3(t)\mathscr{W}_3+b_1(t)\mathscr{S}+$$
$$b_1(t)\mathscr{W}_1\mathscr{S}+b_2(t)\mathscr{W}_2\mathscr{S}+b_3(t)\mathscr{W}_3\mathscr{S}+\mathscr{D} \tag{1.3}$$

是 Noether 算子的充分必要条件是它的标符是非退化的,也就是

$$\det M(t,j)=\det(\boldsymbol{p}(t)+j\boldsymbol{q}(t))\neq 0,\ j=\pm 1, \tag{2.10}$$

Noether 算子 $\mathscr{K}$ 的指数是

$$\operatorname{Ind}\mathscr{K}=\frac{1}{8\pi}\left\{\arg\frac{\det M(t,-1)}{\det M(t,+1)}\right\}_\Gamma. \tag{2.11}$$

证　首先证明条件的**充分性**,根据定义我们知道算子 $\mathscr{K}$ 与对应算子 $\mathscr{M}$ 的标符是完全一致的,因此只要条件(2.10) 满足,那么对应算子 $\mathscr{M}$ 就是 Noether 算子[1]. 又根据引理 1 知道,算子 $\mathscr{M}$ 是算子 $\mathscr{M}_j$,$(j=0,1,2,3)$ 的直接和,从而算子 $\mathscr{M}_0=\mathscr{M}|_{L_p^{4,0}(\Gamma)}$ 也就是 Noether 算子,再根据引理 2 知道算子 $\mathscr{K}$ 也一定是 Noether 算子.

反过来，再证明条件的**必要性**. 这时候，假设算子 $\mathscr{K}$ 是 Noether 算子，于是伴随算子 $\mathscr{T}_j(j=1,2,3)$ 也必定都是 Noether 算子(参看[4]定理7)，再根据引理 2 知道算子 $\mathscr{M}_j(j=0,1,2,3)$ 都是 Noether 算子，从而算子 $\mathscr{M}$ 也是 Noether 算子，从而条件(2.10)必定满足[1].

除此以外，我们知道[4] $\operatorname{Ind}\mathscr{K}=\operatorname{Ind}\mathscr{T}_j(j=1,2,3)$ 而且 $\operatorname{Ind}\mathscr{M}=\operatorname{Ind}\mathscr{K}+\operatorname{Ind}\mathscr{T}_1+\operatorname{Ind}\mathscr{T}_2+\operatorname{Ind}\mathscr{T}_3$，从而有

$$\operatorname{Ind}\mathscr{K}=\frac{1}{8\pi}\left\{\arg\frac{\det M(t,-1)}{\det M(t,+1)}\right\}_{\Gamma}.$$

§ 3.

对于带两个 Carleman 位移的奇异积分方程组来说，只不过是把方程(1.1)中的系数 $a_k(t),b_k(t)(k=0,1,2,3)$ 理解为 $(m\times m)$ 阶方阵，它们的元素属于空间 $H_\mu(\Gamma)$，$\boldsymbol{g}(t)$ 是 m 维向量，它的元素属于 $L_p(\Gamma)$，$p>1$. 而 $K(t,\tau)$ 也应理解为 $(m\times m)$ 阶方阵，它的元素最多只具有弱奇异性. 这样一来，上一段中所有讨论不需要做任何本质上的改变，都可以搬到方程组的情形. 只是对应方程组(1.4)中方程与未知函数的个数由 4 增加到 $4m$.

对于带两个 Carleman 位移的奇异积分方程组来说，如果引用算子记号将有以下结论：

定理 2 为了使算子

$$\mathscr{K}\equiv\boldsymbol{a}_0(t)\mathscr{J}+\boldsymbol{a}_1(t)\mathscr{W}_1+\boldsymbol{a}_2(t)\mathscr{W}_2+\boldsymbol{a}_3(t)\mathscr{W}_3+\boldsymbol{b}_0(t)\mathscr{S}+\boldsymbol{b}_1(t)\mathscr{W}_1\mathscr{S}+\boldsymbol{b}_2(t)\mathscr{W}_2\mathscr{S}+\boldsymbol{b}_3(t)\mathscr{W}_3\mathscr{S}+\mathscr{D}\tag{3.1}$$

是 Noether 算子的充分必要条件是它的标符是非退化的，也就是说

$$\det M(t,j)=\det(\boldsymbol{p}(t)+j\boldsymbol{q}(t))\neq 0,\ j=\pm 1.\tag{3.2}$$

这里 $\boldsymbol{a}_k(t),\boldsymbol{b}_k(t)(k=0,1,2,3)$ 是 $(m\times m)$ 阶方阵，而 $\boldsymbol{p}_k(t),\boldsymbol{q}_k(t)$ 是 $(4m\times 4m)$ 阶方阵，它的具体形式与(1.6)(1.7)完全类似，只不过代替 $a_k(t)$，$b_k(t)$ 的位置，在这里应该换成相应的 $(m\times m)$ 阶方阵.

另外，算子 $\mathscr{K}$ 的指数公式(2.11)也是成立的.

§ 4.

我们考虑以下更为一般的奇异积分方程

$$(\mathcal{K}\varphi)(t)\equiv a_0(t)\varphi(t)+a_1(t)\varphi[\alpha(t)]+a_2(t)\varphi[\beta(t)]+$$
$$a_3(t)\varphi[\gamma(t)]+c_0(t)\overline{\varphi(t)}+c_1(t)\overline{\varphi[\alpha(t)]}+c_2(t)\overline{\varphi[\beta(t)]}+$$
$$c_3(t)\overline{\varphi[\gamma(t)]}+\frac{b_0(t)}{\mathrm{i}\pi}\int_\Gamma\frac{\varphi(\tau)}{\tau-t}\mathrm{d}\tau+b_1(t)\frac{1}{\mathrm{i}\pi}\int_\Gamma\frac{\varphi(\tau)}{\tau-\alpha(t)}\mathrm{d}\tau+$$
$$\frac{b_2(t)}{\mathrm{i}\pi}\int_\Gamma\frac{\varphi(\tau)}{\tau-\beta(t)}\mathrm{d}\tau+\frac{b_3(t)}{\mathrm{i}\pi}\int_\Gamma\frac{\varphi(\tau)}{\tau-\gamma(t)}\mathrm{d}\tau+\boldsymbol{d}_0(t)\overline{\frac{1}{\mathrm{i}\pi}\int_\Gamma\frac{\varphi(\tau)}{\tau-t}\mathrm{d}\tau}+$$
$$\boldsymbol{d}_1(t)\overline{\frac{1}{\mathrm{i}\pi}\int_\Gamma\frac{\varphi(\tau)}{\tau-\alpha(t)}\mathrm{d}\tau}+\boldsymbol{d}_2(t)\overline{\frac{1}{\mathrm{i}\pi}\int_\Gamma\frac{\varphi(\tau)}{\tau-\beta(t)}\mathrm{d}\tau}+$$
$$\boldsymbol{d}_3(t)\overline{\frac{1}{\mathrm{i}\pi}\int_\Gamma\frac{\varphi(\tau)}{\tau-\gamma(t)}\mathrm{d}\tau}+\int_\Gamma K_1(t,\tau)\varphi(\tau)\mathrm{d}\tau+$$
$$\overline{\int_\Gamma K_2(t,\tau)\varphi(\tau)\mathrm{d}\tau}=g(t). \tag{4.1}$$

如果把由方程(4.1)取复共轭值所得到的方程与方程(4.1)联立,并引入新的未知函数

$$\varphi_1(t)=\varphi(t),\ \varphi_2(t)=\overline{\varphi(t)},$$

就可以得到以下方程组

$$a_0(t)\varphi_1(t)+a_1(t)\varphi_1[\alpha(t)]+a_2(t)\varphi_1[\beta(t)]+a_3(t)\varphi_1[\gamma(t)]+$$
$$c_0(t)\varphi_2(t)+c_1(t)\varphi_2[\alpha(t)]+c_2(t)\varphi_2[\beta(t)]+c_3(t)\varphi_2[\gamma(t)]+$$
$$\frac{b_0(t)}{\mathrm{i}\pi}\int_\Gamma\frac{\varphi_1(\tau)}{\tau-t}\mathrm{d}\tau+\frac{b_1(t)}{\mathrm{i}\pi}\int_\Gamma\frac{\varphi_1(\tau)}{\tau-\alpha(t)}\mathrm{d}\tau+\frac{b_2(t)}{\mathrm{i}\pi}\int_\Gamma\frac{\varphi_1(\tau)}{\tau-\beta(t)}\mathrm{d}\tau+$$
$$\frac{b_3(t)}{\mathrm{i}\pi}\int_\Gamma\frac{\varphi_1(\tau)}{\tau-\gamma(t)}\mathrm{d}\tau-\frac{d_0(t)}{\mathrm{i}\pi}\int_\Gamma\frac{\varphi_2(\tau)\overline{\tau'^2(\sigma)}}{\bar\tau-\bar t}\mathrm{d}\tau+\frac{d_1(t)}{\mathrm{i}\pi}\int_\Gamma\frac{\varphi_2(\tau)\overline{\tau'^2(\sigma)}}{\bar\tau-\overline{\alpha(t)}}\mathrm{d}\tau-$$
$$\frac{d_2(t)}{\mathrm{i}\pi}\int_\Gamma\frac{\varphi_2(\tau)\overline{\tau'^2(\sigma)}}{\bar\tau-\overline{\beta(t)}}\mathrm{d}\tau-\frac{d_3(t)}{\mathrm{i}\pi}\int_\Gamma\frac{\varphi_2(\tau)\overline{\tau'^2(\sigma)}}{\bar\tau-\overline{\gamma(t)}}\mathrm{d}\tau+\int_\Gamma K_1(t,\tau)\varphi_1(\tau)\mathrm{d}\tau+$$
$$\int_\Gamma\overline{K_2(t,\tau)}\,\overline{\tau'^2(\sigma)}\varphi_2(\tau)\mathrm{d}\tau=g(t), \tag{4.2}$$
$$\overline{c_0(t)}\varphi_1(t)+\overline{c_1(t)}\varphi_1[\alpha(t)]+\overline{c_2(t)}\varphi_1[\beta(t)]+\overline{c_3(t)}\varphi_1[\gamma(t)]+$$
$$\overline{a_0(t)}\varphi_2(t)+\overline{a_1(t)}\varphi_2[\alpha(t)]+\overline{a_2(t)}\varphi_2[\beta(t)]+\overline{a_3(t)}\varphi_2[\gamma(t)]+$$
$$\frac{\overline{d_0(t)}}{\mathrm{i}\pi}\int_\Gamma\frac{\varphi_1(\tau)}{\tau-t}\mathrm{d}\tau+\frac{\overline{d_1(t)}}{\mathrm{i}\pi}\int_\Gamma\frac{\varphi_1(\tau)}{\tau-\alpha(t)}\mathrm{d}\tau+\frac{\overline{d_2(t)}}{\mathrm{i}\pi}\int_\Gamma\frac{\varphi_1(\tau)}{\tau-\beta(t)}\mathrm{d}\tau+$$
$$\frac{\overline{d_3(t)}}{\mathrm{i}\pi}\int_\Gamma\frac{\varphi_1(\tau)}{\tau-\gamma(t)}\mathrm{d}\tau-\frac{\overline{b_0(t)}}{\mathrm{i}\pi}\int_\Gamma\frac{\varphi_2(\tau)\overline{\tau'^2(\sigma)}}{\bar\tau-\bar t}\mathrm{d}\tau+\frac{\overline{b_1(t)}}{\mathrm{i}\pi}\int_\Gamma\frac{\varphi_2(\tau)\overline{\tau'^2(\sigma)}}{\bar\tau-\overline{\alpha(t)}}\mathrm{d}\tau-$$
$$\frac{\overline{b_2(t)}}{\mathrm{i}\pi}\int_\Gamma\frac{\varphi_2(\tau)\overline{\tau'^2(\sigma)}}{\bar\tau-\overline{\beta(t)}}\mathrm{d}\tau-\frac{\overline{b_3(t)}}{\mathrm{i}\pi}\int_\Gamma\frac{\varphi_2(\tau)\overline{\tau'^2(\sigma)}}{\bar\tau-\overline{\gamma(t)}}\mathrm{d}\tau+\int_\Gamma K_2(t,\tau)\varphi_1(\tau)\mathrm{d}\tau+$$

$$\int_{\Gamma}\overline{K_1(t,\tau)}\,\overline{\tau'^2(\sigma)}\varphi_2(\tau)\mathrm{d}\tau=\overline{g(t)}.$$

这个方程组也叫作方程(4.1)的对应方程组，由于这里要利用取复共轭值的运算，它在复空间 $L^p(\Gamma)$ 中是反线性有界算子. 但是，只要在实数域上来讨论这种算子的话，它在指定空间中就是线性的. 我们把这样的实线性空间用 $\widetilde{L}^p(\Gamma)$ 表示. 以后我们假设算子 $\mathscr{K}$ 是作用在线性空间 $\widetilde{L}^p(\Gamma)$ 上的，也就是说，我们讨论方程(4.1)解的线性无关性是指带实系数的线性无关性. 而讨论方程组(4.2)解的线性无关性，仍然是指带复系数的线性无关性. 这样一来，利用[1]中的结果，可以知道带有两个 Carleman 位移和未知函数复共轭值的奇异积分方程(4.1)和对应奇异积分方程组(4.2)，同时是，或者不是 Noether 方程，而且方程(4.1)和方程组(4.2)的指数将是相等的. 也就是说，我们已经把方程(4.1)转化为第 3 段中讨论的方程组情形，从而不难得到相应的结论.

参考文献

[1] Литвинчук Г С. Краевые задачи и сингулярные интеяралъные уравнения со сдвигом. 1977.

[2] Мусхелишвили Н И. Сингулярные интеяралъные уравнения. 1968.(1962 年版有中译本，上海科技出版社)

[3] Векуа И Н. Системы сингулярных интеяралъных уравнений и некоторые граничные задачи. 1970.(1950 年版有中译本，上海科技出版社)

[4] 赵桢. 带两个 Carleman 位移的奇异积分方程的可解性问题. 北京师范大学学报(自然科学版)，1980，(2)：1-18.

[5] Литвинчук Г С. Об индексе и нормадъной разрешимости одного класса функдионадъных уравнений. ДАН СССР，1963，149(5)：1 029-1 032.

[6] Литвинчук Г С. Теоремы Нетера для одного кдасса сингулярных интегралъных уравнений со сдвигом и сопряжением. ДАН СССР. 1965，162(1)：26-29.

数学研究与评论，
1982,(1):97-108.

关于带两个位移的广义 Hilbert 问题①

On the Generalized Hilbert Problem with Two Carleman's Shifts

Abstract In this paper, the generalized Hilbert problem with two Carleman's shifts is considered for analytic functions under the boundary value condition:

$$\mathrm{Re}\{A(t)\Phi^{+}(t)+B(t)\Phi^{+}[\alpha(t)]+C(t)\Phi^{+}[\beta(t)]\}=h(t), \quad (0.1)$$

where $A(t), B(t), C(t), h(t)$ are given H-continuous functions and $\alpha(t)$, $\beta(t)$ are two different homeomorphisms of boundary Γ onto itself. Γ is a simple closed Lyapunov's curve.

Using the integral representation of analytic functions we establish the equivalent singular integral equation

$$(\mathscr{K}\mu)(t)=g(t). \quad (0.2)$$

Besides the transposed equation

$$(\mathscr{K}'\psi)(t)=0 \quad (0.3)$$

is also obtained.

In order to discuss solvability of problem (0.1) we establish also the adjoint boundary value problem and obtain the necessary and sufficient conditions for solution of problem (0.1). Finally we obtain the formula of index.

---

① 收稿时间:1981-03-06.

在文[1]中提出了带位移的广义 Hilbert 问题,并对这一问题建立了 Noether 性条件. 在文[2]中又利用带位移的奇异积分方程理论得到了这一问题可解的充分必要条件以及指数计算公式. 显然,这些结果推广了一般 Hilbert 问题的相应结论. 本文目的是利用作者在文[4][5]中建立的理论解决带两个位移的广义 Hilbert 问题,并得到了相应的可解条件和指数计算公式.

§1. 问题提法

假设 Γ 是简单的封闭 Ляпунов 曲线,它围出一个有界区域 D^+,用 D^- 代表闭区域 $D^+ + \Gamma$ 到全平面的补集. 我们提出以下边值问题:寻求在区域 D^+ 内解析,在边界 Γ 上满足条件 $H_\mu(\Gamma)$ 的函数 $\Phi(z) = u(x,y) + \mathrm{i}(x,y)$,$z = x + \mathrm{i}y$,它在边界上满足条件

$$a_0(t)u(t) + a_1(t)u[\alpha(t)] + a_2(t)u[\beta(t)] + b_0(t)v(t) + b_1(t)v[\alpha(t)] + b_2(t)v[\beta(t)] = h(t), \tag{1.1}$$

其中 $a_i(t), b_i(t), i = 0,1,2$ 和 $h(t)$ 是满足条件 $H_\mu(\Gamma)$ 的实函数,$\alpha(t)$,$\beta(t)$ 是两个不同的同胚,满足条件 $\alpha[\alpha(t)] = t, \beta[\beta(t)] = t$,它们可以是正位移或反位移. 假设 $\alpha'(t), \beta'(t) \in H_\mu(\Gamma)$,$\alpha'(t) \cdot \beta'(t) \neq 0$. 除此以外,我们还假设 $\alpha[\beta(t)] = \beta[\alpha(t)]$. 为了方便,记 $r(t) = \alpha[\beta(t)] = \beta[\alpha(t)]$,显然 $r(t)$ 也是满足条件 $r[r(t)] = t$ 的位移.

条件(1.1)还可以改写成

$$\mathrm{Re}\{A(t)\Phi^+(t) + B(t)\Phi^+[\alpha(t)] + C(t)\Phi^+[\beta(t)]\} = h(t), \tag{1.2}$$

其中 $A(t) = a_0(t) - \mathrm{i}b_0(t), B(t) = a_1(t) - \mathrm{i}b_1(t), C(t) = a_2(t) - \mathrm{i}b_2(t)$.

§2. 与问题(1.2)(或问题(1.1))等价的奇异积分方程

利用已知的解析函数积分表达式[3]

$$\Phi(z) = \frac{1}{2\mathrm{i}\pi}\int_\Gamma \frac{\mu(\tau)}{\tau - z}\mathrm{d}\tau + \mathrm{i}C, \tag{2.1}$$

其中 $\mu(t)$ 是实函数,它满足条件 $H_\mu(\Gamma)$,而 C 是实常数. 我们取它在边界 Γ 上的边界值 $\Phi^+(t), \Phi^+[\alpha(t)], \Phi^+[\beta(t)]$,然后代入边界条件(1.2),将

得到一个带两个位移的奇异积分方程

$$\mathcal{K}_{\mu} \equiv \mathrm{Re}\{A(t)\mu(t) + B(t)\mu[\alpha(t)] + C(t)\mu[\beta(t)] + \frac{A(t)}{\mathrm{i}\pi}\int_{\Gamma}\frac{\mu(\tau)}{\tau - t}\mathrm{d}\tau + \frac{\nu_1 B(t)}{\mathrm{i}\pi}\int_{\Gamma}\frac{\alpha'(\tau)\mu[\alpha(\tau)]}{\alpha(\tau) - \alpha(t)}\mathrm{d}\tau + \frac{\nu_2 C(t)}{\mathrm{i}\pi}\int_{\Gamma}\frac{\beta'(\tau)\mu[\beta(\tau)]}{\beta(\tau) - \beta(t)}\mathrm{d}\tau\Big\} = g(t), \tag{2.2}$$

其中 $g(t) = 2h(t) + 2C\mathrm{Im}[A(t) + B(t) + C(t)]$，$\nu_1$，$\nu_2$ 取值 1 或者 -1，这依赖于 $\alpha(t)$，$\beta(t)$ 是正位移或者反位移. 问题(1.2)与方程(2.2)在下述意义下是等价的，即如果问题(1.2)有解，那么通过(2.1)知道其密度 $\mu(t)$ 必是方程(2.2)的解；反过来，如果方程(2.2)有解 $\mu(t)$，那么通过(2.1)得到解析函数 $\Phi(z)$，它必为问题(1.2)的解. 根据一般确定相联算子的积分等式[3]

$$\int_{\Gamma}\mu(t)\mathcal{K}'\psi\mathrm{d}t = \int_{\Gamma}\psi(t)\mathcal{K}\mu\mathrm{d}t, \tag{2.3}$$

可以得到方程(2.2)的相联齐次方程

$$\begin{aligned}\mathcal{K}'\psi \equiv\ & [A(t) + \overline{A(t)}]\psi(t) + \nu_1\{B[\alpha(t)] + \overline{B[\alpha(t)]}\}\alpha'(t)\psi[\alpha(t)] + \\ & \nu_2\{C[\beta(t)] + \overline{C[\beta(t)]}\}\beta'(t)\psi[\beta(t)] - \\ & \frac{1}{\mathrm{i}\pi}\int_{\Gamma}\Big[\frac{A(\tau)}{\tau - t} - \frac{\overline{t'^2(s)A(\tau)}}{\bar{\tau} - \bar{t}}\Big]\psi(\tau)\mathrm{d}\tau - \\ & \frac{\nu_1}{\mathrm{i}\pi}\int_{\Gamma}\Big[\frac{B[\alpha(\tau)]}{\tau - t} - \frac{\overline{t'^2(s)B[\alpha(\tau)]}}{\bar{\tau} - \bar{t}}\Big]\alpha'(\tau)\psi[\alpha(\tau)]\mathrm{d}\tau - \\ & \frac{\nu_2}{\mathrm{i}\pi}\int_{\Gamma}\Big[\frac{C[\beta(\tau)]}{\tau - t} - \frac{\overline{t'^2(s)C[\beta(\tau)]}}{\bar{\tau} - \bar{t}}\Big]\beta'(\tau)\psi[\beta(\tau)]\mathrm{d}\tau = 0.\end{aligned} \tag{2.4}$$

我们考虑到以下等式[3]

$$t'(s)\mid_{t=\alpha(t)} = \frac{\alpha'(t)}{\mid\alpha'(t)\mid}t'(s),$$

$$t'(s)\mid_{t=\beta(t)} = \frac{\beta'(t)}{\mid\beta'(t)\mid}t'(s).$$

如果对方程(2.4)的两端同乘上函数 $t'(s)$ 将得到

$$\begin{aligned}t'(s)\mathcal{K}'\psi \equiv\ & [A(t) + \overline{A(t)}]\psi(t)t'(s) + \\ & \nu_1\{B[\alpha(t)] + \overline{B[\alpha(t)]}\}\mid\alpha'(t)\mid\psi[\alpha(t)]t'(s)\mid_{t=\alpha(t)} + \\ & \nu_2\{C[\beta(t)] + \overline{C[\beta(t)]}\}\mid\beta'(t)\mid\psi[\beta(t)]t'(s)\mid_{t=\beta(t)} - \\ & \frac{1}{\mathrm{i}\pi}\int_{\Gamma}\Big[\frac{t'(s)A(\tau)}{\tau - t} - \frac{\overline{t'(s)}\ \overline{A(\tau)}}{\bar{\tau} - \bar{t}}\Big]\psi(\tau)t'(\sigma)\mathrm{d}\sigma -\end{aligned}$$

$$\frac{\nu_1}{\mathrm{i}\pi}\int_\Gamma\left[\frac{t'(s)B[\alpha(\tau)]}{\tau-t}-\frac{\overline{t'(s)}\ \overline{B[\alpha(\tau)]}}{\bar{\tau}-\bar{t}}\right]|\alpha'(\tau)|\psi[\alpha(\tau)]\tau'(\sigma)|_{\tau=\alpha(\tau)}\mathrm{d}\sigma-$$

$$\frac{\nu_2}{\mathrm{i}\pi}\int_\Gamma\left[\frac{t'(s)C[\beta(\tau)]}{\tau-t}-\frac{\overline{t'(s)}\ \overline{C[\beta(\tau)]}}{\bar{\tau}-\bar{t}}\right]|\beta'(\tau)|\psi[\beta(\tau)]\tau'(\sigma)|_{\tau=\beta(\tau)}\mathrm{d}\sigma=0.$$

不难看出，方程 $t'(s)\mathscr{K}'\psi=0$ 对于新未知函数 $\gamma(t)=\psi(t)t'(s)$ 来说，其核与系数都是实的. 这样，我们就可以在实函数类中来求解，也就是说，可以认为 $\gamma(t)$ 是实函数. 方程 $t'(s)\mathscr{K}'\psi=0$ 还可以改写成

$$\mathrm{Re}\left\{A(t)\gamma(t)+\nu_1B[\alpha(t)]|\alpha'(t)|\gamma[\alpha(t)]+\nu_2C[\beta(t)]|\beta'(t)|\gamma[\beta(t)]-\right.$$

$$\left.\frac{t'(s)}{\mathrm{i}\pi}\int_\Gamma\frac{A(\tau)\gamma(\tau)+\nu_1B[\alpha(\tau)]|\alpha'(\tau)|\gamma[\alpha(\tau)]+\nu_2C[\beta(\tau)]|\beta'(\tau)|\gamma[\beta(\tau)]}{\tau-t}\mathrm{d}\sigma\right\}$$

$$=0,\tag{2.5}$$

由于 $t'(s)\neq0$，所以方程(2.5)与(2.4)显然是等价的，为了方便，以后我们把方程(2.5)也叫作方程(2.2)的相联方程. 如果我们用 k 和 k' 分别代表齐次方程(2.2)和(2.5)线性无关解的个数，那么，根据[4]中的结果知道 $\mathrm{lnd}\mathscr{K}=k-k'$. 另外，对于方程(2.2)来说，其特征部分的系数将是

$$a_0(t)=\frac{1}{2}(A(t)+\overline{A(t)}),\ a_1(t)=\frac{1}{2}(B(t)+\overline{B(t)}),$$

$$a_2(t)=\frac{1}{2}(C(t)+\overline{C(t)}),$$

$$\mathrm{i}b_0(t)=\frac{1}{2}(A(t)-\overline{A(t)}),\ \mathrm{i}\nu_1b_1(t)=\frac{\nu_1}{2}(B(t)-\overline{B(t)}),$$

$$\mathrm{i}\nu_2b_2(t)=\frac{\nu_2}{2}(C(t)-\overline{C(t)}).$$

从而，其对应方程组之特征部分的系数矩阵将是

$$\boldsymbol{p}(t)=\begin{pmatrix}a_0(t)&a_1(t)&a_2(t)&0\\a_1[\alpha(t)]&a_0[\alpha(t)]&0&a_2[\alpha(t)]\\a_2[\beta(t)]&0&a_0[\beta(t)]&a_1[\beta(t)]\\0&a_2[\gamma(t)]&a_1[\gamma(t)]&a_0[\gamma(t)]\end{pmatrix},$$

$$\boldsymbol{q}(t)=\begin{pmatrix}b_0(t)&\nu_1b_1(t)&\nu_2b_2(t)&0\\b_1[\alpha(t)]&\nu_1b_0[\alpha(t)]&0&\nu_1\nu_2b_2[\alpha(t)]\\b_2[\beta(t)]&0&\nu_2b_0[\beta(t)]&\nu_1\nu_2b_1[\beta(t)]\\0&\nu_1b_2[\gamma(t)]&\nu_2b_1[\gamma(t)]&\nu_1\nu_2b_0[\gamma(t)]\end{pmatrix},$$

对于方程(2.2)将有以下结论：[4][5]

定理 1　(Noether 性条件) 方程(2.2) 是 Noether 方程的充分必要条件是

$$\det(\boldsymbol{p}(t)+j\boldsymbol{q}(t))\neq 0, j=1 \text{ 或者 } -1. \tag{2.6}$$

定理 2　(指数公式) 当条件(2.6) 满足时,奇异积分方程(2.2) 的指数公式是

$$\begin{aligned}\ln\mathrm{d}\mathscr{K}=k-k'&=\frac{1}{4\pi}\left\{\arg\frac{\det(\boldsymbol{p}(t)-\boldsymbol{q}(t))}{\det(\boldsymbol{p}(t)+q(t))}\right\}_\Gamma\\&=\frac{1}{4\pi}\{\arg\det(\boldsymbol{p}(t)-\boldsymbol{q}(t))\}_\Gamma\end{aligned} \tag{2.7}$$

定理 3　(可解条件)　如果条件(2.6) 满足,那么奇异积分方程(2.2) 可解的充分必要条件是

$$\int_\Gamma\{h(t)+C\mathrm{Im}[A(t)+B(t)+C(t)]\}\gamma_j(t)\mathrm{d}s=0, \tag{2.8}$$

其中$\{\gamma_j(t)\}, j=1,2,\cdots,k'$是相联齐次方程(2.5) 线性无关解的完备系.

§3.　问题(1.2) 的相关问题

假设 $\gamma(t)$ 是相联方程(2.5) 的解,我们做柯西型积分

$$\Psi(z)=\frac{1}{2\mathrm{i}\pi}\int_\Gamma\frac{\psi(\tau)}{\tau-z}\mathrm{d}\sigma, \tag{3.1}$$

其中

$$\psi(\tau)=A(\tau)\gamma(\tau)+\nu_1B[\alpha(\tau)]\mid\alpha'(\tau)\mid\gamma[\alpha(\tau)]+\nu_2C[\beta(\tau)]\mid\beta'(\tau)\mid\gamma[\beta(\tau)]. \tag{3.2}$$

根据方程(2.5),我们将有

$$\mathrm{Re}\{t'(s)\Psi^-(t)\}=0 \tag{3.3}$$

或者

$$t'(s)\Psi^-(t)+\overline{t'(s)\Psi^-(t)}=0. \tag{3.4}$$

我们将证明当满足条件

$$\mathrm{Im}\left\{\int_\Gamma\psi(\tau)\mathrm{d}\sigma\right\}=0 \tag{3.5}$$

时,一定有函数 $\Psi^-(z)\equiv 0, z\in D^-$. 首先,条件(3.5) 可以写成

$$\begin{aligned}\mathrm{Im}\Big\{&\int_\Gamma A(\tau)\gamma(\tau)\mathrm{d}\sigma+\nu_1\int_\Gamma B[\alpha(\tau)]\mid\alpha'(\tau)\mid\gamma[\alpha(\tau)]\mathrm{d}\sigma+\\&\nu_2\int_\Gamma C[\beta(\tau)]\mid\beta'(\tau)\mid\gamma[\beta(\tau)]\mathrm{d}\sigma\Big\}=0,\end{aligned}$$

在上式第二、第三两个积分中分别用 $\alpha(\tau)$，$\beta(\tau)$ 代替 τ 将得到

$$\mathrm{Im}\left\{\int_{\Gamma}[A(\tau)+B(\tau)+C(\tau)]\gamma(\tau)\mathrm{d}\sigma\right\}=0. \tag{3.6}$$

现在来证明 $\Psi^{-}(z)\equiv 0$ 的结论. 由于 $\Psi^{-}(z)$ 可以表示成一个柯西型积分，于是应该有 $\Psi^{-}(\infty)=0$. 这时候，如果我们假设 $\Psi^{-}(z)\neq 0$，那么就将得出矛盾. 事实上，我们用 $N_{D^{-}}$ 和 N_{Γ} 分别代表函数 $\Psi^{-}(z)$ 在区域 D^{-} 内和在边界 Γ 上零点的个数. 根据广义幅角原理，将有

$$\frac{1}{2\pi}\{\arg\Psi^{-}(t)\}_{\Gamma}=-N_{D^{-}}-\frac{1}{2}N_{\Gamma}.$$

考虑到 $\frac{1}{2\pi}\{\arg t'(s)\}_{\Gamma}=1$，所以根据边界条件(3.4)我们得到 $2N_{D^{-}}+N_{\Gamma}=2$，但是，我们知道 $N_{D^{-}}\geqslant 1$，因此，有 $N_{\Gamma}=0$，$N_{D^{-}}=1$，这就是说 $\Psi^{-}(z)$ 只在 ∞ 点具有一阶零点，于是有

$$\int_{\Gamma}\Psi^{-}(t)\mathrm{d}t=-2\pi\mathrm{i}\mathrm{Res}\Psi^{-}(z)\big|_{z=\infty}\neq 0$$

另一方面根据表示式(3.1)，我们得到

$$\begin{aligned}&\mathrm{Res}\Psi^{-}(z)\big|_{z=\infty}\\&=-\frac{1}{2\mathrm{i}\pi}\int_{\Gamma}\{A(\tau)\gamma(\tau)+\nu_1 B[\alpha(\tau)]\mid\alpha'(\tau)\mid\gamma[\alpha(\tau)]+\\&\quad\nu_2 C[\beta(\tau)]\mid\beta'(\tau)\mid\gamma[\beta(\tau)]\}\mathrm{d}\sigma\\&=-\frac{1}{2\mathrm{i}\pi}\int_{\Gamma}[A(\tau)+B(\tau)+C(\tau)]\gamma(\tau)\mathrm{d}\sigma\neq 0.\end{aligned} \tag{3.7}$$

将边界条件(3.4)沿 Γ 关于 σ 进行积分，得到

$$\mathrm{Re}\left\{\int_{\Gamma}\Psi^{-}(t)\mathrm{d}t\right\}=0, \tag{3.8}$$

从而有

$$\mathrm{Re}\left\{\int_{\Gamma}[A(\tau)+B(\tau)+C(\tau)]\gamma(\tau)\mathrm{d}\sigma\right\}=0.$$

因此，

$$\mathrm{Im}\left\{\int_{\Gamma}[A(\tau)+B(\tau)+C(\tau)]\gamma(\tau)\mathrm{d}\sigma\right\}\neq 0.$$

这刚好与(3.6)是矛盾的. 于是应该有 $\Psi^{-}(z)=0$，$z\in D^{-}$. 这样一来，由(3.1)得到等式

$$A(t)\gamma(t)+\nu_1 B[\alpha(t)]\mid\alpha'(t)\mid\gamma[\alpha(t)]+\nu_2 C[\beta(t)]\mid\beta'(t)\mid\gamma[\beta(t)]$$

$$= t'(s)\Psi^{+}(t). \tag{3.9}$$

下面分四种不同情形来进行讨论.

(1) $\alpha(t),\beta(t)$ 都是正位移的情形．这时候,把方程(3.9)中的变量 t 分别换成 $\alpha(t),\beta(t),\gamma(t)$,再把得到的方程与原来的方程联立,我们得到方程组

$$\begin{cases} A(t)\gamma(t)+B[\alpha(t)]\mid\alpha'(t)\mid\gamma[\alpha(t)]+C[\beta(t)]\mid\beta'(t)\mid\gamma[\beta(t)] \\ \quad = t'(s)\Psi^{+}(t), \\ B(t)\gamma(t)+A[\alpha(t)]\mid\alpha'(t)\mid\gamma[\alpha(t)]+C[r(t)]\mid\gamma'(t)\mid\gamma[r(t)] \\ \quad = t'(s)\alpha'(t)\Psi^{+}[\alpha(t)], \\ C(t)\gamma(t)+A[\beta(t)]\mid\beta'(t)\mid\gamma[\beta(t)]+B[r(t)]\mid\gamma'(t)\mid\gamma[r(t)] \\ \quad = t'(s)\beta'(t)\Psi^{+}[\beta(t)], \\ C[\alpha(t)]\mid\alpha'(t)\mid\gamma[\alpha(t)]+B[\beta(t)]\mid\beta'(t)\mid\gamma[\beta(t)]+ \\ \quad A[r(t)]\mid r'(t)\mid\gamma[r(t)]=t'(s)r'(t)\Psi^{+}[r(t)], \end{cases} \tag{3.9$_1$}$$

这个方程组的系数行列式是

$$\Delta_1^{*}(t)=\begin{vmatrix} A(t) & B[\alpha(t)]\mid\alpha'(t)\mid & C[\beta(t)]\mid\beta'(t)\mid & 0 \\ B(t) & A[\alpha(t)]\mid\alpha'(t)\mid & 0 & C[r(t)]\mid r'(t)\mid \\ C(t) & 0 & A[\beta(t)]\mid\beta'(t)\mid & B[r(t)]\mid r'(t)\mid \\ 0 & C[\alpha(t)]\mid\alpha'(t)\mid & B[\beta(t)]\mid\beta'(t)\mid & A[r(t)]\mid r'(t)\mid \end{vmatrix}$$

$$=\mid\alpha'(t)\mid\cdot\mid\beta'(t)\mid\cdot\mid r'(t)\mid\cdot\Delta_1(t),$$

其中

$$\Delta_1(t)=\begin{vmatrix} A(t) & B(t) & C(t) & 0 \\ B[\alpha(t)] & A[\alpha(t)] & 0 & C[\alpha(t)] \\ C[\beta(t)] & 0 & A[\beta(t)] & B[\beta(t)] \\ 0 & C[r(t)] & B[r(t)] & A[r(t)] \end{vmatrix}.$$

利用Noether性条件(2.6)不难知道 $\Delta_1(t)\neq 0$,根据条件 $\mid\alpha'(t)\mid\cdot\mid\beta'(t)\mid\cdot\mid r'(t)\mid\neq 0$. 从而,有 $\Delta_1^{*}(t)\neq 0$,解方程组(3.9)$_1$,可以得到

$$\gamma(t)=\frac{1}{\Delta_1^{*}(t)}\begin{vmatrix} t'(s)\Psi^{+}(t) & B[\alpha(t)]\mid\alpha'(t)\mid & C[\beta(t)]\mid\beta'(t)\mid & 0 \\ t'(s)\alpha'(t)\Psi^{+}[\alpha(t)] & A[\alpha(t)]\mid\alpha'(t)\mid & 0 & C[r(t)]\mid r'(t)\mid \\ t'(s)\beta'(t)\Psi^{+}[\beta(t)] & 0 & A[\beta(t)]\mid\beta'(t)\mid & B[r(t)]\mid r'(t)\mid \\ t'(s)r'(t)\Psi^{+}[r(t)] & C[\alpha(t)]\mid\alpha'(t)\mid & B[\beta(t)]\mid\beta'(t)\mid & A[r(t)]\mid r'(t)\mid \end{vmatrix}$$

$$= \frac{1}{\Delta_1(t)}\{t'(s)\{A_1^*(t)\Psi^+(t) + B_1^*(t)\Psi^+[\alpha(t)]\alpha'(t) +$$

$$C_1^*(t)\Psi^+[\beta(t)]\beta'(t) + D_1^*(t)\Psi^+[r(t)]r'(t)\}\},\qquad(3.10)$$

其中

$$A_1^*(t) = A[\alpha(t)]A[\beta(t)]A[r(t)] - A[\alpha(t)]B[\beta(t)]B[r(t)] - A[\beta(t)]C[r(t)]C[\alpha(t)],$$

$$B_1^*(t) = B[\alpha(t)]B[\beta(t)]B[r(t)] - B[\alpha(t)]A[\beta(t)]A[r(t)] - B[r(t)]C[\alpha(t)]C[\beta(t)],$$

$$C_1^*(t) = C[\alpha(t)]C[\beta(t)]C[r(t)] - C[\beta(t)]A[\alpha(t)]A[r(t)] - C[r(t)]B[\alpha(t)]B[\beta(t)],$$

$$D_1^*(t) = A[\alpha(t)]C[\beta(t)]B[r(t)] + A[\beta(t)]B[\alpha(t)]C[r(t)].$$

考虑到 $\gamma(t)$ 应该是实函数，将(3.10)式右端的分子、分母同乘$\overline{\Delta_1(t)}$将得到边值问题

$$\mathrm{Re}\{\mathrm{i}\,\overline{\Delta_1(t)}t'(s)\{A_1^*(t)\Psi^+(t) + B_1^*(t)\alpha'(t)\Psi^+[\alpha(t)] + C_1^*(t)\beta'(t)\Psi^+[\beta(t)] + D_1^*(t)r'(t)\Psi^+[r(t)]\}\} = 0.\qquad(3.11)$$

问题(3.1)就是当 $\alpha(t),\beta(t)$ 都是正位移时，问题(1.2)的相联问题.

(2) $\alpha(t),\beta(t)$ 都是反位移的情形. 仿照上面情形，把方程(3.9)中的变量t分别换成$\alpha(t),\beta(t),r(t)$，再把得到的方程与原来的方程联立，这时候，为了方便，我们还把第二、第三个方程取共轭值，最后得到方程组

$$\begin{aligned}
&A(t)\gamma(t) - B[\alpha(t)]\,|\alpha'(t)|\,\gamma[\alpha(t)] - C[\beta(t)]\,|\beta'(t)|\,\gamma[\beta(t)] = t'(s)\Psi^+(t),\\
&-\overline{B(t)}\gamma(t) + \overline{A[\alpha(t)]}\,|\alpha'(t)|\,\gamma[\alpha(t)] - \overline{C[r(t)]}\,|r'(t)|\,\gamma[r(t)]\\
&= \overline{t'(s)\alpha'(t)\Psi^+[\alpha(t)]},\\
&-\overline{C(t)}\gamma(t) + \overline{A[\beta(t)]}\,|\beta'(t)|\,\gamma[\beta(t)] - \overline{B[r(t)]}\,|r'(t)|\,\gamma[r(t)]\\
&= \overline{t'(s)\beta'(t)\Psi^+[\beta(t)]},\\
&-C[\alpha(t)]\,|\alpha'(t)|\,\gamma[\alpha(t)] - B[\beta(t)]\,|\beta'(t)|\,\gamma[\beta(t)] +\\
&A[r(t)]\,|r'(t)|\,\gamma[r(t)] = t'(s)r'(t)\Psi^+[r(t)],
\end{aligned}\qquad(3.9)_2$$

这个方程组的系数行列式是

$$\Delta_2^*(t) = \begin{vmatrix}
A(t) & -B[\alpha(t)]\,|\alpha'(t)| & -C[\beta(t)]\,|\beta'(t)| & 0\\
-\overline{B(t)} & \overline{A[\alpha(t)]}\,|\alpha'(t)| & 0 & -\overline{C[r(t)]}\,|r'(t)|\\
-\overline{C(t)} & 0 & \overline{A[\beta(t)]}\,|\beta'(t)| & -\overline{B[r(t)]}\,|r'(t)|\\
0 & C[\alpha(t)]\,|\alpha'(t)| & -B[\beta(t)]\,|\beta'(t)| & A[r(t)]\,|r'(t)|
\end{vmatrix}$$

$$=|\alpha'(t)|\cdot|\beta'(t)|\cdot|r'(t)|\cdot\Delta_2(t),$$

其中

$$\Delta_2(t)=\begin{vmatrix} A(t) & -\overline{B(t)} & -\overline{C(t)} & 0 \\ -B[\alpha(t)] & -\overline{A[\alpha(t)]} & 0 & -C[\alpha(t)] \\ -C[\beta(t)] & 0 & \overline{A[\beta(t)]} & -B[\beta(t)] \\ 0 & -\overline{C[r(t)]} & -\overline{B[r(t)]} & A[r(t)] \end{vmatrix}.$$

利用 Noether 性条件(2.6)仿照情形(1),不难知道 $\Delta_2^*(t)\neq 0$,于是解方程组$(3.9)_2$ 可以得到

$$\gamma(t)=\frac{1}{\Delta_2^*(t)}\cdot$$

$$\begin{vmatrix} t'(s)\Psi^+(t) & -B[\alpha(t)]\,|\alpha'(t)| & -C[\beta(t)]\,|\beta'(t)| & 0 \\ \overline{t'(s)\alpha'(t)\Psi^+[\alpha(t)]} & \overline{A[\alpha(t)]}\,|\alpha'(t)| & 0 & -\overline{C[r(t)]}\,|r'(t)| \\ \overline{t'(s)\beta'(t)\Psi^+[\beta(t)]} & 0 & \overline{A[\beta(t)]}\,|\beta'(t)| & -\overline{B[r(t)]}\,|r'(t)| \\ t'(s)r'(t)\Psi^+[r(t)] & -C[\alpha(t)]\,|\alpha'(t)| & -B[\beta(t)]\,|\beta'(t)| & A[r(t)]\,|r'(t)| \end{vmatrix}$$

$$=\frac{1}{\Delta_2(t)}\{A_2^*(t)t'(s)\Psi^+(t)+B_2^*(t)\overline{t'(s)\alpha'(t)\Psi^+[\alpha(t)]}+$$

$$C_2^*(t)\overline{t'(s)\beta'(t)\Psi^+[\beta(t)]}+D_2^*(t)t'(s)r'(t)\Psi^+[r(t)]\},\quad(3.12)$$

其中

$$A_2^*(t)=\overline{A[\alpha(t)]}\,\overline{A[\beta(t)]}A[r(t)]-\overline{A[\alpha(t)]}B[\beta(t)]\,\overline{B[r(t)]}-$$

$$\overline{A[\beta(t)]}C[\alpha(t)]\,\overline{C[r(t)]},$$

$$B_2^*(t)=\overline{B[\alpha(t)]}\,\overline{A[\beta(t)]}A[r(t)]+\overline{B[r(t)]}C[\beta(t)]C[\alpha(t)]-$$

$$B[\alpha(t)]B[\beta(t)]\,\overline{B[r(t)]},$$

$$C_2^*(t)=\overline{C[\beta(t)]}\,\overline{A[\alpha(t)]}A[r(t)]+B[\alpha(t)]B[\beta(t)]\,\overline{C[r(t)]}-$$

$$C[\alpha(t)]C[\beta(t)]\,\overline{C[r(t)]},$$

$$D_2^*(t)=\overline{A[\alpha(t)]}C[\beta(t)]\,\overline{B[r(t)]}+\overline{A[\beta(t)]}B[\alpha(t)]\,\overline{C[r(t)]}.$$

考虑到$\gamma(t)$应该是实函数,把(3.12)式右端的分子、分母同乘$\overline{\Delta_2(t)}$,将得到边值问题

$$\mathrm{Re}\{\mathrm{i}\,\overline{\Delta_2(t)}\{A_2^*(t)t'(s)\Psi^+(t)+B_2^*(t)\overline{\alpha'(t)t'(s)\Psi^+[\alpha(t)]}+$$

$$C_2^*(t)\overline{\beta'(t)t'(s)\Psi^+[\beta(t)]}+D_2^*(t)r'(t)t'(s)\Psi^+[r(t)]\}\}=0.$$

(3.13)

问题(3.13)就是当$\alpha(t),\beta(t)$都是反位移时,问题(1.2)的相联问题.

(3) 完全类似地可以得到在 $\alpha(t)$ 是正位移，而 $\beta(t)$ 是反位移(或者 $\alpha(t)$ 是反位移，而 $\beta(t)$ 是正位移)的情形下，问题(1.2)的相联问题.

为了对于 $\alpha(t)$，$\beta(t)$ 分别是正位移或反位移情形统一我们的记法，还将引入算子

$$\mathbf{C}_\nu\varphi(t)=\begin{cases}\varphi(t), & \nu=1,\\ \overline{\varphi(t)}, & \nu=-1.\end{cases}$$

如果把方程(3.9)中的变量分别换成 $\alpha(t)$，$\beta(t)$，$r(t)$，并相应地把得到的方程两端再分别作用算子 $\mathbf{C}_{\nu_1}$，$\mathbf{C}_{\nu_2}$，$\mathbf{C}_{\nu_1\nu_2}$ 以后，所得到的方程与原来的方程联立，可以得到方程组

$$\begin{aligned}
&A(t)\gamma(t)+\nu_1 B[\alpha(t)]\,|\alpha'(t)|\,\gamma[\alpha(t)]+\nu_2 C[\beta(t)]\,|\beta'(t)|\,\gamma[\beta(t)]\\
&\qquad=t'(s)\Psi^+(t),\\
&\mathbf{C}_{\nu_1}\{\nu_1 B(t)\gamma(t)+A[\alpha(t)]\,|\alpha'(t)|\,\gamma[\alpha(t)]+\nu_2 C[r(t)]\,|r'(t)|\,\gamma[r(t)]\}\\
&\qquad=\mathbf{C}_{\nu_1}\{t'(s)\alpha'(t)\Psi^+[\alpha(t)]\},\\
&\mathbf{C}_{\nu_2}\{\nu_2 C(t)\gamma(t)+A[\beta(t)]\,|\beta'(t)|\,\gamma[\beta(t)]+\nu_1 B[r(t)]\,|r'(t)|\,\gamma[r(t)]\}\\
&\qquad=\mathbf{C}_{\nu_2}\{t'(s)\beta'(t)\Psi^+[\beta(t)]\},\\
&\mathbf{C}_{v_1\nu_2}\{\nu_2 C[\alpha(t)]\,|\alpha'(t)|\,\gamma[\alpha(t)]+\nu_1 B[\beta(t)]\,|\beta'(t)|\,\gamma[\beta(t)]+\\
&\qquad A[r(t)]\,|r'(t)|\,[r(t)]\}\\
&\qquad=\mathbf{C}_{\nu_1\nu_2}\{t'(s)r'(t)\Psi^+[r(t)]\}.
\end{aligned}\tag{3.9}^*$$

这个方程组的系数行列式是 $\Delta^*(t)=|\alpha'(t)|\cdot|\beta'(t)|\cdot|r'(t)|\cdot\Delta(t)$，其中

$$\Delta(t)=\begin{vmatrix}A(t) & \nu_1\mathbf{C}_{\nu_1}B(t) & \nu_2\mathbf{C}_{\nu_2}C(t) & 0\\ \nu_1 B[\alpha(t)] & \mathbf{C}_{\nu_1}A[\alpha(t)] & 0 & \nu_2\mathbf{C}_{\nu_1\nu_2}C[\alpha(t)]\\ \nu_2 C[\beta(t)] & 0 & \mathbf{C}_{\nu_2}A[\beta(t)] & \nu_2\mathbf{C}_{\nu_1\nu_2}B[\beta(t)]\\ 0 & \nu_2\mathbf{C}_{\nu_1}C[r(t)] & \nu_1\mathbf{C}_{\nu_2}B[r(t)] & \mathbf{C}_{\nu_1\nu_2}A[r(t)]\end{vmatrix}.$$

利用 Noether 性条件(2.6)，不难知道 $\Delta(t)\neq 0$，从而有 $\Delta^*(t)\neq 0$，通过解方程组(3.9)* 可以得到

$$\gamma(t)=\frac{1}{\Delta(t)}\begin{vmatrix}t'(s)\Psi^+(t) & \nu_1 B[\alpha(t)] & \nu_2 C[\beta(t)] & 0\\ \mathbf{C}_{\nu_1}\{t'(s)\alpha'(t)\Psi^+[\alpha(t)]\} & \mathbf{C}_{\nu_1}A[\alpha(t)] & 0 & \nu_2\mathbf{C}_{\nu_1}C[r(t)]\\ \mathbf{C}_{\nu_2}\{t'(s)\beta'(t)\Psi^+[\beta(t)]\} & 0 & \mathbf{C}_{\nu_2}A[\beta(t)] & \nu_1\mathbf{C}_{\nu_2}B[r(t)]\\ \mathbf{C}_{\nu_1\nu_2}\{t'(s)r'(t)\Psi^+[r(t)]\} & \nu_2 C_{\nu_1\nu_2}C[\alpha(t)] & \nu_2\mathbf{C}_{\nu_1\nu_2}B[\beta(t)] & \mathbf{C}_{\nu_1\nu_2}A[r(t)]\end{vmatrix}$$

$$= \frac{1}{\Delta_1(t)}\{A^*(t)t'(s)\Psi^+(t) + B^*(t)\mathbf{C}_{\nu_1}\{t'(s)\alpha(t)\Psi^+[\alpha(t)] + C^*(t)\mathbf{C}_{\nu_2}\{t'(s)\beta'(t)\Psi^+[\beta(t)]\} + D^*(t)\mathbf{C}_{\nu_1\nu_2}[t'(s)r'(t)\Psi^+[r(t)]\}\}. \tag{3.14}$$

其中

$$A^*(t) = \mathbf{C}_{\nu_1}A[\alpha(t)]\mathbf{C}_{\nu_2}A[\beta(t)]\mathbf{C}_{\nu_1\nu_2}A[r(t)] - \mathbf{C}_{\nu_1}C[r(t)]\mathbf{C}_{\nu_2}A[\beta(t)]\mathbf{C}_{\nu_1\nu_2}C[\alpha(t)] - \mathbf{C}_{\nu_1}A[\alpha(t)]\mathbf{C}_{\nu_2}B[r(t)]\mathbf{C}_{\nu_1\nu_2}B[\beta(t)],$$

$$B^*(t) = \nu_1B[\alpha(t)]\mathbf{C}_{\nu_1\nu_2}B[\beta(t)]\mathbf{C}_{\nu_2}B[r(t)] - \nu_1B[\alpha(t)]\mathbf{C}_{\nu_2}A[\beta(t)]\mathbf{C}_{\nu_1\nu_2}A[r(t)] - \nu_1\mathbf{C}_{\nu_2}B[r(t)]C[\beta(t)]\mathbf{C}_{\nu_1\nu_2}C[\alpha(t)],$$

$$C^*(t) = \nu_2C[\beta(t)]\mathbf{C}_{\nu_1}C[r(t)]\mathbf{C}_{\nu_1\nu_2}C[\alpha(t)] - \nu_2B[\alpha(t)]\mathbf{C}_{\nu_1}C[r(t)]\mathbf{C}_{\nu_1\nu_2}B[\beta(t)] - \nu_2C[\beta(t)]\mathbf{C}_{\nu_1}A[\alpha(t)]\mathbf{C}_{\nu_1\nu_2}A[r(t)],$$

$$D^*(t) = \nu_1\nu_2B[\alpha(t)]\mathbf{C}_{\nu_2}A[\beta(t)]\mathbf{C}_{\nu_1}C[r(t)] + \nu_1\nu_2C[\beta(t)]\mathbf{C}_{\nu_1}A[\alpha(t)]\mathbf{C}_{\nu_2}B[r(t)].$$

考虑到 $\gamma(t)$ 是实函数,把(3.14)式右端的分子、分母同乘$\overline{\Delta(t)}$,我们将得到边值问题

$$\mathrm{Re}\{\mathrm{i}\,\overline{\Delta(t)}\{A^*(t)t'(s)\Psi^+(t) + B^*(t)\mathbf{C}_{\nu_1}\{t'(s)\alpha'(t)\Psi^+[\alpha(t)]\} + C^*(t)\mathbf{C}_{\nu_2}\{t'(s)\beta(t)\Psi^+[\beta(t)]\} + D^*(t)\mathbf{C}_{\nu_1\nu_2}\{t'(s)r'(s)\Psi^+[r(t)]\}\}\} = 0 \tag{3.15}$$

这就是问题(1.2)的相联问题的统一形式.

为此,我们还必须讨论以下两种情形:

① 方程(2.5)的任何解都满足条件(3.6);

② 方程(2.5)的解中至少有一个不满足条件(3.6).

根据定理3知道,只要满足条件(2.6),那么,方程(2.2)可解的充要条件就是(2.8).这样一来,对于情形①,即相联方程(2.5)的任何解都满足条件(3.6),从而,常数 C 可以保持是任意的,而条件(2.8)将可以写成

$$\int_\Gamma h(t)\gamma_j(t)\mathrm{d}s = 0,\quad j = 1,2,\cdots,k. \tag{3.16}$$

但是,对于情形②,即相联方程(2.5)的线性无关解 $\gamma_1(t),\gamma_2(t),\cdots,\gamma_k(t)$ 中至少有一个(无妨假设就是 $\gamma_{k1}(t)$)不满足条件(3.6).这时候,我们做函数

$$\gamma_j^*(t) = \gamma_j(t) - \beta_j\gamma_{k1}(t),\ j = 1,2,\cdots,k'-1,$$

其中

$$\beta_j = \frac{\mathrm{Im}\left\{\int_\Gamma [A(t)+B(t)+C(t)]\gamma_j(t)\mathrm{d}s\right\}}{\mathrm{Im}\left\{\int_\Gamma [A(t)+B(t)+C(t)]\gamma_{k'}(t)\mathrm{d}s\right\}}.$$

不难验证,这些函数是相联方程(2.5)满足附加条件(3.6)的解. 事实上,

$$\mathrm{Im}\int_\Gamma [A(t)+B(t)+C(t)]\gamma_j^*(t)\mathrm{d}s = \mathrm{Im}\int_\Gamma [A(t)+B(t)+C(t)]\gamma_j(t)\mathrm{d}s -$$

$$\beta_j \mathrm{Im}\int_\Gamma [A(t)+B(t)+C(t)]\gamma_{k'}(t)\mathrm{d}s = 0.$$

对于上述的 $k'-1$ 个解 $\gamma_j^*(t)$ 来说,可解条件(2.8)仍可以表示成(3.16)的形式,而对于 $\gamma_{k'}(t)$ 可以得到

$$\int_\Gamma h(t)\gamma_{k'}(t)\mathrm{d}s = -2C\int_\Gamma \mathrm{Im}[A(t)+B(t)+C(t)]\gamma_{b'}(t)\mathrm{d}s. \qquad (3.17)$$

由于上式右端的积分部分不等于零,只要适当地选择常数 C 就可以使得(3.17)是满足的.

由于函数 $\gamma_j(t)$ 和 $h(t)$ 都是实的,那么,条件(3.16)还可以写成

$$\mathrm{Re}\left\{\int_\Gamma h(t)\gamma_j(t)\overline{t'(s)}\mathrm{d}t\right\} = 0. \qquad (3.18)$$

从而,有如下定理成立:

定理 4 如果 Norther 条件(2.6)成立,那么广义 Hilbert 问题(1.2)可解的充分必要条件具有以下形式

$$\mathrm{Re}\left\{\int_\Gamma \left\{\frac{h(t)A^*(t)}{\Delta(t)} + \nu_1 \frac{h[\alpha(t)]\mathbf{C}_{\nu_1}B^*[\alpha(t)]}{\mathbf{C}_{\nu_1}\Delta[\alpha(t)]} + \nu_2 \frac{h[\beta(t)]\mathbf{C}_{\nu_2}C^*[\beta(t)]}{\mathbf{C}_{\nu_2}\Delta[\beta(t)]} +\right.\right.$$

$$\left.\left.\nu_1\nu_2 \frac{h[r(t)]\mathbf{C}_{\nu_1\nu_2}D^*[r(t)]}{\mathbf{C}_{\nu_1\nu_2}\Delta[r(t)]}\right\}\Psi_j^+(t)\mathrm{d}t\right\} = 0,$$

其中 $\Psi_j^+(t)$ 是相联问题(3.15)的线性无关解.

证 只要把相联方程(2.5)的解 $\gamma_j(t)$ 利用表示式(3.14)代入(3.18)就可以得到

$$\mathrm{Re}\left\{\int_\Gamma h(t)\frac{1}{\Delta(t)}\{A^*(t)t'(t)\Psi_j^+(t) + B^*(t)\mathbf{C}_{\nu_1}\{t'(s)\alpha'(t)\Psi_j^+[\alpha(t)]\} +\right.$$

$$\left.C^*(t)\mathbf{C}_{\nu_2}\{t'(s)\beta'(t)\Psi_j^+[\beta(t)]\} + D^*(t)\mathbf{C}_{\nu_1\nu_2}\{t'(s)r'(t)\Psi_j[r(t)]\}\}\mathrm{d}s\right\}$$

$$= \mathrm{Re}\left\{\int_\Gamma h(t)\left\{\frac{1}{\Delta(t)}A^*(t)t'(s)\Psi_j^+(t) + \frac{\mathbf{C}_{\nu_1}B^*(t)}{\mathbf{C}_{\nu_1}\Delta(t)}t'(s)\alpha'(t)\Psi_j^+[\alpha(t)] +\right.\right.$$

$$\frac{\mathbf{C}_{\nu_2}C^*(t)}{\mathbf{C}_{\nu_2}\Delta(T)}t'(s)\beta'(t)\Psi_j^+[\beta(t)]+\frac{\mathbf{C}_{\nu_1\nu_2}D^*(t)}{\mathbf{C}_{\nu_1\nu_2}\Delta(t)}t'(s)r'(t)\Psi_j^+[r(t)]\}\mathrm{d}s\Big\}$$

$$=\mathrm{Re}\Big\{\int_\Gamma\Big\{\frac{h(t)}{\Delta(t)}A^*(t)+\nu_1\frac{\mathbf{C}_{\nu_1}B^*[\alpha(t)]}{\mathbf{C}_{\nu_1}\Delta[\alpha(t)]}h[\alpha(t)]+\nu_2\frac{\mathbf{C}_{\nu_2}C^*[\beta(t)]}{\mathbf{C}_{\nu_2}\Delta[\beta(t)]}h[\beta(t)]+$$

$$\nu_1\nu_2\frac{\mathbf{C}_{\nu_1\nu_2}D^*[r(t)]}{\mathbf{C}_{\nu_1\nu_2}\Delta[r(t)]}h[r(t)]\Big\}\Psi_j^+(t)\mathrm{d}s\Big\}=0.$$

§ 4.　广义 Hilbert 问题(1.2)的 Noether 性条件与指数公式

定义　如果齐次问题(1.2)与相联问题(3.15)的线性无关解的个数 l 和 l' 都是有限的，而且满足定理 4 中的可解条件，我们就把问题(1.2)叫作 Noether 问题．这样一来，我们得到如下定理．

定理 5　如果 Noether 条件(2.6)满足，那么广义 Hilbert 问题(1.2)就是 Noether 问题.

除此以外，关于广义 Hilbert 问题的指数，将有以下定理成立.

定理 6　如果 Noether 条件(2.6)满足，那么广义 Hilbert 问题(1.2)的指数公式是

$$\text{æ}=l-l'=\frac{1}{4\pi}\{\arg\det(\boldsymbol{p}(t)-\boldsymbol{q}(t))\}_\Gamma+1,\tag{4.1}$$

其中 $\boldsymbol{p}(t)$，$\boldsymbol{q}(t)$ 是在第 2 段中提到的系数矩阵.

证　(1) 如果相联齐次方程(2.5)的解都满足条件(3.6)，那么相联齐次方程(2.5)的每一个解都对应着相联问题(3.15)的一个解，从而有 $l'=k'$. 而方程(2.2)右端中的常数 C 保持是任意的，于是应该有 $l=k+1$，最后，得到

$$\text{æ}=l-l'=k-k'+1=\frac{1}{4\pi}\{\arg\det(\boldsymbol{p}(t)-\boldsymbol{q}(t))\}_\Gamma+1.$$

(2) 如果相联齐次方程(2.5)的解中只要有一个不满足条件(3.6). 这时候，方程(2.5)满足条件(3.6)的解刚好应该有 $k'-1$ 个，从而，有 $l'=k'-1$，但是，这时候，方程(2.5)右端中的常数 C 是由公式(3.17)唯一确定的. 于是应该有 $l=k$，最后我们仍将得到

$$\text{æ}=l-l'=k-k'+1=\frac{1}{4\pi}\{\arg\det(\boldsymbol{p}(t)-\boldsymbol{q}(t))\}_\Gamma+1,$$

参考文献

[1] Литвинчук Г С, Хасаъов Э Г. О Краевой задаче Гилъбѐрта со сдвигом. ДАН СССР,1962,142(2):274-277.

[2] Литвинчук Г С, Хасабов Е Г. Один класс сингулярных интегралъных уравнений и оо̄об шенная краевая задача типа задачи карлемана. Сибирс. мат. ж. 1964,5(4):858-880.

[3] Литвинчук Г С. Краевые задачи и сингулярные интегралъные уравнення со сдвигом. 1977.

[4] 赵桢. 带两个 Carleman 位移的奇异积分方程的可解性问题. 北京师范大学学报(自然科学版),1980,(2):1-18.

[5] 赵桢. 带两个 Carleman 位移的奇异积分方程 Noether 可解的充分必要条件. 数学年刊,1981,2A(1):91-100.

应用数学与计算数学，
1982,(1):49-54.

关于带位移的奇异积分方程与边值问题

On the Singular Integral Equation with Shifts and Boundary Value Problems

众所周知，奇异积分方程理论对很多实际问题都具有重要意义. 在20世纪四五十年代，苏联学者就系统地研究了奇异积分方程和方程组的理论，从而大大地推动了解析函数与广义解析函数边值问题理论的研究[1～4]. 在20世纪初 Hilbert，Haseman 以及晚些时候 Carleman 都研究过带位移的边值问题. 但是对这一类问题进行系统地研究还是 Н. И. Муселишвнлн 在20世纪40年代开始的. 由于这一类边值问题理论发展的需要，最近十多年来带位移的奇异积分方程与边值问题理论也有了相应的发展. 在专著[5]中系统地总结了这方面的工作，并且给出了大量参考文献. 文章[6][10]系统地介绍了带位移的奇异积分方程的 Noether 理论与带位移的边值问题. 本文目的是综述带位移的奇异积分方程与边值问题的理论，重点是介绍有关带两个 Carleman 位移的奇异积分方程 Noether 理论以及某些多元素边值问题的一些结果.

§1. 带位移的奇异积分方程的 Noether 理论

一般讲，所谓带位移的奇异积分方程是指以下类型的积分方程

$$(\mathscr{K}\varphi)(t) \equiv \sum_{k=0}^{n-1}\left\{\alpha_k(t)\varphi[\alpha_k(t)] + \frac{C_k(t)}{\mathrm{i}\pi}\int_\Gamma \frac{\varphi(\tau)}{\tau-\alpha_k(t)}\mathrm{d}\tau\right\} + \int_\Gamma K(t,\tau)\varphi(\tau)\mathrm{d}\tau = g(t), \tag{1.1}$$

其中 $K(t,\tau)$ 在 Γ 上只具有弱奇异性，而系数 $a_k(t),c_k(t)\in C(\Gamma)$，而右端 $g(t)\in L^p(\Gamma)$，$p>1$. 有时候，为了方便，还可以假设系数和右端都属于空间 $H_\mu(\Gamma)$. 这里，$\alpha(t)$ 是满足条件(K_n) 的位移，而 $\alpha_k(t)=\alpha[\alpha_{k-1}(t)]$，$\alpha_0(t)=t,k=1,2,\cdots,n-1$. 对于这一类奇异积分方程已经建立了系统的 Noether 理论[5][6]，利用它还解决了不少带位移的边值问题[5][10]. 我们的工作主要是把这一类奇异积分方程理论对带有两个位移的方程进行了推广，从而建立了相应的 Noether 可解性条件和指数公式[7][8][17]，并研究了某些边值问题[9][11][12][14～16][18][19][21].

首先我们考虑位移满足条件(K_2) 的情形. 假设 $\alpha(t)$，$\beta(t)$ 是两个不同的位移（它们可以是正位移或反位移）满足条件(K_2)，除此以外，假设 $\alpha[\beta(t)]=\beta[\alpha(t)]$. 为了方便，我们记 $\gamma(t)=\alpha[\beta(t)]=\beta[\alpha(t)]$，显然，$\gamma(t)$ 也是满足条件(K_2) 的位移，还要求 $\alpha'(t),\beta'(t)\in H_\mu(\Gamma)$，而且 $\alpha'(t)\cdot\beta'(t)\neq 0$. 以后如果不做特殊声明，就认为所有这些条件都是满足的.

我们考虑带两个 Carleman 位移的奇异积分方程

$$(\mathscr{K}\varphi)(t)\equiv a_0(t)\varphi(t)+a_1(t)\varphi[\alpha(t)]+a_2(t)\varphi[\beta(t)]+a_3(t)\varphi[\gamma(t)]+\frac{b_0(t)}{\mathrm{i}\pi}\int_\Gamma\frac{\varphi(t)}{\tau-t}\mathrm{d}\tau+\frac{b_1(t)}{\mathrm{i}\pi}\int_\Gamma\frac{\varphi(t)}{\tau-\alpha(t)}\mathrm{d}\tau+\frac{b_2(t)}{\mathrm{i}\pi}\int_\Gamma\frac{\varphi(t)}{\tau-\beta(t)}\mathrm{d}\tau+\frac{b_3(t)}{\mathrm{i}\pi}\int_\Gamma\frac{\varphi(t)}{\tau-\gamma(t)}\mathrm{d}\tau+\int_\Gamma K(t,\tau)\varphi(\tau)\mathrm{d}\tau=g(t),\tag{1.2}$$

它还可以写成算子形式

$$\mathscr{K}\equiv a_0(t)\mathscr{I}+a_1(t)\mathscr{W}_1+a_2(t)\mathscr{W}_2+a_3(t)\mathscr{W}_3+b_0(t)\mathscr{S}+b_1(t)\mathscr{W}_1\mathscr{S}+b_2(t)\mathscr{W}_2\mathscr{S}+b_3(t)\mathscr{W}_3\mathscr{S}+\mathscr{D},\tag{1.3}$$

其中 $\mathscr{I}$ 是恒等算子，$\mathscr{W}_k$ 是位移算子，$\mathscr{S}$ 是奇异积分算子，而 $\mathscr{D}$ 是完全连续算子.

我们可以引入所谓的对应方程组和伴随方程[7]，其对应方程组的特征部分的系数矩阵是

$$\boldsymbol{p}(t)=\begin{bmatrix} a_0(t) & a_1(t) & a_2(t) & a_3(t)\\ a_1[\alpha(t)] & a_0[\alpha(t)] & a_3[\alpha(t)] & a_2[\alpha(t)]\\ a_2[\beta(t)] & a_3[\beta(t)] & a_0[\beta(t)] & a_1[\beta(t)]\\ a_3[\gamma(t)] & a_2[\gamma(t)] & a_1[\gamma(t)] & a_0[\gamma(t)]\end{bmatrix},\tag{1.4}$$

$$\boldsymbol{q}(t)=\begin{bmatrix} b_0(t) & \gamma_1 b_1(t) & \gamma_2 b_2(t) & \gamma_1\gamma_2 b_3(t) \\ b_1[\alpha(t)] & \gamma_1 b_0[\alpha(t)] & \gamma_2 b_3[\alpha(t)] & \gamma_1\gamma_2 b_2[\alpha(t)] \\ b_2[\beta(t)] & \gamma_1 b_3[\beta(t)] & \gamma_2 b_0[\beta(t)] & \gamma_1\gamma_2 b_1[\beta(t)] \\ b_3[\gamma(t)] & \gamma_1 b_2[\gamma(t)] & \gamma_2 b_1[\gamma(t)] & \gamma_1\gamma_2 b_0[\gamma(t)] \end{bmatrix}, \tag{1.5}$$

这里，γ_1,γ_2 分别等于 $+1$ 或者 -1，这依赖于 $\alpha(t),\beta(t)$ 是正位移或者反位移而定. 按照常用的方法，还可以引入方程(1.2)的相联方程以及相联方程的对应方程组，我们知道它刚好是方程(1.2)之对应方程组的相联方程组[7]. 从而得到以下结论：[7][8]

定理 1 为了使算子 $\mathscr{K}$ 是 Noether 算子的充分必要条件是

$$\det(\boldsymbol{p}(t)\pm\boldsymbol{q}(t))\neq 0, \tag{1.6}$$

这里 $\boldsymbol{p}(t),\boldsymbol{q}(t)$ 分别由公式(1.4)(1.5)确定.

定理 2 Noether 算子 $\mathscr{K}$ 的指数是

$$\operatorname{lnd}\mathscr{K}=\frac{1}{8\pi}\left\{\arg\frac{\det(\boldsymbol{p}(t)-\boldsymbol{q}(t))}{\det(\boldsymbol{p}(t)+\boldsymbol{q}(t))}\right\}_\Gamma. \tag{1.7}$$

另外，对于(1.2)类型的奇异积分方程组以及带有未知函数复共轭值的方程

$$\begin{aligned}(\mathscr{K}\varphi)(t)\equiv{}& a_0(t)\varphi(t)+a_1(t)\varphi[\alpha(t)]+a_2(t)\varphi[\beta(t)]+a_3(t)\varphi[\gamma(t)]+\\ & c_0(t)\overline{\varphi(t)}+c_1(t)\overline{\varphi[\alpha(t)]}+c_2(t)\overline{\varphi[\beta(t)]}+c_3(t)\overline{\varphi[\gamma(t)]}+\\ & \frac{b_0(t)}{\mathrm{i}\pi}\int_\Gamma\frac{\varphi(\tau)}{\tau-t}\mathrm{d}\tau+\frac{b_1(t)}{\mathrm{i}\pi}\cdot\\ & \int_\Gamma\frac{\varphi(\tau)}{\tau-\alpha(t)}\mathrm{d}\tau+\frac{b_2(t)}{\mathrm{i}\pi}\int_\Gamma\frac{\varphi(\tau)}{\tau-\beta(t)}\mathrm{d}\tau+\frac{b_3(t)}{\mathrm{i}\pi}\int_\Gamma\frac{\varphi(\tau)}{\tau-\gamma(t)}\mathrm{d}\tau+\\ & d_0(t)\overline{\frac{1}{\mathrm{i}\pi}\int_\Gamma\frac{\varphi(\tau)}{\tau-t}\mathrm{d}\tau}+d_1(t)\overline{\frac{1}{\mathrm{i}\pi}\int_\Gamma\frac{\varphi(\tau)}{\tau-\alpha(t)}\mathrm{d}\tau}+d_2(t)\overline{\frac{1}{\mathrm{i}\pi}\int_\Gamma\frac{\varphi(\tau)}{\tau-\beta(t)}\mathrm{d}\tau}+\\ & d_3(t)\overline{\frac{1}{\mathrm{i}\pi}\int_\Gamma\frac{\varphi(\tau)}{\tau-\gamma(t)}\mathrm{d}\tau}+\int_\Gamma K_1(t,\tau)\varphi(\tau)\mathrm{d}\tau+\overline{\int_\Gamma K_2(t,\tau)\varphi(\tau)\mathrm{d}\tau}=g(t),\end{aligned} \tag{1.8}$$

也将有类似的结论成立.[8]

对于位移 $\alpha(t),\beta(t)$ 分别满足条件(K_n)，(K_m)的情形，我们讨论奇异积分方程

$$\begin{aligned}(\mathscr{K}\varphi)(t)\equiv{}&\sum_{k=0}^{m-1}\sum_{i=0}^{n-1}\left\{a_{kj}(t)\varphi[\beta_k(\alpha_j(t))]+\frac{b_{kj}(t)}{\mathrm{i}\pi}\int_\Gamma\frac{\varphi(\tau)\mathrm{d}\tau}{\tau-\beta_k(\alpha_j(t))}\right\}+\\ &\int_\Gamma K(t,\tau)\varphi(\tau)\mathrm{d}\tau=g(t),\end{aligned} \tag{1.9}$$

也得到了类似的可解充分必要条件[17].

研究奇异积分方程的重要方法之一是把它正则化. 对于某些带位移的奇异积分方程在[20]中讨论了它们的等价正则化问题,并且给出了正则化算子的构造方法.

关于带有非 Carleman 位移的积分方程,一直到目前为止成果还不多,在[5]中建立了带一个非 Carleman 位移的奇异积分方程的 Noether 性充分条件. 在[13]中利用类似的方法,对带两个非 Carleman 位移的奇异积分方程,建立了 Noether 性充分条件.

§2. 某些带位移的边值问题

在20世纪70年代初期,苏联学者就讨论过带两个位移的三元素边值问题,并得到了某些三元素边值问题的 Noether 性条件和指数公式. 我们利用 §1 中的结果,最近又研究了以下边值问题.

2.1 某些三元素边值的问题

假设 Γ 是简单封闭的 Lyapunoff 曲线,它把整个平面分成为两部分:有界区域 D^+ 和无界区域 D^-,不妨假设原点 $O\in D^+$.

(1) 要求根据边界条件

$$\Phi^+(t)=G_1(t)\Phi^-[\alpha(t)]+G_2(t)\Phi^+[\beta(t)]+g(t),t\in\Gamma \qquad (2.1)$$

来寻求分片解析函数 $\Phi(z)$,其中 $G_1(t)\neq 0,G_2(t)\neq 0,G_1(t),G_2(t)\in H_\mu(\Gamma),0<\mu<1,g(t)\in L^2(\Gamma),\alpha(t),\beta(t)$ 是两个不同的移位[14]. 只要利用柯西型积分和 Сохоцкий-Plemelj 公式就可以把上述问题转化为 §1 中所讨论的奇异积分方程.

定理 3 (i) 如果 $\alpha(t),\beta(t)$ 都是正位移,那么边值问题(2.1)的 Noether 条件是 $1-G_2(t)G_2[\beta(t)]\neq 0$,而问题的指数是

$$k=\frac{1}{4\pi}\left\{\arg\frac{G_1(t)G_1[\beta(t)]}{1-G_2(t)G_2[\beta(t)]}\right\}_\Gamma;$$

(ii) 如果 $\alpha(t)$ 是正位移,$\beta(t)$ 是反位移,那么 Noether 条件自然成立,(因为 $G_1(t)\neq 0$),而问题的指数是

$$x=\frac{1}{2\pi}\{\arg G_1(t)\}_\Gamma;$$

(iii) 如果 $\alpha(t),\beta(t)$ 都是反位移,那么 Noether 条件也是自然成立的(因为 $G_1(t)\neq 0,G_2(t)\neq 0$),而问题的指数是

$$k = \frac{1}{2\pi}\{G_1[\beta(t)]G_2(t)\}_\Gamma.$$

如果再附加一个不等式条件还可以得到齐次问题线性无关解个数 l 和非齐次问题可解条件个数 p 的统一公式:[14]

$$l = \max(0, x), p = \max(0, -x).$$

(2) 求在区域 D^+ 内解析,在闭区域 $D^+ + \Gamma$ 上 Hölder 连续的函数 $\Phi(z)$,在 Γ 上满足边界条件

$$\Phi^+(t) = A(t)\Phi^+[\alpha(t)] + B(t)\Phi^+[\beta(t)] + H(t), \tag{2.2}$$

其中 $A(t), B(t), H(t)$ 都属于空间 $H_\mu(\Gamma)$[15]. 为了消除超定性还要求满足一组等式条件(条件 S). 解决问题的方法主要引入了与问题(2.2)在一定意义下等价的辅助问题(是对于解析函数对应的边值问题),然后,就不难归结为§1中所讨论的奇异积分方程,从而得到 Noether 可解的充分必要条件和指数公式. 应该指出当 $\alpha(t), \beta(t)$ 均为正位移的情形辅助问题将不是 Noether 问题[15].

(3) 当区域 D^+ 是多连通情形,文[18]还讨论了关于解析函数和解析函数对应的如下边值问题

$$\begin{cases} \Phi^+(t) = A(t)\Phi^+[\alpha(t)] + B(t)\overline{\Phi^+[\beta(t)]} + H(t), \\ \Phi^-(t) = A(t)\Phi^+[\alpha(t)] + B(t)\Phi^+[\beta(t)] + H(t), \\ \Phi_1^+(t) = A(t)\Phi_2^+[\alpha(t)] + B(t)\overline{\Phi_2^+[\beta(t)]} + H(t), \\ \Phi_1^-(t) = A(t)\Phi_2^+[\alpha(t)] + B(t)\Phi_2^+[\beta(t)] + H(t); \end{cases} \tag{2.3}$$

其中 $A(t), B(t), H(t)$ 都属于空间 $H_\mu(\Gamma)$,并满足一组消除超定性的条件. 利用§1中的奇异积分方程理论,对于 $\alpha(t), \beta(t)$ 分别是正、反位移的情形,也得到了相应的 Noether 条件和指数公式,并指出了 $\beta(t)$ 是反位移,而 $\alpha(t)$ 是正位移时,问题(2.3)将不是 Noether 问题的结论[18].

2.2　广义 Hilbert 问题

问题的提法是:要求在区域 D^+ 内解析,在边界 Γ 上满足 Hölder 条件的函数 $\Phi(z)$,它在边界 Γ 上满足条件

$$\mathrm{Re}\{A(t)\Phi^+(t) + B(t)\Phi^+[\alpha(t)] + C(t)\Phi^+[\beta(t)]\} = h(t), \tag{2.4}$$

这里,$A(t), B(t), C(t), h(t)$ 都属于空间 $H_\mu(\Gamma)$. 这个问题显然是一般 Hilbert 问题的自然推广. 利用§1中建立的奇异积分方程理论,不难得到与问题相应的积分方程可解的 Noether 性条件的指数公式[9]. 为了得

到问题(2.4)的可解条件，文[9]还引入了它的相联问题

$$\begin{aligned}&\mathrm{Re}\{\overline{\mathrm{i}\Delta(t)}\{A^*(t)t'(s)\Psi^+(t)+B^*(t)C_{\gamma_1}\{t'(s)\alpha'(t)\Psi^+[\alpha(t)]\}+\\&\quad e^*(t)C_{\gamma_2}\{t'(s)\beta'(t)\Psi^+[\beta(t)]\}+\\&\quad D^*(t)C_{\gamma_1\gamma_2}\{t'(s)\gamma'(t)\Psi^+[\gamma(t)]\}\}\}=0,\end{aligned}\tag{2.5}$$

这里 $\Delta(t),A^*(t),B^*(t),C^*(t),D^*(t)$ 都是通过已知系数 $A(t),B(t),C(t)$ 确定的函数，而 C_ν 是共轭算子

$$C_\nu[\varphi(t)]=\begin{cases}\varphi(t), & \gamma=1,\\ \overline{\varphi(t)}, & \gamma=-1.\end{cases}$$

只要相应积分方程的 Noether 条件成立，对于广义 Hilbert 问题将有以下结论：[9]

定理 4 广义 Hilbert 问题(2.4)可解的充分必要条件是

$$\mathrm{Re}\left\{\int_\Gamma\left\{\frac{h(t)A^*(t)}{\Delta(t)}+\nu_1\frac{h[\alpha(t)]\boldsymbol{C}_{\nu_1}B^*[\alpha(t)]}{\boldsymbol{C}_{\nu_1}\Delta[\alpha(t)]}+\nu_2\frac{h[\beta(t)]\boldsymbol{C}_{\nu_1}B^*[\beta(t)]}{\boldsymbol{C}_{\nu_2}\Delta[\beta(t)]}+\right.\right.$$
$$\left.\left.\nu_1\nu_2\frac{h[\gamma(t)]\boldsymbol{C}_{\nu_1\nu_2}D^*[\gamma(t)]}{\boldsymbol{C}_{\nu_1\nu_2}\Delta[\gamma(t)]}\right\}\cdot\Psi_j^+(t)\mathrm{d}t\right\}=0,$$

其中 $\Psi_j^+(t)$ 是相联问题(2.5)的线性无关解. 问题(2.4)的指数公式是

$$x=\frac{1}{4\pi}\{\arg\det(\boldsymbol{p}(t)-\boldsymbol{q}(t))\}_\Gamma+1,$$

这里 $\boldsymbol{p}(t),\boldsymbol{q}(t)$ 是(1.4)(1.5)形式的矩阵.

如果 $\alpha(t),\beta(t)$ 分别满足更一般的条件(K_m)和(K_n)，在文[19]中讨论了以下形式的边值问题

$$\mathrm{Re}\left\{\sum_{p=0}^{m-1}\sum_{q=0}^{n-1}A_{pq}(t)\Phi^+[\alpha_p(\beta_q(t))]\right\}=h(t),\tag{2.6}$$

并得到了与上面完全类似的结论. 另外，还借助问题(2.6)的系数引入了三类函数 $\Delta(t),\theta(t),V(t)$，从而得到了几种退化情形的进一步结论[19].

2.3 基本边值问题

求分片解析函数 $\Phi(t)$，它在边界上分别满足条件：

(1) $a(t)\Phi^+(t)+b(t)\Phi^+[\alpha(t)]+c(t)\Phi^-[\beta(t)]+d(t)\Phi^-[\gamma(t)]=g(t)$；

(2) $a(t)\Phi^+(t)+b(t)\Phi^+[\alpha(t)]+c(t)\overline{\Phi^-[\beta(t)]}+d(t)\overline{\Phi^-[\gamma(t)]}=g(t)$；

(3) $a(t)\Phi^+(t)+b(t)\overline{\Phi^+[\alpha(t)]}+c(t)\Phi^-[\beta(t)]+d(t)\overline{\Phi^-[\gamma(t)]}=g(t)$.

由于这些问题的 Noether 可解条件与带一个位移的基本边值问题类似，所以我们把它们叫作带两个位移的基本边值问题. 在[12] 中以(1) 为例引入了在一定意义下等价的辅助问题及其相关问题，从而得到以下结论：

定理 5　齐次问题$(1)^{\circ}(g(t)=0)$ 与相联齐次问题的线性无关解的个数之差为 $l-l'=\frac{x}{2}$. 而非齐次问题(1) 的可解条件为

$$\int_{\Gamma}\frac{g(\tau)a[\alpha(\tau)]-g[\alpha(\tau)]b(\tau)}{\Delta_1(\tau)}\Psi^{+}(\tau)\mathrm{d}\tau=0,$$

其中 $\Psi^{+}(\tau)$ 为相联齐次问题的解.

另外，对于退化情形还可以得到 $l=\max\left(0,\frac{x}{2}\right)$, $p=\max\left(0,-\frac{x}{2}\right)$的结论[12].

2.4　复合边值问题

在[11][21] 中还在多连通区域情形下，对解析函数的各种复合边值问题得到了可解性的相应结论，并对齐次问题非零解的个数得到了相应的估计式.

参考文献

[1] 穆斯黑利什维利. 奇异积分方程. 上海：上海科技出版社，1962.

[2] Векуа И Н. Системы сингулярных интеяралъных уравнений и некоторые граничные задачи. 1970.

[3] Гахов Ф Д. Краевые задачи. 1963.

[4] 维库阿. 广义解析函数. 北京：人民教育出版社，1960.

[5] Литвинчук Г С. Краевые задачи и сингулярных интеяралъные уравнения со сдвигом. 1977.

[6] 赵桢. 带位移的奇异积分方程的 Noether 理论. 应用数学与计算数学，1979，(6)：53-62.

[7] 赵桢. 带两个 Carleman 位移的奇异积分方程的可解性问题. 北京师范大学学报(自然科学版)，1980，(2)：1-18.

[8] 赵桢. 带两个 Carleman 位移的奇异积分方程 Noether 可解的充分必要条件. 数学年刊，1981，2A(1)：91-100.

[9] 赵桢. 关于带两个位移的广义 Hilbert 问题. 数学研究与评论，1982，(1)：97-108.

[10] 刘来福.带位移的边值问题.应用数学与计算数学,1979,(6):82-89.

[11] 刘来福.解析函数带位移韵复合边值问题.数学年刊,1981,2(3):325-337.

[12] 刘来福.解析函数带两个位移的基本边值问题.1981 年全国奇异积分方程与边值问题会议报告.

[13] 陈方权.带两个非 Carleman 位移的奇异积分方程.1981 年全国奇异积分方程与边值问题会议报告.

[14] 陈方权.一类带两个 Carleman 位移的三元素边值问题.北京师范大学学报(自然科学版),1981,(2):1-10.

[15] 蒋绍惠.带两个位移的广义 Carleman 边值问题.北京师范大学学报(自然科学版),1981,(1):23-33.

[16] 蒋绍惠,李跃堂.多连通区域内关于解析函数对的带两个位移的广义 Carleman 边值问题.桂林电子工业学院学报,1986,(1):109-121.

[17] 赵达夫.一类带 Carleman 位移的奇异积分方程 Noether 可解条件.北京师范大学数学系研究生毕业论文,1981.

[18] 楚泽甫.多连通区域内一类三元素边值问题的 Noether 理论.北京师范大学数学系研究生毕业论文,1981.

[19] 李正吾.带两个 Carleman 位移的希尔伯特边值问题 Noether 理论.北京师范大学数学系研究生毕业论文,1981.

[20] 林益.带 Carleman 位移的奇异积分方程的正则化问题.北京师范大学数学系研究生毕业论文,1981.

[21] 黄海洋.解析函数带位移的复合边值问题.北京师范大学数学系研究生毕业论文,1981 .

应用数学与计算数学，
1984，(3)：76-79.

关于非线性奇异积分方程

On the Nonlinear Singular Integral Equation

线性奇异积分方程理论已经得到了系统的发展，在这方面特别应该提出的是苏联学者 Н. И. Муселишвили，Н. П. Векуа，С. Г. Михлин 等人的工作．这种理论已经有了系统的专著，并有大量的文献目录．关于非线性奇异积分方程的理论只是在不久前才开始发展，因此，在这些人的专著中都没有涉及，只是在 Pogorgelski 的专著中对于非线性奇异积分方程发展初期的某些基本理论问题有所反映．近年来，非线性奇异积分方程理论发展得很快，在 А. И. Гусейнов，Х. Ш. Мухтаров 的专著[1]中系统地总结了这方面的理论，并且给出了大量的参考文献．本文的主要目的就是要综述有关非线性奇异积分方程的某些理论结果．

§ 1.

А. И. Гусейнов 研究单位圆到某一确定区域的保角映射函数时，归结为解一类非线性奇异积分方程，1974 年[2]，他研究了非线性方程

$$u(x)=\int_{-\pi}^{\pi}\Phi[x,s,u(s),\lambda]\cot\frac{s-x}{2}\mathrm{d}s \equiv Su, \tag{1.1}$$

首先建立了算子 S 在空间 $H_\delta(0<\delta<1)$ 中的有界性和在 H_δ 中某一球体上的一致连续性，对 $\Phi[x,s,u,\lambda]$ 做了一些自然的限制以后，根据 Schauder 原理，证明了方程(1.1) 在 H_δ 中解的存在与唯一性．以后这种理论在不可压缩理想流体中圆柱体的绕流问题中找到了应用．1950 年 А. И. Гусейнов 的结果还进一步对于方程

$$u(x)=\lambda\Phi\left[x,u(x),-\frac{1}{2\pi}\int_{-\pi}^{\pi}u(s)\cot\frac{s-x}{2}\mathrm{d}s\right] \tag{1.2}$$

和方程组

$$u_i(x)=\lambda\int_{-\pi}^{\pi}\Phi_i[x,s,u_1(s),\cdots,u_n(s)]\cot\frac{s-x}{2}\mathrm{d}s, i=1,2,\cdots,n, \tag{1.3}$$

进行了推广.

1955 年[3] Б. И. Гехт 在函数类 $H_{k_1,k_2,\delta}$(即 $\|u\|_C\leqslant K_1$,而且 $|u(x_1)-u(x_2)|\leqslant K_2|x_1-x_2|^{\delta}$) 中研究了方程(1.2),证明了在 Гусейнов 的条件下,当 $|\lambda|$ 适当小时,方程(1.2) 在 L^2 中有唯一解,而且解可以通过迭代法求得. 除此以外,他证明了这个解也仍然属于类 $H_{k_1,k_2,\delta}$. 这种手法后来还用来解决各种不同类型的非线性奇异积分方程. 例如,方程

$$u(x)=f(x)+\Phi\left[\lambda,x,u(x),-\frac{1}{2\pi}\int_{-\pi}^{\pi}u(s)\cot\frac{s-x}{2}\mathrm{d}s\right] \tag{1.2$'$}$$

或

$$u^{(n)}(x)=\lambda\Phi\Big[x,u(x),u'(x),\cdots,u^{(n)}(x),$$
$$-\frac{1}{2\pi}\int_{-\pi}^{\pi}u(s)\cot\frac{s-x}{2}\mathrm{d}s,\cdots,-\frac{1}{2\pi}\int_{-\pi}^{\pi}u(s)\cot\frac{s-x}{2}\mathrm{d}s\Big] \tag{1.4}$$

以及方程组

$$u_i(x)=\lambda\Phi_i\Big[x,u_1(x),\cdots,u_n(x),-\frac{1}{2\pi}\int_{-\pi}^{\pi}u_1(s)\cot\frac{s-x}{2}\mathrm{d}s,\cdots,$$
$$-\frac{1}{2\pi}\int_{-\pi}^{\pi}u_n(s)\cot\frac{s-x}{2}\mathrm{d}s\Big],\quad i=1,2,\cdots,n. \tag{1.5}$$

§ 2.

非线性奇异积分方程与解析函数的非线性边值问题之间有着紧密的联系. 一般提法下,非线性边值问题是根据边界条件 $F[s,u(s),v(s)]=0$,s 是弧坐标,来确定区域内解析函数 $f(z)=u(x,y)+\mathrm{i}v(x,y)$. 当然,要解决这种一般提法下的边值问题是很困难的.

但是,如果是在圆周上,而且边界条件能够写成

$$l(s)u(s)+\lambda\Phi[s,u(s),v(s)]=0,\quad l(s)\neq 0, \tag{2.1}$$

那么,只需要 $v(0,0)=0$,它就可以归结为研究方程(1.2) 的可解性. 这时候,边界条件中线性部分的指标是含影响问题(2.1) 解的个数的. ①

① 也可以讨论更一般的边界条件 $a(s)u(s)+b(s)v(s)=\Phi[s,u(s),v(s),\lambda]$.

与这种非线性希尔伯特问题相联系还可以讨论非线性奇异积分方程

$$a(x)u(x)-\frac{b(x)}{2\pi}\int_{-\pi}^{\pi}u(s)\cot\frac{s-t}{2}\mathrm{d}s$$

$$=f(x)+\Phi\left[x,u(x),-\frac{1}{2\pi}\int_{-\pi}^{\pi}u(s)\cot\frac{s-t}{2}\mathrm{d}s,\lambda\right]. \tag{2.2}$$

只要利用正则化因子的方法[4]，方程(2.2)的可解性就可以与方程(1.2)′的可解性问题联系起来.

应该指出在研究非线性奇异积分方程时，某些不变的泛函空间将起重要作用，例如，空间 $H_\varphi, H_{R,K,\delta}, H_{K1,K2}(\omega_1,\omega_2,\omega)$ 等[1]. 关于非线性函数 $\Phi(x,s,u)$ 当然需要附加某些适当的光滑性条件[5][6]才能解决问题.

§3.

对于带有柯西型核的非线性奇异积分方程

$$u(t)=\frac{1}{2i\pi}\int_\Gamma\frac{f(t,\tau,u(\tau))}{t-\tau}\mathrm{d}t \tag{3.1}$$

以及

$$u(t)=\lambda\mu\left[t,u(t),\rho\int_\Gamma\frac{K(t,\tau,u(\tau))}{\tau-t}\mathrm{d}\tau\right], \tag{3.2}$$

其中 Γ 是封闭的 Ляпунов 曲线，只需利用 Schauder 原理就可以建立方程(3.1)(3.2)解的存在定理.[2]

对于方程组

$$u_i(t)=\lambda\int_\Gamma\frac{f_i[t,\tau,u_1(\tau),u_2(\tau),\cdots,u_n(\tau)]}{t-\tau}\mathrm{d}\tau \tag{3.3}$$

还可以归结为对分片解析向量 $\Phi(z)=[\Phi_1(z),\Phi_2(z),\cdots,\Phi_n(z)]$ 所提的 Riemann 型边值问题

$$\Phi_\gamma^+(t)=G_\gamma(t)\Phi_\gamma^-(t)+\lambda f[t,\Phi_1^-(t),\cdots,\Phi_n^-(t),\Phi_1^+(t),\cdots,\Phi_n^+(t)],$$

$$\gamma=1,2,\cdots,n. \tag{3.4}$$

已经有不少作者讨论过这种类型的边值问题，从而非线性奇异积分方程(3.1)～(3.3)的问题也可以得到解决. 除此以外，利用 И. Н. Веяуа 的积分表示式，我们可以把非线性 Riemann 边值问题

$$\mathrm{Re}\sum_{k=1}^m\left[a_k(t)\varphi^{(k)}(t)+\int_\Gamma\frac{G_k(t,\tau)\varphi^{(k)}(\tau)}{\tau-t}\mathrm{d}\tau\right]$$

$$= F\left[t, \varphi(t), \varphi^{(1)}(t), \cdots, \varphi^{(m)}(t), \int_{\Gamma} \frac{K_1(t,\tau)\varphi(\tau)}{\tau - t} d\tau, \cdots, \int_{\Gamma} \frac{K_m(t,\tau)\varphi^{(m)}(\tau)}{\tau - t} d\tau\right] \tag{3.5}$$

归结为非线性奇异积分方程,然后在一般的假设条件下,再利用 Schauder 原理,也可以证明相应方程的解是存在的.[7]

§ 4.

讨论非线性奇异积分方程

$$u(x) = \lambda \int_a^b \frac{f[x, s, u(s)]}{s - x} ds, \tag{4.1}$$

如果假设 $f(x,s,t)$ 在区域 D 中满足条件

$$\begin{cases} f(a,a,t) = f(b,b,t) \equiv 0, \\ | f(x_1,s_1,t_1) - f(x_2,s_2,t_2) | \\ \leqslant A_1\omega^*(| x_1 - x_2 |) + A_2\omega(| s_1 - s_2 |) + A_3 | t_1 - t_2 |, \end{cases} \tag{4.2}$$

其中 $\omega^*(\sigma), \omega(\sigma) \in \Phi, \omega^*(\sigma)\ln\dfrac{b-a}{\sigma} \leqslant A_4\omega(\sigma), A_1, A_2, A_3, A_4$ 是常数.

那么,在空间 $H_{R,K_0}(\omega)$ 中方程(4.1)有唯一解,这个解可以用 Picard 逐次逼近法求得,并且对近似解还得到了相应的估计式.[8]

另外,А. И. Гусейнов 还对方程组

$$u_i(x) = \lambda \sum_{j=1}^{n} \int_a^b \frac{K_{ij}[x, s, u_1(s), \cdots, u_n(s)]}{s - x} ds + F_i(s), i = 1, 2, \cdots, n \tag{4.3}$$

进行了推广.

类似地还可以在空间 $H_{R,K}(\omega,0), I_{\varphi,P}^{R,K}, H_{K1,K2}(\omega_1,\omega_2,\omega)$ 中来讨论方程(4.1)的解,从而得到类似的结论.[1]

还可以利用牛顿 - 康特洛维奇方法来求解非线性奇异积分方程. 这时,如果 $f(s,u), f'_u(s,u), f''_{uu}(s,u)$ 在 $D = \{a \leqslant s \leqslant b, -\infty < u < +\infty\}$ 上有定义,而且满足条件

$$\begin{cases} f(a,u)=f(b,u)\equiv 0, \\ f'_u(a,u)=f'_u(b,u)\equiv 0, \\ f''_{uu}(a,u)=f''_{uu}(b,u)\equiv 0, \\ |f(s_1,u_1)-f(s_2,u_2)|\leqslant A_0[|s_1-s_2|^\delta+|u_1-u_2|], \\ |f'_u(s_1,u_1)-f'_u(s_2,u_2)|\leqslant A_1[|s_1-s_2|^\delta+|u_1-u_2|], \\ |f''_{uu}(s_1,u_1)-f''_{uu}(s_2,u_2)|\leqslant A_2[|s_1-s_2|^\delta+|u_1-u_2|], \end{cases} \tag{4.4}$$

只要算子 $P(u)\equiv u(x)-\lambda\int_a^b \frac{f[s,u(s)]}{s-u}\mathrm{d}s$ 在 Fréchet 意义下可微，并且在 H_δ^α 中的每一点上都存在着有界的逆算子，从而牛顿－康特洛维奇法就是可以实现的.[9]

再介绍利用泛函修正值的方法来求解非线性奇异积分方程问题. 讨论方程

$$u(x)-\lambda F\left(x,\int_a^b \frac{f[x,s,u(s)]}{s-x}\mathrm{d}s\right)=g(x), \tag{4.5}$$

这里 $g(x)\in L^2[a,b]$，λ 是数值参数.

所谓泛函修正值方法就是作为初始近似可以取任意元素 $u_0(x)\in L^2[a,b]$，而以后的近似应该选取

$$u_n(x)=\lambda F\left(x,\int_a^b \frac{f[x,s,u_{n-1}(s)]}{s-x}\mathrm{d}s\right)+\lambda a_n(x)+g(x), \tag{4.6}$$

其中 $a_n(x)$ 是所谓的泛函修正值[1]. 不难证明，这种迭代过程是收敛的，从而方程(4.5)的解在 $L^2[a,b]$ 中是存在的，而且是唯一的，(只要 $|\lambda|$ 适当小) 并且有适当的估计式[1].

最后还可以讨论方程

$$u(x)-\lambda r(x)\int_a^b \frac{f[s,u(s)]}{r(s)(s-x)}\mathrm{d}s=g(x) \tag{4.7}$$

和

$$u(x)=\lambda\left[r_1(x)\int_a^b \frac{f[s,u(s)]}{r_1(s)(s-x)}\mathrm{d}s+g(x)\right], \tag{4.8}$$

从而得到关于方程(4.7)(4.8) 解存在的相应结论.

参考文献

[1] Гусейнов А И, Мухтаров Х Ш. Введение в Теорию нелинейных сингулярных интур. Москва,1980.

[2] Гусейнов А И, Матем С. новая серия,1947,20(2):293-309.

[3] Гехт Б И. Труды новочеркасского полит. ин-та,1955,26:436-454.

[4] Гахов Ф Д. Краевые задачи. М-Наука,1977.

[5] Берколайко М З,Рутидкий Я Б. ДАН СССР,1970,192(6):1 199-1 201.

[6] Сиб. Матем Ж. 1971,12(5):1 015-1 025.

[7] Zdzislaw R. Bull. Wajsk. Akad. Techn. I. Dabranskiego,1968,17(6):53-62.

[8] Гусейнов А И. Изв. АН СССР,Сер.《Мат. Н. 》,1984,12(2):193-212.

[9] Верттейм Б А. ДАН СССР,1956,110(5):719-722.

四川师范大学学报(自然科学版),
1994,17(2):114-116.

双解析函数的某些性质①

Some Properties of Bianalytic Functions

我们知道,解析函数具有非常好的性质,因此,它在力学、数学物理中有着广泛和重要的应用.例如,研究无源和无旋的物理场时,解析函数就显示了非常巨大的威力,但是,研究有源或有旋的物理场时,解析函数这一重要工具却是无能为力的.[1]中提出了半解析函数的概念,并研究了有关的一些性质,只是由于这类函数对于 Cauchy-Riemann 方程组中的某一个方程完全不做要求,致使在应用时,在唯一性问题上又出现了很大困难.为了解决这种困难,作者在本文中试图对于半解析函数附加某些控制条件,从而,提出了一类新的函数,即双解析函数,并给出了关于双解析函数的一些定理.

众所周知,柯西-黎曼方程组的复形式是

$$\frac{\partial w}{\partial \bar{z}}=0, \tag{0.1}$$

这里$\frac{\partial}{\partial \bar{z}}=\frac{1}{2}\left(\frac{\partial}{\partial x}+\mathrm{i}\frac{\partial}{\partial y}\right)$, $w=u+\mathrm{i}v$, $z=x+\mathrm{i}y$.

我们考虑在平面上某一区域 G 内定义的函数 $w(z)$,并假设它具有关于 $\bar{z}$ 的二阶偏导数,这时候,如果 $w(z)$ 满足二阶复微分方程

$$\frac{\partial^2 w}{\partial \bar{z}^2}=0,\ z\in G \tag{0.2}$$

我们将称之为双解析函数.

根据[2]知道

① 北京市自然科学基金资助项目.

$$\frac{\partial w}{\partial \bar{z}} = f,$$

这里，f 是解析函数，这样，将有

$$w(z) = -\frac{1}{\pi}\iint_G \frac{f(\xi,\eta)}{\xi - z}\mathrm{d}\xi\mathrm{d}\eta + \Phi(z), \tag{0.3}$$

其中 $\xi = \xi + \mathrm{i}\eta \in G$，而 $\Phi(z)$ 是 G 内的任意解析函数.

§1.　双解析函数的唯一性

我们考虑以下边值问题：

问题 A　寻求复方程(2)在区域 G 内的正则解，并要求它在边界 Γ 上满足条件

$$\frac{\partial w}{\partial \bar{z}}\,|_{\Gamma} = 0, \tag{1.1}$$

$$w\,|_{\Gamma} = 0, \tag{1.2}$$

定理 1　问题 A 在 G 内只有零解.

事实上，由(0.2)式可知 $\frac{\partial \omega}{\partial \bar{z}}$ 必是 G 内的解析函数，这样，再根据条件(1.1)可知 $\frac{\partial w}{\partial \bar{z}} \equiv 0, z \in G$，从而，$w(z)$ 又必是区域 G 内的解析函数，再根据条件(1.2) $w(z)\,|_{\Gamma} = 0$，必定还有 $w(z) \equiv 0$.

定理 2　如果不计一个解析函数加项，那么对于双解析函数 $w(z)$ 来说，只要规定 $\frac{\partial w}{\partial \bar{z}}\,|_{\Gamma} = r(t), t \in \Gamma$，这里 $r(t)$ 是给定的连续函数，$w(z)$ 就是唯一确定的.

事实上，由(0.3)式可知，如果不计一个解析函数加项 $\Phi(z)$，那么有

$$w(z) = \frac{-1}{\pi}\iint_G \frac{f(\xi,\eta)}{\xi - z}\mathrm{d}\xi d\eta \equiv T_G(f), \tag{0.3$'$}$$

这里 $f(z) = \frac{\partial w}{\partial z}$ 是解析函数，显然，只要知道 $\frac{\partial w}{\partial z}\,|_{\Gamma} = f\,|_{\Gamma} = r(t), f(z)$ 就是唯一确定的，也就是说 $w(z) = T_G(f)$ 是唯一确定的.

下面为了叙述方便，我们把忽略解析函数加项 $\Phi(z)$ 的双解析函数 $w(z)$. 将简称为双解析函数.

§2.　双解析函数的基本函数系

我们考虑下列函数组

$$\begin{aligned} &w_{2n,0}(z) = T_G(z^n), \\ &w_{2n+1,0}(z) = T_G(\mathrm{i}z^n), \quad n \in \mathbf{N}, \\ &w_{2n,k}(z) = T_G[(z - z_u)^{-n}], \\ &w_{2n+1,k}(z) = T_G[\mathrm{i}(z - z_k)^{-n}], \quad n \in \mathbf{N}^*;\ k = 1,2,\cdots,m. \end{aligned} \tag{2.1}$$

其中 $z_1, z_2, \cdots, z_m$ 分别是在 $\Gamma_1, \Gamma_2, \cdots, \Gamma_m$ 内即多连通区域的补域中的定点，如果是单连通区域（即 $m = 0$）的情形，我们就仅有函数 $w_{n,0}(z)$，这时候，简单地记作 $w_n(z)$.

不难知道，任意双解析函数，可以用函数组(2.1)的（具有实系数的）线性组合在 G 内一致逼近.

我们将把函数组(2.1)叫作双解析函数的基本函数系.

定理 3 如果 G 是圆域：$|z| < R$，$w(z)$ 是 G 内的双解析函数，它在 $G+\Gamma$ 上连续，那么在 G 内 $w(z)$ 可以表示成一致收敛的级数（广义泰勒级数）

$$w(z) = \sum_{n=0}^{+\infty} C_n w_n(z), \tag{2.2}$$

这里

$$\begin{aligned} &w_n(z) = w_{n,0}(z), \\ &C_{2n} = \mathrm{Re}\left[\frac{1}{2\pi_{\mathrm{i}}}\int_\Gamma \frac{\frac{\partial \mathrm{w}}{\partial t}}{t^{n+1}}\mathrm{d}t\right], \quad n \in \mathbf{N}, \\ &C_{2n+1} = \mathrm{Im}\left[\frac{1}{2\pi_{\mathrm{i}}}\int_\Gamma \frac{\frac{\partial \mathrm{w}}{\partial t}}{t^{n+1}}\mathrm{d}t\right], \end{aligned} \tag{2.3}$$

事实上，由于 $\frac{\partial w}{\partial \bar{z}} = f$ 在 G 内解析，在 $G+\Gamma$ 上连续，从而，有

$$\frac{\partial w}{\partial \bar{z}} = \sum_{n=0}^{+\infty} a_n z^n,$$

这里 $a_n = \frac{1}{2\pi \mathrm{i}}\int_\Gamma \frac{\frac{\partial w}{\partial \bar{t}}}{t^{n+1}}\mathrm{d}t$，它在 G 内是一致收敛的.

只需要再把这一表示式代入(0.3)′就得到表示式(2.2). 显然，这里

$$C_{2n}=\operatorname{Re}\left[\frac{1}{2\mathrm{i}\pi}\int_{\Gamma}\frac{\frac{\partial \mathrm{w}}{\partial t}}{t^{n+1}}\mathrm{d}t\right]=\operatorname{Re}a_n,\qquad n\in\mathbf{N},$$
$$C_{2n+1}=\operatorname{Im}\left[\frac{1}{2\mathrm{i}\pi}\int_{\Gamma}\frac{\frac{\partial \mathrm{w}}{\partial t}}{t^{n+1}}\mathrm{d}t\right]=\operatorname{Im}a_n,$$

定理 4 假设G是圆环域$R_1<|z|<R_2$,我们考虑G内的双解析函数$w(z)$,还要求它在$G+\Gamma$上连续,则在G内$w(z)$可以表示成一致收敛的级数(广义洛朗敛数):

$$w(z)=\sum_{n=-\infty}^{+\infty}C_n w_n(z),\tag{2.4}$$

其中

$$\begin{aligned}&w_n=w_{n,0},\\&w_{-2n}=T_G(z^{-n}),\qquad n\in\mathbf{N}^*,\\&w_{-2n+1}=T_G(\mathrm{i}z^{-n}),\end{aligned}\tag{2.5}$$

$$C_{2n}=\operatorname{Re}\left[\frac{1}{2\mathrm{i}\pi}\int_{\Gamma}\frac{\frac{\partial \mathrm{w}}{\partial t}}{t^{n+1}}\mathrm{d}t\right],\qquad n\in\mathbf{N},\Gamma_2:|z|=R_2$$
$$C_{2n+1}=\operatorname{Im}\left[\frac{1}{2\mathrm{i}\pi}\int_{\Gamma}\frac{\frac{\partial \mathrm{w}}{\partial t}}{t^{n+1}}\mathrm{d}t\right],$$
$$C_{-2n}=-\operatorname{Re}\left[\frac{1}{2\mathrm{i}\pi}\int_{\Gamma_1}\frac{\partial w}{\partial t}t^{n-1}\mathrm{d}t\right],\qquad n\in\mathbf{N}^*,\Gamma_1:|z|=R_1,$$
$$C_{-2n-1}=-\operatorname{Im}\left[\frac{1}{2\mathrm{i}\pi}\int_{\Gamma_1}\frac{\partial w}{\partial t}t^{n}\mathrm{d}t\right]$$

今后为了方便,我们也可以把级数(2.4)中的$\sum_{n=0}^{+\infty}C_n w_n(z)$叫作该级数的双解析部分,而$\sum_{n=-1}^{-\infty}C_n w_n(z)$,叫作它的主要部分.

类似地,还可以考虑关于$(z-z_0)$的广义泰勒级数和广义洛朗级数.

参考文献

[1]王见定.半解析函数、共轭解析函数.北京:北京工业大学出版社,1988.

[2] Вскуа И Н.广义解析函数.北京:人民教育出版社,1960.

北京师范大学学报(自然科学版),
1995,31(2):175-179.

双解析函数与复调和函数以及它们的基本边值问题[①]

Bianalytic Functions, Complex Harmonic Functions and Their Basic Boundary Value Problems

摘要 讨论双解析函数和复调和函数的某些性质,借助于这两类函数还可以讨论一类半解析函数的问题,最后,还讨论了两类基本边值问题.

关键词 双解析函数;复调和函数;基本边值问题;半解析函数.

解析函数具有非常绝妙的性质,在力学、数学物理中有着广泛和重要的应用.例如,研究无源和无旋的物理场时,解析函数就显示了巨大的威力,但是当研究有源或有旋的物理场时,解析函数这一重要工具却是无能为力的.文献[1]曾提出了半解析函数的概念,并研究了有关的一些性质,只是由于这类函数定义中对于柯西-黎曼方程组(C-R 方程组)中的某一个可以完全不做要求,致使当应用这一理论时,在唯一性问题上遇到了难以克服的困难.

为了解决这个问题,本文试图对半解析函数附加某些控制条件,从而又提出了两类新的函数,即双解析函数和复调和函数,并给出了一系列关于双解析函数性质的定理.最后还讨论了对这两类函数所提的基本边值问题.

我们指出:C-R 方程组的复形式是

① 北京市自然科学基金资助项目,
收稿日期:1994-12-07.

$$\frac{\partial W}{\partial \bar{z}} = 0, \tag{0.1}$$

这里，$W(z)=u(x,y)+\mathrm{i}v(x,y)$，$z=x+\mathrm{i}y$，$\frac{\partial}{\partial \bar{z}}=\left(\frac{\partial}{\partial x}+\mathrm{i}\frac{\partial}{\partial y}\right)$. 以后，我们还将用到导数$\frac{\partial}{\partial z}=\left(\frac{\partial}{\partial x}-\mathrm{i}\frac{\partial}{\partial y}\right)$和 $\Delta\equiv\frac{\partial^2}{\partial z\partial \bar{z}}=\frac{\partial^2}{\partial \bar{z}\partial z}$.

定义 1 假设 G 是平面上的区域，在 G 上给定了复函数 $W(z)$，要求它具有关于 $\bar{z}$ 的二阶导数$\frac{\partial^2 W}{\partial \bar{z}^2}$. 如果给定的函数 $W(z)$满足以下方程式

$$\frac{\partial^2 W}{\partial \bar{z}^2} = 0, \tag{0.2}$$

那么称 $W(z)$是 G 上的双解析函数. 以后将用 $\mathscr{D}_2(G)$来代表所有双解析函数构成的集合.

定义 2 假设 G 是平面上的区域，在 G 上给定了复函数 $W(z)$，要求它具有关于 $\bar{z}$ 和 z 的二阶混合导数$\frac{\partial^2 W}{\partial \bar{z}\partial z}=\frac{\partial^2 W}{\partial z\partial \bar{z}}$. 如果给定的函数 $W(z)$满足以下方程式

$$\Delta \equiv \frac{\partial^2 W}{\partial \bar{z}\partial z} = 0, \tag{0.3}$$

那么称 $W(z)$是 G 上的复调和函数. 以后将用 $\mathscr{H}_2(G)$来代表所有复调和函数构成的集合.

§1.　双解析函数

根据方程(0.2) 我们得到[2] $\frac{\partial W}{\partial \bar{z}}=f$，这里，$f(z)=f_1(x,y)+\mathrm{i}f_2(x,y)$是任意解析函数. 从而有

$$W(z) = -\frac{1}{\pi}\iint_G \frac{f(\zeta)}{\zeta - z}\mathrm{d}\xi\mathrm{d}\eta + \Phi(z), \tag{1.1}$$

这里，$\zeta=\xi+\mathrm{i}\eta\in G$，并且 $\Phi(z)$也是任意解析函数.

定理 1 （唯一性定理） 假设 $W(z)$是 G 上的双解析函数，如果

$$\left.\frac{\partial W}{\partial \bar{z}}\right|_{\partial G} = 0, \tag{1.2}$$

$$W\,|_{\partial G} = 0, \tag{1.3}$$

那么 $W(z)\equiv 0$，$z\in G$，这里，∂G 是 G 的边界.

证　因为$\frac{\partial W}{\partial \bar{z}}=f(z)$是解析函数，一般认为它不是常数，从而，根据条件(1.2)，将有$\frac{\partial W}{\partial \bar{z}}\equiv 0, z\in G$；再根据(1.1)(1.3)，还有$W(z)\equiv 0, z\in G$.

定理 2　（双解析函数第一表示式）　如果$W(z)\in \mathscr{D}_2(G)$，那么有以下表示式成立：

$$W(z)=-\frac{1}{\pi}\iint_G \frac{\frac{\partial W}{\partial \bar{\zeta}}}{\zeta - z}\mathrm{d}\zeta\mathrm{d}\eta+\Phi(z), \tag{1.4}$$

这里，$\zeta=\zeta+\mathrm{i}\eta\in G$，$\Phi(z)$是任意的解析函数. 根据(1.1)，定理 2 显然是成立的.

定理 3　（双解析函数第二表示式）　如果$W(z)\in \mathscr{D}_2(G)$，那么有以下表示式成立：

$$W(z)=\bar{z}\varphi(z)+\Phi(z), \tag{1.5}$$

这里，$\varphi(z)$也是任意的解析函数.

证　取$\varphi(z)=\frac{\partial W}{\partial \bar{z}}$，显然知道，$W_1(z)=\bar{z}\varphi(z)$是双解析函数，并且有$\frac{\partial W_1}{\partial \bar{z}}=\varphi(z)$，于是，有$\frac{\partial (W-W_1)}{\partial \bar{z}}=0$. 最后得到$W(z)=\bar{z}\varphi(z)+\Phi(z)$.

为了简单，我们把忽略不计一个解析函数加项$\Phi(z)$的函数$W(z)$，有时候也叫作双解析函数.

根据第二表示式我们知道，下面的一些定理也是成立的.

定理 4　（关于双解析函数零点的定理）　如果$W(z)\in \mathscr{D}_2(G)$，那么它的零点必是孤立的，并且零点个数一定是有限的.

定理 5　（关于双解析函数奇点的定理）　如果$W(z)\in \mathscr{D}_2(G)$，那么它的奇点是孤立的，并且奇点个数也一定是有限的.

定理 6　（双解析函数泰勒展开定理）　如果在圆盘G上，G：$|z|<R$，$W(z)$是双解析函数，那么在G上有以下的展开式成立：

$$W(z)=\sum_{k=0}^{+\infty} c_k \bar{z} z^k, \tag{1.6}$$

这里，

$$c_k=\frac{1}{2\pi \mathrm{i}}\int_{\partial G}\frac{\frac{\partial W}{\partial \bar{t}}}{t^{k+1}}\mathrm{d}t, k\in \mathbf{N} \tag{1.7}$$

因为$W=\bar{z}\frac{\partial W}{\partial \bar{z}}$并且$\frac{\partial W}{\partial z}$是解析函数. 于是，定理的结论显然是成立的.

定理 7 （双解析函数洛朗展开定理） 假设在圆环域 G 上，G：$R_1 < |z| < R_2$，$0 \leqslant R_1 < R_2 < +\infty$，$W(z)$ 是双解析函数，则在 G 上有以下的展开式成立：

$$W(z) = \sum_{k=-\infty}^{+\infty} c_k \bar{z} z^k, \tag{1.8}$$

这里，系数 c_k 可以由以下公式给出

$$c_k = \frac{1}{2\pi \mathrm{i}} \int_\gamma \frac{\dfrac{\partial W}{\partial t}}{t^{k+1}} \mathrm{d}\bar{t},$$

其中 γ 是圆周 $|z| = r$，$R_1 < r < R_2$，r 是任意的，而且这种展开式是唯一的.

证 只要把 $\varphi(z) = \dfrac{\partial W}{\partial \bar{z}}$ 展开成洛朗级数，再利用第二表示式，就可以得到(1.8)

应该指出，在任意圆环 $G^* \subset G$ 上，G^*：$r_1 < |z| < r_2$，只要 $R_1 < r_1 < r_2 < R_2$，于是，(1.8)就都是绝对和一致收敛的.

以后，我们也将把(1.8)中的级数 $\sum\limits_{k=0}^{+\infty} c_k \bar{z} z^k$ 和 $\sum\limits_{k=-1}^{-\infty} c_k \bar{z} z^k$ 分别叫作函数 $W(z)$ 的双解析部分和主要部分.

类似地，我们也可以利用洛朗展开式对双解析函数的孤立奇点进行分类：

(1) $z=0$ 是可去奇点的充分必要条件是(1.8)中不含主要部分.

(2) $z=0$ 是 m 阶极点的充分必要条件是(1.8)中的主要部分只含有 m 项，而且 $c_m \neq 0$.

(3) $z=0$ 是本性奇点的充分必要条件是(1.8)中的主要部分含有无穷多项.

我们也可以讨论双解析函数关于 $(z-z_0)^k$ 的泰勒级数和洛朗级数.这里 z_0 是区域 G 内的任意点.

定理 8 如果 $W(z) = u(x,y) + \mathrm{i}v(x,y) \in \mathscr{D}_2(G)$，那么它的实部 $u(x,y)$ 和虚部 $v(x,y)$ 都是双调和函数.

证 由定义 1，$\dfrac{\partial^2 W}{\partial \bar{z}^2} = 0$. 于是，

$$\Delta^2 W = \frac{\partial^4 W}{\partial z^2 \partial \bar{z}^2} = 0,\ W = u + \mathrm{i}v,$$

从而有，$\Delta^2 u = 0$，$\Delta^2 v = 0$.

§ 2. 复调和函数

根据方程(0.3),对于复调和函数 $W(z)$,有[2] $\frac{\partial W}{\partial \bar{z}}=\bar{f}$,这里,$f(z)=f_1(x,y)+\mathrm{i}f_2(x,y)$是任意解析函数.从而有

$$W(z)=-\frac{1}{\pi}\iint_G \frac{\overline{f(\zeta)}}{\zeta-z}\mathrm{d}\zeta\mathrm{d}n+\Phi(z). \tag{2.1}$$

其中 $\zeta=\zeta+\mathrm{i}\eta\in G$,而且 $\Phi(z)$是 G 上的任意解析函数.

定理 9(唯一性定理) 假设 $W(z)$是 G 上的复调和函数,如果

$$\left.\frac{\partial W}{\partial \bar{z}}\right|_{\partial G}=0, \tag{2.2}$$

$$W\,|_{\partial G}=0, \tag{2.3}$$

那么 $W(z)\equiv 0, z\in G$,这里∂G 是 G 的边界.

证 $\frac{\partial W}{\partial \bar{z}}=\overline{f(z)}$,$f(z)$是解析函数,于是,根据(2.2)我们有$\frac{\partial W}{\partial \bar{z}}\equiv 0$,$z\in G$,然后,再根据 (2.1)(2.3)还有 $W(z)\equiv 0, z\in G$.

我们再讨论非齐次 C-R 方程组

$$\frac{\partial u}{\partial x}-\frac{\partial v}{\partial y}=p(x,y),\qquad \frac{\partial u}{\partial y}+\frac{\partial v}{\partial x}=0,$$

其中 $p(x,y)$是任意调和函数.它的复形式是

$$\frac{\partial W}{\partial \bar{z}}=p(x,y),\quad W=u+\mathrm{i}v, \tag{2.4}$$

而它的一般解将是

$$\begin{aligned}W(z)&=-\frac{1}{\pi}\iint_G \frac{p(x,y)}{\zeta-z}\mathrm{d}\zeta\mathrm{d}\eta+\Phi(z)\\&=-\frac{1}{\pi}\iint_G \frac{\varphi(\zeta)}{\zeta-z}\mathrm{d}\zeta\mathrm{d}\eta-\frac{1}{\pi}\iint_G \frac{\overline{\varphi(\zeta)}}{\zeta-z}\mathrm{d}\zeta\mathrm{d}\eta+\Phi(z),\end{aligned} \tag{2.5}$$

其中 $\varphi(z)$是任意解析函数.这里,我们看出

$$W_1(z)=-\frac{1}{\pi}\iint_G \frac{\varphi(\zeta)}{\zeta-z}\mathrm{d}\zeta\mathrm{d}\eta$$

是一个双解析函数,而

$$W_2(z)=-\frac{1}{\pi}\iint_G \frac{\overline{\varphi(\zeta)}}{\zeta-z}\mathrm{d}\zeta\mathrm{d}\eta$$

是一个复调和函数.

根据定理 1 和定理 8,我们知道,对于函数(2.5)唯一性定理也是成立的. 从而,只要对半解析函数附加这样的条件,即,使得(2.4)中的 $p(x,y)$ 是调和函数. 那么,借助于双解析函数和复调和函数,我们就已经可以讨论带有源(或者旋)的物理场了.

§ 3. 基本边值问题

首先我们讨论对于双解析函数的边值问题(问题 A):

寻求复方程

$$\frac{\partial^2 W}{\partial \bar{z}^2} = 0,\ z \in G \tag{3.1}$$

的解,要求它在边界 ∂G 上满足以下条件:

$$\left.\frac{\partial W}{\partial \bar{z}}\right|_{\partial G} = g_1(t),\ W|_{\partial G} = g_2(t),\ t \in \partial G \tag{3.2}$$

这里,$g_1(t)$,$g_2(t)$ 都是在边界上给定的连续函数.

定理 10　问题 A 一定有解,而且解是唯一的.

证　由于 $\frac{\partial W}{\partial \bar{z}}$ 是解析函数,根据条件(3.2)知道,可以唯一地确定 $\frac{\partial W}{\partial \bar{z}} = \varphi(z)$,$z \in G$,从而,$W_1(z) = -\frac{1}{\pi}\iint_G \frac{\varphi(\zeta)}{\zeta - z}\mathrm{d}\zeta\mathrm{d}\eta$ 也随之而唯一确定. 然后,根据双解析函数第一表示式,得到

$$W(z) = W_1(z) + \Phi(z) = -\frac{1}{\pi}\iint_G \frac{\varphi(\zeta)}{\zeta - z}\mathrm{d}\zeta\mathrm{d}\eta + \Phi(z),$$

这里 $W_1(z)$ 已经唯一确定,而 $\Phi(z)$ 是任意解析函数. 再由条件(3.2)可以知道

$$W(z)|_{\partial G} = W_1(z)|_{\partial G} + \Phi(z)|_{\partial G} = g_2(t),$$

这里,$W_1(z)|_{\partial G} = -\frac{1}{\pi}\iint_G \frac{\varphi(\zeta)}{\zeta - t}\mathrm{d}\zeta\mathrm{d}\eta = W_1(t)$ 是确定的已知函数. 于是问题就转化为对于解析函数的边值问题

$$\Phi(z)|_{\partial G} = g_2(t) - W_1(t) = \gamma(t). \tag{3.3}$$

再由解析函数的唯一性定理,就可以得到问题(3.3)的唯一解.

完全类似地,可以讨论对于复调和函数的边值问题(问题 B):

寻求复方程

$$\frac{\partial^2 W}{\partial \bar{z}\partial z} = 0,\ z \in G \tag{3.4}$$

的解，要求它在边界∂G上满足以下条件：

$$\left.\frac{\partial W}{\partial \bar{z}}\right|_{\partial G}=g_1(t),\ W|_{\partial G}=g_2(t),\ t\in\partial G, \tag{3.5}$$

这里，$g_1(t)$，$g_2(t)$都是在边界上给定的连续函数.

定理 11 问题 B 的解一定存在，而且解是唯一的.

证 与定理 10 的证明完全类似，这里，只是需要先根据条件(3.5)，考虑$\left.\overline{\frac{\partial W}{\partial z}}\right|_{\partial G}=\overline{g_1(t)}$，而$\overline{\frac{\partial W}{\partial z}}$是解析函数，也就可以唯一确定$\overline{f(z)}$.

最后，类似地，对于附加了控制条件的半解析函数，也可以讨论相应的边值问题. 再根据 $W|_{\partial G}=g_2(t)$，又可以唯一确定 $\Phi(z)$.

参考文献

[1] 王见定. 半解析函数、共轭解析函数. 北京：北京工业大学出版社，1988.
[2] 维库阿 · 依 · 涅. 广义解析函数. 北京：人民教育出版社，1960.
[3] 赵桢. 双解析函数的某些性质. 四川师范大学学报(自然科学版)，1994，17(2)：14.

Abstract Some properties of bianalytic and complex harmonic functions are considered. With the help of them, one kind of semi-analytic functions can be also considered. The basic boundary value problems for bianalytic and complex harmonic functions are established.

Keywords bianalytic function; complex harmonic function; basic boundary value problem; semi-analytic function.

宁夏大学学报(自然科学版),
1998,19(1):6-8.

一类三阶复偏微分方程的Schwarz问题①

Schwarz's Problem for a Class of Complex Partial Differential Equations of Third Order

摘要 研究一类三阶复偏微分方程$\frac{\partial^3 W}{\partial \bar{z}^2 \partial z}=0, z\in G$解的性质,同时讨论它的Schwarz问题的可解性.

关键词 复偏微分方程;双解析函数;复调和函数;复正则函数;Schwarz问题.

解析函数理论在力学和数学物理中有着广泛的应用,特别是在研究无源和无旋的物理场时它具有非常重要的意义.但是如果研究有源或有旋的物理场时,解析函数理论却显得无能为力.文[1]曾提出半解析函数的概念,但是由于半解析函数的定义中对于C-R方程组的某一个方程完全不做要求,从而在解决唯一性问题时遇到了难以克服的困难.文[2]中提出:对半解析函数类添加适当的控制条件就可以使唯一性问题得以解决.为此,对双解析函数和复调和函数的性质及其边值问题进行了一系列研究,并得到了一些重要的结果[2~4].本文的目的是研究一类三阶复偏微分方程解的性质,并讨论它的Schwarz问题的可解性.

§1. 复正则函数的性质

首先讨论如下的偏微分方程

---

① 国家自然科学基金资助项目.
收稿日期:1997-12-08.

$$\frac{\partial^3 W}{\partial \bar{z}^2 \partial z} = 0,\ z \in G. \tag{1.1}$$

这里，$\frac{\partial}{\partial \bar{z}} = \frac{1}{2}\left(\frac{\partial}{\partial x} + \mathrm{i}\frac{\partial}{\partial y}\right)$, $\frac{\partial}{\partial z} = \frac{1}{2}\left(\frac{\partial}{\partial x} - \mathrm{i}\frac{\partial}{\partial y}\right)$, $\Delta \equiv \frac{\partial^2 W}{\partial \bar{z}\partial z} = \frac{\partial^2 W}{\partial z \partial \bar{z}}$.

以后把这类方程的解叫作复正则函数，记作 $D_3(G)$. 显然，双解析函数类 $D_2(G)$是它的一个真子类，即 $D_2(G) \subset D_3(G)$. 容易看出，方程(1.1)与下列两个方程中任何一个都等价.

$$\frac{\partial W}{\partial \bar{z}} = x(z),\ z \in G; \tag{1.2}$$

$$\Delta W = \varphi(z),\ z \in G. \tag{1.3}$$

这里，$x(z)$和 $\varphi(z)$分别是任意的复调和函数和解析函数. 复调和函数 $x(z) = \varphi_1(z) + \overline{\varphi_2(z)}$，而 $\varphi_i(z)$，$i=1,2$ 是解析函数[4].

1.1 复正则函数的第一表示式

根据(1.2)我们知道

$$\begin{aligned} W(z) &= -\frac{1}{\pi}\iint_G \frac{x(\zeta)}{\zeta - z}\mathrm{d}T_\zeta + \varphi(z) \\ &= -\frac{1}{\pi}\iint_G \frac{\varphi_1(\zeta)}{\zeta - x}\mathrm{d}T_\zeta - \frac{1}{\pi}\iint_G \frac{\overline{\varphi_2(\zeta)}}{\zeta - z}\mathrm{d}T_\zeta + \varphi(z). \end{aligned} \tag{1.4}$$

其中 $\varphi(z)$，$\varphi_i(z)$，$i=1,2$ 都是任意的解析函数，而 $x(z) = \varphi_1(z) + \overline{\varphi_2(z)}$ 是任意的复调和函数.

1.2 复正则函数的第二表示式

根据(1.3)我们知道[5]

$$\begin{aligned} W(z) &= \frac{2}{\pi}\iint_G \ln|\zeta - z|\varphi(\zeta)\mathrm{d}T_\zeta + x(z) \\ &= \frac{2}{\pi}\iint_G \ln|\zeta - z|\varphi(\zeta)\mathrm{d}T_\zeta + \varphi_1(z) + \overline{\varphi_2(z)}. \end{aligned}$$

其中 $\varphi(z)$，$\varphi_2(z)$，$i=1,2$ 都是任意的解析函数，而 $x(z) = \varphi_1(z) + \overline{\varphi_2(z)}$ 是任意的复调和函数.

总之，$D_3(G)$类中的函数一般来说应依赖于三个任意的解析函数，或者说应依赖于一个复调和函数和一个解析函数.

1.3 复正则函数的唯一性

假设 Γ 是区域 G 的边界，Γ 上的点以后将用 t 来表示.

根据复正则函数的第一表示式可以看出，只要$\left.\frac{\partial W}{\partial \bar{z}}\right|_\Gamma = 0$，就可以唯一

确定复调和函数 $x(z)=0$;再根据实条件 $\mathrm{Re}\,W|_\Gamma=0$,将有 $\mathrm{Re}\,\varphi(z)|_\Gamma=\mathrm{Re}\,Tx(z)|_\Gamma\equiv 0$. 这样就可以确定解析函数 $\varphi(z)=\mathrm{i}C$,只要再规定 $\varphi(0)=0$,将有 $C=0$,从而 $W(z)\equiv 0$.

根据复正则函数的第二表示式还可看出,只要实条件 $\mathrm{Re}\,\Delta W|_\Gamma=0$,就可以唯一确定解析函数 $\varphi(z)=\mathrm{i}C$. 规定 $\varphi(0)=0$,将有 $C=0$,从而 $\varphi(z)\equiv 0$. 再根据条件 $W|_\Gamma=0$,将有 $x(z)|_\Gamma=\dfrac{2}{\pi}\iint_G \ln|\zeta-z|\varphi(\zeta)\mathrm{d}T_\zeta|_\Gamma\equiv 0$,这就可以唯一确定复调和函数 $x(z)\equiv 0$,从而 $W(z)\equiv 0$.

定理 1 如果 $W(z)\in D_3(G)$,那么满足下列两组条件

$$\left.\frac{\partial W}{\partial \bar z}\right|_\Gamma=0,\ \mathrm{Re}\,W(z)|_\Gamma=0;\ \mathrm{Re}\,\Delta W(z)|_\Gamma=0,\ W(z)|_\Gamma=0$$

中的任何一组,规定 $\varphi(0)=0$,则 $W(z)\equiv 0$.

§2. 复正则函数的 Schwarz 问题

Schwarz 问题(问题 S) 根据下列边界条件

$$\left.\frac{\partial W}{\partial \bar z}\right|_\Gamma=\gamma_1(t)+\mathrm{i}\gamma_2(t),\ \mathrm{Re}\,\varphi(z)|_\Gamma=\gamma_3(t);\tag{2.1}$$

$$\mathrm{Re}\,\Delta W(z)|_\Gamma=\gamma_3(t),\ W(z)|_\Gamma=\gamma_1(t)+\mathrm{i}\gamma_2(t)\tag{2.2}$$

中的任何一组(这里 $\gamma_i(t)$,$i=1,2,3$ 是任意给定的连续函数)来寻求三阶复方程(1.1)在区域 $G+\Gamma$ 上的连续解(需要再规定 $\varphi(0)=0$).

定理 2 问题 S 的解一定存在而且唯一.

证 首先考虑满足条件(2.1)的情形:根据复正则函数的第一表示式可以看出,只要$\left.\dfrac{\partial W}{\partial \bar z}\right|_\Gamma=\gamma_1(t)+\mathrm{i}\gamma_2(t)$,就可以唯一确定复调和函数$x(z)$;再根据实条件 $\mathrm{Re}\,W(z)|_\Gamma=\gamma_3(t)$,将有 $\mathrm{Re}\,\varphi(z)|_\Gamma=\mathrm{Re}\,Tx(z)|_\Gamma$(这是确定的已知函数). 这样,只要再规定 $\varphi(0)=0$,就可以唯一确定解析函数 $\varphi(z)$,于是 $W(z)$也将是唯一确定的.

再考虑满足条件(2.2)的情形:根据复正则函数的第二表示式还可以看出,只要根据实条件 $\mathrm{Re}\,\Delta W|_\Gamma=\gamma_3(t)$,规定 $\varphi(0)=0$ 就可以唯一确定解析函数 $\varphi(z)$. 再根据条件 $W|_\Gamma=\gamma_1(t)+\mathrm{i}\gamma_2(t)$,将有 $x(z)|_\Gamma=\dfrac{2}{\pi}\iint_G \ln|\zeta-z|\varphi(\zeta)\mathrm{d}T_\zeta|_\Gamma$(这是确定的已知函数). 这样就可以唯一确定复调和函数 $x(z)$,从而 $W(z)$也将是唯一确定的.

§3.　复正则函数的一个子类 $D_3^*(G)$

根据复正则函数的第一表示式(1.4)，如果知道 $\varphi_1(z)\equiv\varphi_2(z)$，那么得到表示式

$$W^*(z)=-\frac{1}{\pi}\iint_G\frac{\varphi_1(\zeta)}{\zeta-z}\mathrm{d}T_\zeta-\frac{1}{\pi}\iint_G\frac{\overline{\varphi_1(\zeta)}}{\zeta-z}\mathrm{d}T_\zeta+\varphi(z)$$

$$=-\frac{1}{\pi}\iint_G\frac{2\mathrm{Re}[\varphi_1(\zeta)]}{\zeta-z}\mathrm{d}T_\zeta+\varphi(z).\tag{3.1}$$

这里 $\mathrm{Re}[\varphi_1(z)]=p(x,y)$是任意的(实)调和函数.

由此看出，这一类函数 $W^*(z)$将满足一阶复方程$\frac{\partial W^*(z)}{\partial\bar{z}}=p(x,y)$，它刚好是第一类半解析函数[1]，只是附加了 $p(x,y)$是调和函数的条件. 以后，我们将把这一子类称为类 $D_3^*(G)$，记作 $W^*(z)\in D_3^*(G)$.

对于类 $D_3^*(G)$我们也研究它的 Schwarz 问题.

问题 S*　寻求类 $D_3^*(G)$中在 $G+\Gamma$ 上连续的函数，使它满足下列边界条件

$$\left.\frac{\partial W^*}{\partial\bar{z}}\right|_\Gamma=\gamma_1(t);\tag{3.2}$$

$$\mathrm{Re}\,W^*(z)\,|_\Gamma=\gamma_2(t),\ t\in\Gamma.\tag{3.3}$$

这里 $\gamma_i(t)$，$i=1,2$ 是给定的实连续函数.

定理 3　如果规定 $\varphi(0)=0$，那么问题 S^* 一定有唯一解。

证　因为$\frac{\partial W^*(z)}{\partial\bar{z}}$是实调和函数，所以根据(3.2)可以唯一确定调和函数 $p(x,y)=\mathrm{Re}[\varphi_1(z)]$；再根据(3.3)我们得到

$$\mathrm{Re}\,\varphi(z)\,|_\Gamma=\gamma_2(t)+\frac{1}{\pi}\iint_G\frac{2p(\zeta,\eta)}{\zeta-z}\mathrm{d}T_\zeta\,|_\Gamma,\ \zeta=\xi+\mathrm{i}\eta.$$

上式右端是完全确定的已知函数，这正好是对解析函数所提的 Schwarz 问题，再根据条件 $\varphi(0)=0$，得到 $\varphi(z)$是唯一确定的. 剩下的工作只需要把 $p(x,y)$和 $\varphi(z)$代入表示式(3.1)就可以得到问题 S^* 的解.

参考文献

[1] 王见定. 半解析函数、共轭解析函数. 北京：北京工业大学出版社，1988.

[2] 赵桢. 双解析函数，复调和函数和它们的基本边值问题. 北京师范大学学报

（自然科学版），1995，31(2)：175.

[3] Zhao Zhen. Riemann-Hilbert's problem for bianalytic functions. 北京师范大学学报（自然科学版），1996，32(3)：316.

[4] Zhao Zhen. Schwarz's problems for complex partial differential equations of second order. Beijing Mathematics，1996，1(2)：132.

[5] Vekua I N. Generalized analytic functions. Oxford：Pergamon，1962.

[6] Mushkelishvili N I. Singular integral equations. Groningen：Noordhoof，1953.

Abstract In this paper the solution of a class of complex partial differential equations of third order as follows

$$\frac{\partial^3 W}{\partial \bar{z}^2 \partial z}=0,\ z\in G,$$

where $\frac{\partial}{\partial \bar{z}}=\frac{1}{2}\left(\frac{\partial}{\partial x}+\mathrm{i}\frac{\partial}{\partial y}\right)$, $\frac{\partial}{\partial z}=\frac{1}{2}\left(\frac{\partial}{\partial x}-\mathrm{i}\frac{\partial}{\partial y}\right)$, $\Delta\equiv\frac{\partial^2 W}{\partial \bar{z}\partial z}=\frac{\partial^2 W}{\partial z\partial \bar{z}}$, is considered and the conditions of uniqueness are obtained. Furthermore, the Schwarz problem for them is also considered and some corresponding theorems are obtained.

Keywords complex partial differential equation; bianalytic function; complex harmonic function; complex regular function; Schwarz problem.

Mathematica Pannonica,
1991,2(1):49-61.

一类非线性复合型三元方程组的初边值问题①

An Initial and Boundary Value Problem for Nonlinear Composite Type Systems of Three Equations

Abstract Boundary value problems for systems of composite type were investigated by A. Dzhuraev, see [1]. Using the theory of singular integral equations in [1] linear problems for linear systems of three and of four equations having one and two real characteristics, respectively are treated. Here a nonlinear problem for a nonlinear system of three equations is studied by utilizing a method from the theory of elliptic systems (see e. g. [3][4]) based on Schauder imbedding. The case of three equations is important in particular because every elliptic second order equation in two independent variables may be reduced to a first order composite type system of three equations.

Keywords composite type systems; elliptic systems; a priori estimate; Schauder imbedding.

---

① This work was done while the Wen G C and Zhao Z visited the Free University Berlin in fall 1988 on exchange programs between the Educational State Commission of P. R. China and DFG and DAAD, respectively.

本文与 Begehr H 和闻国椿合作.

Received:1989-12.

§ 1. Formulation of the initial and boundary value problem

In this paper, we consider the nonlinear system of first order composite type equations

$$\begin{cases} w_{\bar z} = F(z, w, w_z, s), \\ F = Q_1 w_z + Q_2 \overline{w}_{\bar z} + A_1 w + A_2 \overline{w} + A_3 s + A_4, \end{cases} \tag{1.1}$$

$$\begin{cases} s_y = G(z, w, s), \\ G = B_1 w + B_2 \overline{w} + B_3 s + B_4, \end{cases} \tag{1.2}$$

in a bounded simply connected domain D, where

$$Q_j = Q_j(z, w, w_z, s),\ j = 1,2,$$

$$A_j = A_j(z, w, s), B_j = B_j(z, w, s),\ j = 1,2,3,4,$$

and $w(z)$, $Q_j, A_j, B_j\ (j=1,2), A_4$ are complex valued functions, $B_2 = \overline{B}_1$, $s(z)$, A_3, $B_j\ (j=3,4)$ are real valued functions. For the sake of convenience, we may assume that D is the unit disk and the lower boundary of D is $\gamma = \{|z|=1,\ y \leqslant 0\}$. We suppose that system (1.1) and (1.2) satisfy the following condition.

Condition C

(1) $Q_j(z, w, U, s), j=1,2$, $A_j(z, w, s)$, $j=1,2,3,4$ are measurable in $z \in D$ for all continuous functions $w(z)$, $s(z)$ and all measurable functions $U(z)$ on $\overline{D}$, satisfying

$$L_p[A_j(z, w(z), s(z)), \overline{D}] \leqslant k_0 < +\infty,\ j = 1,2,4,$$

$$L_p[A_3(z, w(z), s(z)), \overline{D}] \leqslant \varepsilon, \tag{1.3}$$

where $p(>2)$, $k_0(>0)$ and $\varepsilon(>0)$ are positive constants.

(2) The above mentioned functions are continuous in $w \in \mathbf{C}$ (the complex plane) and $s \in \mathbf{R}$ (the real axis) for almost every point $z \in D$ and $U \in \mathbf{C}$.

(3) The complex equation (1.1) satisfies the uniform ellipticity condition

$$|F(z, w, U_1, s) - F(z, w, U_2, s)| \leqslant q_0 |U_1 - U_2|, \tag{1.4}$$

for almost every point $z \in D$ and w, U_1, $U_2 \in \mathbf{C}$, $s \in \mathbf{R}$, in which $q_0(<1)$ is a non-negative constant.

(4) $B_j(z, w, s)\ (j=1,2,3,4)$, $G(z, w, s)$ are continuous for $z \in \overline{D}$

for all Hölder continuous functions $w_j(z), s_j(z) \in C_\beta(\bar{D})$ $(j=1,2)$ satisfying

$$\begin{cases} C_\beta[B_j(z,w_1,s_1),\ \bar{D}] \leqslant k_0 < +\infty,\ j=1,2,3,4, \\ G(z,w_1,s_1) - G(z,w_2,s_2) = B_1^*(w_1-w_2) + B_2^*(\overline{w_1-w_2}) + \\ \qquad B_3^*(s_1-s_2), \end{cases} \tag{1.5}$$

in which $C_\beta[B_j^*, \bar{D}] \leqslant k_0$, $\beta(0<\beta<1)$ is real, for $j=1,2,3$.

For system (1.1) and (1.2) we discuss the following nonlinear initial and boundary value problem.

Problem A

$$\operatorname{Re}[\overline{\lambda(t)}w(t)] = P(t,w,s),\ t \in \Gamma = \partial D, \tag{1.6}$$

$$a(t)s(t) = Q(t,w,s),\ t \in \gamma. \tag{1.7}$$

Here $\lambda(t), P(t,w,s)$ are Hölder continuous functions, $|\lambda(t)|=1$, and $\lambda(t), P_0(t)=P(t,0,0), P(t,w,s)$ satisfy

$$\begin{cases} C_\alpha[\lambda[t(\zeta)],L] \leqslant k_0,\ C_\alpha[P_0[t(\zeta)],L] \leqslant k_1,\ L=\zeta(\Gamma), \\ C_\alpha[P(t(\zeta),w_1,s_1) - P(t(\zeta),w_2,s_2),L] \leqslant \\ \varepsilon\{C_\alpha[w_1-w_2,L] + C_\alpha[s_1-s_2,l]\},\ l=\zeta(\gamma), \end{cases} \tag{1.8}$$

for all $w_j[t(\zeta)] \in C_\alpha(L), s_j(t) \in C_\alpha(l)$, $j=1,2$, where $\zeta(z)$ is the homeomorphic solution to the Beltrami equation $\zeta_{\bar{z}} = q(z)\zeta_z$ with a proper $q(|q(z)| \leqslant q_0 < 1)$ which maps D onto the unit disk H such that $\zeta(0)=0$, $\zeta(1)=1$; $z(\zeta)$ is the inverse function of $\zeta(z)$, k_1 and ε are positive constants. Moreover, $|a(t)|=1$, $Q_0(t)=Q(t,0,0)$ and $Q(t,w,s)$ satisfy

$$\begin{cases} C_\beta[Q_0(t),\gamma] \leqslant k_2, \\ C_\beta[Q(t,w_1,s_1) - Q(t,w_2,s_2),\gamma] \leqslant \\ \quad k_2 C_\beta(w_1-w_2,\gamma) + \varepsilon C_\beta(s_1-s_2,\gamma), \end{cases} \tag{1.9}$$

in which k_2 is a positive constant. Obviously Problem A is not necessarily solvable. Hence we consider the modified initial-boundary value problem (**Problem B**) where (1.6) is replaced by

$$\operatorname{Re}[\overline{\lambda(t)}w(t)] = P(t,w,s) + h(t),\ t \in \Gamma, \tag{1.10}$$

with

$$h(t) = \begin{cases} 0,\ t \in \Gamma,\ \text{if } K \geqslant 0,\ K = \dfrac{1}{2\pi}\Delta_\Gamma \arg \lambda(t), \\ h_0 + \operatorname{Re}\displaystyle\sum_{m=1}^{-k-1}(h_m^+ + \mathrm{i}h_m^-)t^m,\ t \in \Gamma,\ \text{if } K < 0, \end{cases} \tag{1.11}$$

where h_0, $h_m^{\pm}(m=1,2,\cdots,-K-1)$ are unknown real constants to be determined appropriately. If $K \geqslant 0$, we assume that the solution $w(z)$ to **Problem A** satisfies the side conditions

$$\operatorname{Im}[\overline{\lambda(a_j)}w(a_j)] = b_j,\ j = 1,2,\cdots,2K+1, \tag{1.12}$$

where $a_j(j=1,2,\cdots,2K+1)$ are distinct points on Γ, and $b_j(j=1,2,\cdots,2K+1)$ are real constants with the condition $|b_j|<k_1$.

In the following, we first give an a priori estimate of solutions to Problem B. Afterwards, we prove **Problem B** and **Problem A** to be solvable by using the Schauder fixed-point theorem. Under some more restrictions, we can discuss the uniqueness of the solution to **Problem B.**

§ 2. A proper estimate of solutions to the initial and boundary value problem

First of all, we discuss the system of first order composite type equations

$$\begin{cases} w_{\bar z} = F^*(z,w,w_z,s), \\ F^* = Q_1 w_z + Q_2 \overline{w_{\bar z}} + A_1 w + A_2 \overline{w} + A, \end{cases} \tag{2.1}$$

$$\begin{cases} s_y = G^*(z,w,s) \\ G^* = B_3 s + B, \end{cases} \tag{2.2}$$

together with the following linear initial and boundary value problem.

Problem B*

$$\operatorname{Re}[\overline{\lambda(t)}w(t)] = P_0(t) + h(t),\ t \in \Gamma, \tag{2.3}$$

$$\operatorname{Im}[\overline{\lambda(a_j)}w(a_j)] = b_j,\ j = 1,2,\cdots,2K+1, K \geqslant 0, \tag{2.4}$$

$$a(t)s(t) = Q_0(t),\ t \in \gamma, \tag{2.5}$$

where $Q_j, A_j(j=1,2)$, B_3, λ, P_0, h, b_j, a, Q_0 are defined as in section 1, and $A=A(z, w, s)$, $B=B(z, w, s)$ are similar to A_4, B_4, but satisfying the conditions

$$L_p[A,\overline{D}] \leqslant k_3,\ C_\beta[B,\overline{D}] \leqslant k_4, \tag{2.6}$$

for any $w(z)$, $s(z) \in C_\beta(\overline{D})$, in which k_3, k_4 are non-negative constants.

Lemma 2.1 If $[w(z),s(z)]$ is a solution to **Problem B*** for the system (1.1) (2.2), then $[w(z),s(z)]$ satisfies the estimates

$$C_\beta[w,\overline{D}]\leqslant M_1(k_1+k_3),L_{p_0}[|w_z|+|w_{\bar z}|,\overline{D}]\leqslant M_2(k_1+k_3) \tag{2.7}$$

$$C_\beta^*[s,\overline{D}]:=C_\beta[s,\overline{D}]+C[s_y,\overline{D}]\leqslant M_3(k_2+k_4), \tag{2.8}$$

where $M_j=M_j(q_0,p_0,k_0,\alpha,k,K)$, $j=1,2,3$, $k=(k_1,k_2,k_3,k_4)$, $\beta=\min(\alpha,\ 1-\frac{2}{p_0})$, $p_0=\min\ (p,\frac{1}{1-\alpha})$.

Proof Substituting the solution $[w,\ s]$ to **Problem B*** into the complex system (2.1) (2.2), and assuming that $k'=\max(k_1,k_3)>0$, $k''=\max\ (k_2,k_4)>0$, we put

$$W(z)=\frac{w(z)}{k'},\ S(z)=\frac{s(z)}{k''}. \tag{2.9}$$

It is clear that $W(z)$ is a solution to the boundary value problem

$$W_{\bar z}=Q_1W_z+Q_2\overline{W}_{\bar z}+A_1W+A_2\overline{W}+\frac{A}{k'}, \tag{2.10}$$

$$\mathrm{Re}[\overline{\lambda(t)}W(t)]=\frac{P_0(t)+h(t)}{k'},\ t\in\Gamma, \tag{2.11}$$

$$\mathrm{Im}[\overline{\lambda(a_j)}W(a_j)]=\frac{b_j}{k'},\ j=1,2,\cdots,2K+1,K\geqslant 0. \tag{2.12}$$

Noting that

$$L_p\left[\frac{A}{k'},\overline{D}\right]\leqslant 1,\ C_\alpha\left[\frac{P_0(t(\zeta))}{k'},L\right]\leqslant 1,\ \left|\frac{b_j}{k'}\right|\leqslant 1, \tag{2.13}$$

and according to Theorem 5.6 of Chapter 5 in [3] or Theorem 4.3 of Chapter 2 in [4], we know that $W(z)$ satisfies the estimate

$$C_\beta[W,\overline{D}]\leqslant M_1,\ L_{p_0}[|W_{\bar z}|+|W_z|,\overline{D}]\leqslant M_2. \tag{2.14}$$

Moreover, $S(z)$ is a solution to the initial value problem

$$S_y=B_3S+\frac{B}{k''}, \tag{2.15}$$

$$a(t)S(t)=\frac{Q_0(t)}{k''},\ t\in\gamma, \tag{2.16}$$

where $C_\beta\left[\frac{B}{k''},\overline{D}\right]\leqslant 1$, $C_\beta\left[\frac{Q_0}{k''},\gamma\right]\leqslant 1$. On the basis of Theorem 2.4 in [2], $S(z)$ can be seen to satisfy the estimate

$$C_\beta^*[S,\overline{D}]\leqslant M_3. \tag{2.17}$$

From (2.14)(2.17) it follows that (2.7) (2.8) for $k'>0$ and $k''>0$ are true. If $k'=0$ or $k''=0$, then (2.7) (2.8) for $k'=\varepsilon>0$ or $k''=\varepsilon>0$ hold. Letting ε tend to 0, we obtain (2.7) (2.8) for $k'=0$ or $k''=0$. □

Theorem 2.2 Let the complex system (1.1) and (1.2) satisfy Condition C and the constant ε in (1.3) (1.8) and (1.9) be small enough. Then the solution $[w(z),s(z)]$ to **Problem B** for (1.1) (1.2) satisfies the estimate

$$U=C_\beta[W,\overline{D}]+L_{p_0}[|w_{\bar z}|+|w_z|,\overline{D}]\leqslant M_4, \tag{2.18}$$

$$V=C_\beta^*[s,\overline{D}]\leqslant M_5, \tag{2.19}$$

where $M_j=M_j(q_0,p_0,k_0,a,k,K)$, $j=4,5$.

Proof Let the solution $[w(z),s(z)]$ be inserted into the complex system (1.1) (1.2), the boundary condition (1.10), the side condition (1.12) and the initial condition (1.7). We see that $A=A_3s+A_4$, $B=B_1w+B_2\overline{w}+B_4$, $P(t,w,s)$, $Q(t,w,s)$, b_j satisfy

$$L_p[A,\overline{D}]\leqslant\varepsilon C[s,\overline{D}]+L_p[A_4,\overline{D}]\leqslant\varepsilon C[s,\overline{D}]+k_0, \tag{2.20}$$

$$C_\beta[B,\overline{D}]\leqslant C_\beta[B_1w+B_w\overline{w},\overline{D}]+C_\beta[B_4,\overline{D}]\leqslant 2k_0C_\beta[w,\overline{D}]+k_0, \tag{2.21}$$

$$\begin{aligned}C_\alpha[P,L]&\leqslant C_\alpha[P_0(t(\zeta)),L]+C_\alpha\{[P[t(\zeta),w,s]\}-P_0[t(\zeta),L]\\&\leqslant k_1+\varepsilon\{C_\alpha[w,L]+C_\beta[s,l]\},\end{aligned} \tag{2.22}$$

$$|b_j|\leqslant k_1,\ j=1,2,\cdots,2K+1,K\geqslant 0, \tag{2.23}$$

$$\begin{aligned}C_\beta[Q,\gamma]&\leqslant C_\beta[Q_0(t),\gamma]+k_0C_\beta[w,\gamma]+\varepsilon C_\beta[s,\gamma]\\&\leqslant k_2+k_2C_\beta[w,\overline{D}]+\varepsilon C_\beta[s,\overline{D}].\end{aligned} \tag{2.24}$$

Using (2.7) and (2.8) we have

$$\begin{aligned}U&\leqslant(M_1+M_2)\{\varepsilon C[s,\overline{D}]+k_0+k_1+\varepsilon C_\alpha(w,L)+C_\alpha[s,l]\}\\&\leqslant(M_1+M_2)[k_0+k_1+\varepsilon C_\beta(w,\overline{D})+\varepsilon C_\beta(s,\overline{D})]\\&\leqslant(M_1+M_2)(k_0+k_1+\varepsilon U+\varepsilon V),\end{aligned} \tag{2.25}$$

$$\begin{aligned}V&\leqslant M_3[2k_0C_\beta(w,\overline{D})+k_0+k_2+k_2C_\beta(w,\overline{D})+\varepsilon C_\beta(s,\overline{D})]\\&\leqslant M_3[k_0+k_2+(2k_0+k_2)U+\varepsilon V].\end{aligned} \tag{2.26}$$

Choosing the constant ε so small that

$$(M_1+M_2)\varepsilon\leqslant\frac{1}{2},\ M_3[1+2(2k_0+k_2)(M_1+M_2)]\varepsilon\leqslant\frac{1}{2},$$

one can show

$$U \leqslant \frac{(M_1+M_2)(k_0+k_1+\varepsilon V)}{1-(M_1+M_2)\varepsilon} < 2(M_1+M_2)(k_0+k_1+\varepsilon V), \quad (2.27)$$

$$\begin{aligned} V &\leqslant M_3[k_0+k_2+2(2k_0+k_2)(M_1+M_2)(k_0+k_1+\varepsilon V)+\varepsilon V] \\ &\leqslant \frac{M_3[k_0+k_2+2(2k_0+k_2)(k_0+k_1)(M_1+M_2)]}{1-M_3[1+2(2k_0+k_2)(M_1+M_2)]\varepsilon} \quad (2.28) \\ &\leqslant 2M_3[k_0+k_2+2(2k_0+k_2)(k_0+k_1)(M_1+M_2)] = M_5, \end{aligned}$$

$$U \leqslant 2(M_1+M_2)(k_0+k_1+\varepsilon M_5) = M_4. \square \quad (2.29)$$

§ 3. Solvability of the initial and boundary value problem

Firstly we prove the existence of solutions to **Problem B** for the system

$$\begin{cases} w_{\bar z} = F(z,w,w_z), F = Q_1 w_z + Q_2 \overline{w}_{\bar z} + A_1 w + A_2 \overline{w} + A_3, \\ Q_j = Q_j(z,w_z), j=1,2, \quad A_j = A_j(z), j=1,2,3 \end{cases} \quad (3.1)$$

and (1.2) by using the parameter extension method, and then verify the existence of solutions to **Problem B** for the system (1.1) and (1.2) by using Theorem 2.2 and the Schauder fixed point theorem. Finally, we give conditions for **Problem A** for (1.1), (1.2) to be solvable.

Theorem 3.1 Let the system (3.1) (1.2) satisfy Condition C and the constant ε be small enough, then **Problem B** for (3.1), (1.2) is solvable.

Proof We consider the following initial boundary value problem with parameter t ($0 \leqslant t \leqslant 1$).

Problem B′

$$w_{\bar z} = tF(z,w,w_z) + A(z) \text{ in } D,\ A \in L_{p_0}(\overline{D}), \quad (3.2)$$

$$\operatorname{Re}[\overline{\lambda(z)}w(z)] = tP(z,w,s) + p(z) + h(z), \text{ on } \Gamma,\ p \in C_\beta(\Gamma), \quad (3.3)$$

$$\operatorname{Im}[\overline{\lambda(a_j)}w(a_j)] = b_j,\ j=1,2,\cdots,2K+1, K \geqslant 0, \quad (3.4)$$

$$s_y = tG(z,w,s) + B(z) \text{ in } D,\ B \in C_\beta(\overline{D}), \quad (3.5)$$

$$a(z)s(z) = tQ(z,w,s) + q(z) \text{ on } \gamma,\ q \in C_\beta(\gamma). \quad (3.6)$$

When $t=0$, **Problem B$'$** has a unique solution $[w(z),s(z)]$ with $w\in C_\beta(\overline{D})$, $s\in C_\beta^*(\overline{D})$—see [2]~[4].

Assuming that **Problem B$'$** for $t_0\,(0<t_0<1)$ is solvable, we will prove that there exists a positive constant δ such that **Problem B$'$** on

$$E=\{t\mid |t-t_0|\leqslant\delta,\ 0\leqslant t\leqslant 1\} \tag{3.7}$$

for any $A\in L_{p_0}(\overline{D})$, $B\in C_\beta(\overline{D})$, $p\in C_\beta(\Gamma)$ and $q\in C_\beta(\gamma)$ has a unique solution $[w(z),s(z)]$, $w\in C_\beta(\overline{D})\cap W_{p_0}^1(D)$, $s\in C_\beta^*(\overline{D})$.

We rewrite (3.2)~(3.6) as

$$w_{\bar z}-t_0F(z,w,w_z)=(t-t_0)F(z,w,w_z)+A(z), \tag{3.8}$$

$$\mathrm{Re}[\overline{\lambda(z)}w(z)]-t_0P(z,w,s)=(t-t_0)P(z,w,s)+p(z)+h(z), \tag{3.9}$$

$$\mathrm{Im}\,[\overline{\lambda(a_j)}w(a_j)]=b_j,\ j=1,2,\cdots,2K+1,K\geqslant 0, \tag{3.10}$$

$$s_y-t_0G(z,w,s)=(t-t_0)G(z,w,s)+B(z), \tag{3.11}$$

$$a(z)s(z)-t_0Q(z,w,s)=(t-t_0)Q(z,w,s)+q(z). \tag{3.12}$$

Choosing arbitrary functions $w_0\in C_\beta(\overline{D})\cap W_{p_0}^1(D)$, $s_0\in C_\beta^*(\overline{D})$, for instance $w_0(z)\equiv 0$, $s_0(z)\equiv 0$, we substitute $w_0(z),s_0(z)$ into the corresponding positions of the right hand sides in (3.8)~(3.12). By assumption, for t_0 the initial-boundary value problem (3.8)~(3.12) has a unique solution $[w_1(z),s_1(z)]$, $w_1\in C_\beta(\overline{D})\cap W_{p_0}^1(D)$, $s_1\in C_\beta^*(\overline{D})$. Let us substitute $w_1(z),s_1(z)$ into the right hand sides of (3.8)~(3.12) and find unique solution $[w_2(z),s_2(z)]$, $w_2\in C_\beta(\overline{D})\cap W_{p_0}^1(D)$, $s_2\in C_\beta^*(\overline{D})$ to this system. Thus, we obtain $[w_n(z),s_n(z)]$, $n\in N^*$, satisfying

$$w_{n+1,\bar z}-t_0F(z,w_{n+1},w_{n+1,z})=(t-t_0)F(z,w_n,w_{nz})+A(z), \tag{3.13}$$

$$\mathrm{Re}[\bar\lambda w_{n+1}]-t_0P(z,w_{n+1},s_{n+1})=(t-t_0)P(z,w_n,s_n)+p(z)+h(z), \tag{3.14}$$

$$\mathrm{Im}\,[\overline{\lambda(a_j)}w_{n+1}(a_j)]=b_j,\ j=1,2,\cdots,2K+1,K\geqslant 0, \tag{3.15}$$

$$s_{n+1y}-t_0G(z,w_{n+1},s_{n+1})=(t-t_0)G(z,w_n,s_n)+B(z), \tag{3.16}$$

$$a(z)s_{n+1}-t_0Q(z,w_{n+1},s_{n+1})=(t-t_0)Q(z,w_n,s_n)+q(z). \tag{3.17}$$

Setting $W_{n+1}=w_{n+1}-w_n$, $S_{n+1}=s_{n+1}-s_n$ from (3.13)~(3.17), we have

$$W_{n+1,\bar{z}}-t_0[F(z,W_{n+1},W_{n+1,z})-F(z,W_n,W_{nz})]=$$
$$(t-t_0)F(z,w_n,w_{nz})-F(z,W_{n-1}W_{n-1,z}), \tag{3.18}$$
$$\mathrm{Re}[\bar{\lambda}w_{n+1}]-t_0[P(z,w_{n+1},s_{n+1})-P(z,w_n,s_n)]=$$
$$(t-t_0)P(z,w_n,s_n)-P(z,w_{n-1},s_{n-1})+h(z), \tag{3.19}$$
$$\mathrm{Im}\,[\overline{\lambda(a_j)}w_{n+1}(a_j)]=0,\ j=1,2,\cdots,2K+1,K\geqslant 0, \tag{3.20}$$
$$S_{n+1,y}-t_0[G(z,w_{n+1},s_{n+1})-G(z,w_n,s_n)]=$$
$$(t-t_0)G(z,w_n,s_n)-G(z,w_{n-1},s_{n-1}), \tag{3.21}$$
$$a(z)S_{n+1}-t_0[Q(z,w_{n+1},s_{n+1})-Q(z,w_n,s_n)]=$$
$$(t-t_0)[Q(z,w_n,s_n)-Q(z,w_{n-1},s_{n-1})]. \tag{3.22}$$

By Condition C

$$L_{p_0}[F(z,W_n,W_{nz})-F(z,W_{n-1},W_{n-1,z}),\bar{D}]\leqslant$$
$$L_{p_0}[W_{nz},\bar{D}]+2k_0C_\beta[W_n,\bar{D}], \tag{3.23}$$
$$C_\alpha\{P[z(\zeta),w_n(z(\zeta)),s_n(z(\zeta))]-$$
$$P[z(\zeta),w_{n-1}(z(\zeta)),s_{n-1}(z(\zeta))],L\}\leqslant$$
$$\varepsilon\{C_\alpha[W_n(z(\zeta)),L],C_\alpha[S_n(z(\zeta)),l]\}, \tag{3.24}$$
$$C_\beta[G(z,w_n,s_n)-G(z,w_{n-1},s_{n-1}),\bar{D}]\leqslant$$
$$2k_0C_\beta[W_n,\bar{D}]+k_0C_\beta[S_n,\bar{D}], \tag{3.25}$$
$$C_\beta[Q(z,w_n,s_n)-Q(z,w_{n-1},s_{n-1}),\gamma]\leqslant$$
$$k_2C_\beta[W_n,\gamma]+\varepsilon C_\beta[S_n,\gamma] \tag{3.26}$$

can be obtained.

According to the method in the proof of Theorem 2.2, we can conclude that

$$U_{n+1}:=C_\beta[W_{n+1},\bar{D}]+L_{p_0}[|W_{n+1,\bar{z}}|+|W_{n+1,z}|,\bar{D}]\leqslant|t-t_0|M_6U_n, \tag{3.27}$$
$$V_{n+1}:=C_\beta^*[S_{n+1},\bar{D}]\leqslant|t-t_0|M_6V_n, \tag{3.28}$$

where $M_6=M_6(q_0,p_0,k_0,\alpha,k,K,\varepsilon)\geqslant 0$.

Choosing $\delta=\frac{1}{2(M_6+1)}$, then for $|t-t_0|\leqslant\delta,0\leqslant t\leqslant 1$, and $n>N+1>1$, we can derive the inequality

$$U_{n+1}\leqslant\frac{1}{2}U_n\leqslant\frac{1}{2^N}U_1,\ V_{n+1}\leqslant\frac{1}{2^N}V_1. \tag{3.29}$$

Moreover, if $n,m>N+1$, then

$$C_\beta[w_n - w_m, \overline{D}] + L_{p_0}[|(w_n - w_m)_{\bar z}| + |(w_n - w_m)_z|, \overline{D}] \leqslant$$

$$\frac{1}{2^N}\sum_{j=0}^{+\infty}\frac{1}{2^j}U_1 = \frac{1}{2^N - 1}U_1, \tag{3.30}$$

$$C_\beta^*[s_n - s_m, \overline{D}] \leqslant \frac{1}{2^N - 1}C_\beta^*[s_1, \overline{D}].$$

This shows that $C_\beta[w_n - w_m, \overline{D}] + L_{p_0}[|(w_n - w_m)_{\bar z}| + |(w_n - w_m)_z|, \overline{D}] \to 0$, $C_\beta^*[s_n - s_m, \overline{D}] \to 0$, if n, $m \to +\infty$. Hence there exist $w_* \in C_\beta(\overline{D}) \cap W_{p_0}^1(D)$, $s_* \in C_\beta^*(\overline{D})$, such that $C_\beta[w_n - w_*, \overline{D}] + L_{p_0}[|(w_n - w_*)_{\bar z}| + |(w_n - w_*)_z|, \overline{D}] \to 0$, $C_\beta^*[s_n - s_*, \overline{D}] \to 0$, as $n \to +\infty$, and $[w_n(z), s_n(z)]$ is just a solution to **Problem B′** on E for (3,2)~(3.6). Thus, we know that when $t \in N$, $\left[\frac{1}{\delta}\right]\delta$, 1, **Problem B′** *for* (3.2)~(3.6) is solvable. In particular, when $t=1$, $A=0$, $p=0$, $B=0$, $q=0$, **Problem B′** i. e. **Problem B** for (3.1), (1.2) is solvable. □

Theorem 3.2 Under the same hypotheses as in Theorem 2.2, **Problem B** for (1.1)(1.2) has a solution.

Proof We introduce a bounded and closed convex set B_M in the Banach space $C(\overline{D}) \times C(\overline{D})$, the elements of which are vectors of functions $w = [w, s]$ satisfying the conditions

$$C[w, \overline{D}] \leqslant M_4, \quad C[s, \overline{D}] \leqslant M_5, \tag{3.31}$$

where M_4, M_5 are the constants stated in (2.18) (2.19). We choose an arbitrary vector of functions $\Omega = [W, S] \in B_M$ and insert $W(z)$, $S(z)$ into the appropriate positions of the complex equation (1.1). Following Theorem 3.1, there exists a solution $[w(z), s(z)]$ to the initial boundary value **Problem B′**:

$$\begin{aligned} w_{\bar z} &= f(z, w, W, s, w_z), \\ f &= Q_1(z, W, w_z, s)w_z + Q_2(z, W, w_z, s)\overline{w_{\bar z}} + \\ &\quad A_1(z, W, s)w + A_2(z, W, s)\overline{w} + A_3(z, W, s), \end{aligned} \tag{3.32}$$

and (1.2) (1.6) (1.10) (1.12) (1.7).

According to Theorem 2.2, the solution $[w(z), s(z)]$ satisfies the estimates (2.18) and (2.19), obviously $w = [w, s] \in B_M$. Denoting this mapping from $\Omega \in B_M$ onto w by $w = \underset{\sim}{S}[\Omega]$, it is clear that $\underset{\sim}{S}$ is an operator which maps B_M onto a compact set in B_M.

To prove that $\underset{\sim}{S}$ is continuous in B_M, we select a sequence of vectors $[W_n, S_n](n\in\mathbf{N})$ satisfying the conditions.

$$C[W_n - W_0, \overline{D}] \to 0, C[S_n - S_0, \overline{D}] \to 0 \text{ as } n \to +\infty \quad (3.33)$$

and consider the difference $w_n - w_0 = \underset{\sim}{S}(\Omega_n) - \underset{\sim}{S}(\Omega_0)$. We have

$$[w_n - w_0]_{\bar{z}} = f(z, w_n, W_n, w_{nz}) - f(z, w_0, W_0, w_{0z}), \quad (3.34)$$

$$\operatorname{Re}[\overline{\lambda(t)}(w_n - w_0)] = P(z, w_n, s_n) - P(z, w_0, s_0) + h(t), t \in \Gamma, \quad (3.35)$$

$$\operatorname{Im}[\overline{\lambda(a_j)}(w_n(a_j) - w_0(a_j))] = 0,\ j = 1,2,\cdots,2K+1,\ K \geqslant 0, \quad (3.36)$$

$$(s_n - s_0)y = G(z, w_n, s_n) - G(z, w_0, s_0), \quad (3.37)$$

$$a(t)[s_n - s_0] = Q(z, w_n, s_n) - Q(t, w_0, s_0), t \in \gamma. \quad (3.38)$$

The complex equation (3.34) can be written as

$$[w_n - w_0]_{\bar{z}} - [f(z, w_n, W_n, w_{nz}) - f(z, w_0, W_n, w_{0z})] = c_n,$$
$$c_n = f(z, w_0, W_n, w_{0z}) - f(z, w_0, W_0, w_{0z}). \quad (3.39)$$

Using the method in the proof of Theorem 2.2 of Chapter 4 in [3] or Theorem 2.6 of Chapter 2 in [4], we can verify that $L_{p_0}[c_n, \overline{D}] \to 0$ as $n \to +\infty$. Hence, applying the method used in the proof of Theorem 2.1,

$$C_\beta[w_n - w_0, \overline{D}],\ C_\beta[s_n - s_0, \overline{D}] \leqslant M_7 L_{p_0}[c_n, \overline{D}] \quad (3.40)$$

can be concluded where M_7 is a non-negative constant. If $n \to +\infty$, then $C[w_n - w_0, \overline{D}] \to 0, C[s_n - s_0, \overline{D}] \to 0$. Hence, $w = S(\Omega)$ is a continuous mapping from B_M onto a compact set in B_M. On the basis of the Schauder fixed point theorem, there exists a vector $w = [w, s] \in B_M$, so that $w = S(w)$, and $w = [w, s]$ is just a solution to **Problem B** for the system (1.1) and (1.2). □

Theorem 3.3 Suppose that the system (1.1) (1.2) satisfies the same conditions as in Theorem 2.2, then the following statement holds

(1) If $K \geqslant 0$, **Problem A** for (1.1) (1.2) is solvable.

(2) If $K < 0$, there are $-2K-1$ conditions for **Problem A** to be solvable.

Proof Let us substitute the solution $[w(z), s(z)]$ to **Problem B** into the boundary condition (1.10). If $h(z)=0$, $z\in\Gamma$, then $[w(z),s(z)]$ is also a solution to **Problem A** for (1.1) (1.2). The total number of real equalities in $h(z)=0$ is just the total number of conditions stated in the theorem. □

Finally, in order to discuss the uniqueness of the solution to **Problem B** and **Problem A** for (1.1), (1.2) the following additional condition is imposed.

There exist $A_1^*, A_2^* \in L_{p_0}(\overline{D})$, with $L_{p_0}[A_2^*, \overline{D}]$ small enough, such that

$$F(z,w_1,U,s_1)-F(z,w_2,U,s_2)=A_1^*(w_1-w_2)+A_2^*(s_1-s_2), \tag{3.41}$$

for any functions $w_j, s_j\in C_\beta(\overline{D})$, $j=1,2$, and $U\in L_{p_0}(\overline{D})$ $(2<p_0<p)$.

Theorem 3.4 (1.1) (1.2) satisfies **Condition C** and (3.41), and the constant ε in (1.3) (1.8) (1.9) is small enough, then the solutions to **Problem B** are unique.

Proof Let $[w_1(z),s_1(z)]$, $[w_2(z),s_2(z)]$ be two solutions to **Problem B** for (1.1) (1.2). It is clear that $[w,s]=[w_1-w_2,s_1-s_2]$ is a solution to the initial-boundary value problem

$$w_{\bar z}=Qw_z+A_1^*w+A_2^*s,$$

$$Q=\begin{cases}\dfrac{F(z,w_1,w_{1z},s_1)-F(z,w_1,w_{2z},s_2)}{w_z}, & w_z\neq 0,\\ 0, & w_z=0,\end{cases}$$

$$s_y=B_1^*w+B_2^*\overline{w}+B_3^*s,$$

$$\mathrm{Re}[\overline{\lambda(t)}w(t)]=P(t,w_1,s_1)-P(t,w_2,s_2)+h(t),\ t\in\Gamma,$$

$$\mathrm{Im}\,[\overline{\lambda(a_j)}w(a_j)]=0,\ j=1,2,\cdots,2K+1,\ 0\leqslant K;$$

$$a(t)s(t)=Q(t,w_1,s_1)-Q(t,w_2,s_2),\ t\in\gamma.$$

With the method used in the proof of Theorem 2.2, we can show

$$C_\beta[w,\overline{D}]+L_{p_0}[|w_{\bar z}|+|w_z|,\overline{D}]=0,$$

$$C_\beta^*(s,\overline{D})=0,$$

so that $w(z)\equiv 0$, $s(z)\equiv 0$, i.e. $w_1(z)\equiv w_2(z)s_1(z)\equiv s_2(z)$ in $\overline{D}$. □

References

[1] Dzhuraev A. Systems of equations of composite type. Nauka Moscow, 1972 (Russian); Longman, Essex, 1989. (English translation)

[2] Ross S L. Differential equations. Blaisdell, New York etc. , 1965.

[3] Wen G C. Linear and nonlinear elliptic complex equations. Science Techn. Publ. House, Shanghai, 1986. (in Chinese)

[4] Wen G C, Begehr. H. Boundary value problems for elliptic equations and systems. Longman, Essex, 1990.

Wen Guochun, Zhao Zhen. eds. Integral Equations and Boundary Value Problems, World Scientific,
1991:281-288.

平面上复椭圆型方程和复合型奇异积分方程的边值问题①

Boundary Value Problems for Complex Elliptic Equations on the Plane and Singular Integral Equation of Composite Type

Abstract In this paper problem A, problem B and one kind of special singular integral equations are considered. For them the conditions of solvability are also obtained.

§0. Introduction

Obviously, the solutions of some elliptic equations on the plane can be represented by T operator. For example, the solution of equation as follows:

$$\frac{\partial W}{\partial \bar{z}} = AW + B\overline{W}$$

can be represented by the sum of T operator and an arbitrary analytic function $\Phi(z)$. [1]

Therefore a lot of mathematicians considered characters of T operator and obtained some useful results.

Method of singular integral equations used to solve the boundary

① Supported by the National Natural Science Foundation of China.

value problems is very effective in obtaining the conditions of solvability and the representation of solutions.[1][2]

A few years ago Soviet mathematicians considered the theory of two-dimensional singular integral equations. They effectively generalized the method of singular integral equations and solved a lot of problems in mathematical physics.

Due to the results of book[3], I suggest a new method to solve some boundary value problems (problem A and problem B), and obtain the conditions of solvability for them.

§ 1. Problem A

We consider the function as follows:

$$W(z)=\Phi(z)-\frac{1}{\pi}\iint_G \frac{f(\zeta)}{\zeta-z}\mathrm{d}\zeta, \tag{1.0}$$

where $\Phi(z)$ is an arbitrary analytic function in G, $f(z)$ is a Holder continuous function in G, here G is an arbitrary $(p+1)$ connected domain on the complex plane, its boundary Γ satisfies a certain smooth conditions (see Ch. 1 in [1]).

Under these hypotheses we know that $\frac{\partial W}{\partial \bar{z}}=f(z)$, and $\frac{\partial W}{\partial \bar{z}}=-\frac{1}{\pi}\iint_G \frac{f(\zeta)}{(\zeta-z)^2}\mathrm{d}\zeta$ satisfy a Holder condition on the closed domain $G+\Gamma$.

In this paper we denote the class of all functions represented in the form (1.0) and in the form of T operator by X and X^* respectively.

The following fact is well-known: for each given continuous function $h(t)$ on Γ, there is a continuous function $h(z)$ in G such that it takes $h(t)$ on Γ, thus, if $h(z)\neq 0$, $z\in G$, then such a function $h(t)$ is called a standard function given on Γ.

We first consider the following boundary value problem.

Problem A To define the function $W(z)$ in class X such that the following boundary condition:

$$\operatorname{Re}\left[h(t)\frac{\partial W}{\partial \bar{z}}+\lambda(t)W\right]=\gamma(t) \tag{1.1}$$

is satisfied on Γ, where $\lambda(t)$, $h(t)$, $\gamma(t)$ are given Holder continuous functions on Γ, $|\lambda(t)|=1$, and $h(t)$ is a standard function given on Γ.

Particularly, if we define the solution of problem A only in class X^*, then such a problem will be called problem A^*.

According to the representation of functions in class X, we have

$$W(z) = \Phi(z) - \frac{1}{\pi}\iint_G \frac{f(\zeta)}{\zeta - z}\mathrm{d}\zeta = \Phi(z) + W^*(z), \tag{1.2}$$

and thus
$$\frac{\partial W}{\partial \bar{z}} = \frac{\partial W^*}{\partial \bar{z}} = f(z).$$

Suppose that $h(z)$, $\lambda(z)$ are continuous functions in G, they take the boundary values $h(t)$, $\lambda(t)$ on Γ respectively. We consider two-dimensional singular integral equation as follows:

$$h(z)f(z) - \lambda(z)\frac{1}{\pi}\iint_G \frac{f(\zeta)}{\zeta - z}\mathrm{d}\zeta = 0. \tag{1.3}$$

By assumption we have $|h(z)|>0$, thus equation (1.3) is one of the equations considered in Dzhuraev's book[3], and therefore equation (1.3) possesses the Fredholm properties. In other words, equation (1.3) has exactly N (a finite number) linearly independent solutions.

Let $z\to t$ and use the Sokhotski-Plemelj formulas, we take the boundary value in equation (1.3), and get

$$h(t)f(t) - \lambda(t)\frac{1}{\pi}\iint_G \frac{f(\zeta)}{\zeta - t}\mathrm{d}\zeta = h(t)\frac{\partial W^*}{\partial \bar{z}} + \lambda(t)W^* = 0,$$

thus we have
$$\operatorname{Re}\left[h(t)\frac{\partial W^*}{\partial \bar{z}} + \lambda(t)W^*\right] = 0. \tag{1.4}$$

The following theorem will be obtained:

Theorem 1 If $h(t)$ is a standard function given on Γ, then problem A must be solvable, and its solution is

$$W^*(z) = -\frac{1}{\pi}\iint_G \frac{f(\zeta)}{\zeta - z}\mathrm{d}\zeta,$$

where $f(z)$ is a general solution of equation (1.3), or $f(z) = \sum_{i=1}^{N} c_j f_j(z)$, here $f_j(z), j=1,2,\cdots,N$ are a complete system of linearly independent solutions of equation (1.3), $c_j, j=1,2,\cdots,N$ are arbitrary constants.

According to representation (1.2) we have

$$\operatorname{Re}\left[h(t)\frac{\partial W}{\partial \bar{z}}+\lambda(t)W\right]=\operatorname{Re}\left[h(t)\frac{\partial W^{*}}{\partial \bar{z}}+\lambda(t)W^{*}\right]+$$

$$\operatorname{Re}\left[\lambda(t)\Phi(t)\right]=\gamma(t).$$

Due to theorem 1 without any difficulties we can reduce problem A to a Riemann Hilbert problem:

$$\operatorname{Re}\left[\lambda(t)\Phi(t)\right]=\gamma(t) \tag{1.5}$$

for analytic function $\Phi(z)$.

Theorem 2 If $h(t)$ is a standard function given on Γ, then problem A and problem (1.5) are solvable simultaneously. (of course we are interested only in obtaining the solution of problem A by the solution of problem (1.5)).

Suppose $x=\frac{1}{2\pi}[\arg\overline{\lambda(t)}]_{\Gamma}$, it will be called the index of problem A.

Using integral representation of analytic functions as follows:

$$\Phi(z)=\frac{1}{\pi i}\int_{\Gamma}\frac{u(t)}{t-z}dt+i\int_{\Gamma_p}u(t)ds, \tag{1.6}$$

where $\mu(t)$ is perfectly defined on Γ_p, and is admitted with an additive constant k_j on the other $\Gamma_j(j=1,2,\cdots,p-1)$.

Let $z\to t$ in formula (1.6), use the Sokhotski-Plemelj formulas, and then substitute it into (1.5), after a simple computation we can get the following (real) integral equation:

$$K\mu\equiv\alpha(t)\mu(t)+\operatorname{Re}\left\{\frac{\lambda(t)}{\pi i}\int_{\Gamma}\frac{\mu(t)}{\tau-t}d\tau\right\}+\beta(t)\int_{\Gamma_p}\mu(\gamma)ds=\gamma(t), \tag{1.7}$$

where $\alpha(t)-i\beta(t)=\lambda(t)$. Equation (1.7) is a singular integral equation of canonical type, which is equivalent to problem (1.5).

Substituting $\mu(t)$ into (1.6) we can get the solution of problem (1.5). The following theorem holds:

Theorem 3 Suppose that $h(t)$ is a standard function given on Γ, if $x>p-1$, thus problem A is solvable for arbitrary $\gamma(t)$, and its solution will be:

$$W(z)=\frac{1}{\pi \mathrm{i}}\int_{\Gamma}\frac{\mu(t)}{t-z}\mathrm{d}t+\mathrm{i}\int_{\Gamma_p}\mu(t)\mathrm{d}s-\frac{1}{\pi}\iint_{G}\frac{f(\zeta)}{\zeta-z}\mathrm{d}\zeta, \qquad (1.8)$$

where $f(z)=\sum_{j=1}^{N}c_j f_j(z)$, and $f_j(z)$, $j=1,2,\cdots,N$, are a complete system of linearly independent solutions of equation (1.3), c_j, $j=1,2,\cdots,N$ are arbitrary constants, and $\mu(t)$ is the solution of equation (1.7) (it includes $2x-p+1$ arbitrary constants).

When $x<0$, problem A is solvable if and only if $-2x+p-1$ conditions:

$$\int_{\Gamma}\gamma(t)\lambda(t)\Psi_j(t)\mathrm{d}(t)=0,\ j=1,2,\cdots,-2x+p-1$$

are satisfied for $\gamma(t)$, where $\Psi_j(t)$ are a complete system of linearly independent solutions of transposed problem of (1.5).

Therefore the solution of problem A will be (1.8) too, and $\mu(t)$ will be uniquely defined there.

§ 2. Problem B

In this section we consider another boundary value problem.

Firstly we introduce the concept of condition B. For each two continuous functions $a(t)$, $b(t)$ given on Γ, there are two continuous functions $a(t)$, $b(z)$ in G such that, they take $a(t)$, $b(t)$ on Γ respectively, if $|a(z)|>|b(z)|$, $z\in G$, then we say functions $a(t)$, $b(t)$ satisfy condition B on Γ.

Problem B To define the functions $W(z)$ in class X such that the following boundary condition:

$$\operatorname{Re}\left[a(t)\frac{\partial W}{\partial \bar{z}}+b(t)\frac{\partial W}{\partial \bar{z}}+c(t)W\right]=\gamma(t) \qquad (2.1)$$

is satisfied on Γ, where $a(t)$, $b(t)$, $c(t)$ and $\gamma(t)$ are Holder continuous functions given on Γ, $|b(t)|\neq 0$, $|a(t)|>|b(t)|$,

$$\frac{\partial}{\partial \bar{z}}=\frac{1}{2}\left(\frac{\partial}{\partial x}+\mathrm{i}\frac{\partial}{\partial y}\right),\quad \frac{\partial}{\partial z}=\frac{1}{2}\left(\frac{\partial}{\partial x}-\mathrm{i}\frac{\partial}{\partial y}\right),$$

in addition, we assume that $a(t)$, $b(t)$ satisfy condition B on Γ.

In particular, if we define the solution of problem B only in class X^*, then such a problem will be called problem B^*.

In this time for $W^*(z)=-\frac{1}{\pi}\iint_G \frac{f(\zeta)}{\zeta-z}d\zeta$ we have

$$\frac{\partial W^*}{\partial \bar{z}}=f(z),\quad \frac{\partial W^*}{\partial z}=-\frac{1}{\pi}\iint_G \frac{f(\zeta)}{(\zeta-z)^2}d\zeta. \tag{2.2}$$

We consider the two-dimensional singular integral equation:

$$a(z)f(z)-b(z)\frac{1}{\pi}\iint_G \frac{f(\zeta)}{(\zeta-z)^2}d\zeta-c(z)\frac{1}{\pi}\iint_G \frac{f(\zeta)}{\zeta-z}d\zeta=0, \tag{2.3}$$

where $a(z)$, $b(z)$, $c(z)$ are continuous functions in G, they take the values $a(t)$, $b(t)$, $c(t)$ on Γ respectively. By assumption we know that $|a(z)|>|b(z)|$, $z\in G$, and thus equation (2.3) is one of equations considered in Dzhuraev's book[3]. Therefore, the Fredholm theorem holds for equation (2.3), in other words equation (2.3) has exactly N (a finite number) linearly independent solutions.

Let $z\to t$ and use the Sokhotski-Plemelj formulas, we take the boundary value in equation (2.3) and obtain the following:

$$a(t)f(t)-b(t)\frac{1}{\pi}\iint_G \frac{f(\zeta)}{(\zeta-t)^2}d\zeta-c(t)\frac{1}{\pi}\iint_G \frac{f(\zeta)}{\zeta-t}d\zeta$$

$$=a(t)\frac{\partial W^*}{\partial \bar{z}}+b(t)\frac{\partial W^*}{\partial z}+c(t)W^*=0,$$

furthermore, we have $\operatorname{Re}\left[a(t)\frac{\partial W^*}{\partial \bar{z}}+b(t)\frac{\partial W^*}{\partial z}+c(t)W^*\right]=0.$

$$\tag{2.4}$$

The following theorem holds:

Theorem 4 If $a(t)$, $b(t)$ satisfy condition B on Γ, then homogeneous problem B^* is always solvable, and its solution can be represented in the form:

$$W^*(z)=-\frac{1}{\pi}\iint_G \frac{f(\zeta)}{\zeta-z}d\zeta,$$

where $f(z)$ is a general solution of equation (2.3),

or $\qquad f(z)=\sum_{j=1}^{N}c_jf_j(z)$, here $f_j(z)$, $j=1,2,\cdots,N$

are a complete system of linearly independent solutions of equation (2.3), c_j, $j=1,2,\cdots,N$, are arbitrary constants.

Using representation (1.2) we have

$$\operatorname{Re}\left[a(t)\frac{\partial W}{\partial \bar{z}}+b(t)\frac{\partial W}{\partial \bar{z}}+c(t)W\right]$$

$$=\operatorname{Re}\left[a(t)\frac{\partial W^*}{\partial \bar{z}}+b(t)\frac{\partial W^*}{\partial z}+c(t)W^*\right]+\operatorname{Re}\left[b(t)\frac{\partial \Phi}{\partial z}+c(t)\Phi\right]$$

$$=\gamma(t).$$

According to theorem 1, we can without difficulty reduce problem B to the problem with oblique derivative:

$$\operatorname{Re}\left[b(t)\frac{\partial \Phi}{\partial z}+c(t)\Phi\right]=\gamma(t) \tag{2.5}$$

for analytic functions $\Phi(z)$.

This problem has been considered by Bojarski and Danilyuk.[4][5] They established the necessary and sufficient conditions of solvability with different methods.

Theorem 5 If functions $a(t)$, $b(t)$ satisfy condition B on Γ, then problem B and problem (2.5) are solvable simultaneously. (of course, we are interested only in obtaining the solution of problem B by the solution of problem (2.5)).

Using the integral representation of analytic functions as follows:[2]

$$\Phi(z)=\int_\Gamma \mu(t)\ln\left(1-\frac{z}{t}\right)\mathrm{d}s+\int_\Gamma \mu(t)\mathrm{d}s+\mathrm{i}C, \tag{2.6}$$

where $\mu(t)$ is a (real) unknown function (Holder continuous function), C is an arbitrary (real) constant.

Obviously, we have

$$\Phi'(z)=-\int_\Gamma \frac{\mu(t)}{t-z}\mathrm{d}t=\frac{1}{2\pi\mathrm{i}}\int_\Gamma -\frac{2\pi\mathrm{i}\,\overline{t'}\mu(t)}{t-z}\mathrm{d}t. \tag{2.7}$$

Let $z\to t$ in formulas (2.6) (2.7), use Sokhotski-Plemelj formulas and then substitute them into (2.5), we can get the (real) integral equation as follows:

$K\mu\equiv\operatorname{Re}\{-\pi\mathrm{i}\,\overline{t'}b(t)\}\mu(t)+$

$$\int_{\Gamma}\mu(\tau)\mathrm{Re}\left\{c(t)\left[1+\ln\left(1-\frac{t}{\tau}\right)\right]+\frac{b(t)}{t-\tau}\right\}\mathrm{d}s=\gamma(t)+C\sigma(t). \tag{2.8}$$

where $\sigma(t)=-\mathrm{Im}\,c(t)$. After a simple computation we know that index of equation (2.8) will be $x=2(n+1)$, here $n=\frac{1}{2\pi}[\arg\overline{b(t)}]_{\Gamma}$.

For further application we consider also the transposed homogeneous equation:

$$K'v\equiv$$

$$\mathrm{Re}\{-\pi\mathrm{i}\,\bar{t}'b(t)\}v(t)+\int_{\Gamma}v(\tau)\mathrm{Re}\left\{c(\tau)\left[1+\ln\left(1-\frac{\tau}{t}\right)\right]+\frac{b(\tau)}{\tau-t}\right\}\mathrm{d}s=0. \tag{2.9}$$

With the help of (2.6) we know that integral equation (2.8) is equivalent to problem (2.5). Substituting $\mu(t)$ into (2.6) we can obtain the solution of problem (2.5). Consequently we have:

Theorem 6 Problem (2.5) is always solvable for each given $\gamma(t)$ if and only if the transposed homogeneous equation (2.9) has no solutions distinct from zero.

But if the transposed homogeneous equation (2.9) has exactly k' linearly independent solutions $v_j(t)$, $j=1,2,\cdots,k'$, then the necessary and sufficient conditions of solvability for problem (2.5) will be

$$\int_{\Gamma}v_j(t)\gamma(t)\mathrm{d}s=0,\ j=1,2,\cdots,k. \tag{2.10}$$

In this case homogeneous problem (2.5) has exactly $k=x+k'$ linearly independent solutions. Obviously homogeneous problem (2.5) has unique solution only if $x+k'=0$.

Theorem 7 Suppose that functions $a(t)$, $b(t)$ satisfy condition B on Γ, and equation (2.9) has only zero-solution ($k'=0$), then problem B is always solvable for each given $\gamma(t)$, and its solutions can be represented in the form:

$$W(z)=\int_{\Gamma}\mu(t)\ln\left(1-\frac{z}{t}\right)\mathrm{d}s+\int_{\Gamma}\mu(t)\mathrm{d}s+\mathrm{i}C-\frac{1}{\pi}\iint_{G}\frac{f(\zeta)}{\zeta-z}\mathrm{d}\zeta, \tag{2.11}$$

where $f(z) = \sum_{j=i}^{N} c_j f_j(z)$ and $f_j(z)$ $j=1,2,\cdots,N$ are a complete system of linearly independent solutions of equation (2.3), c_j, $j=1,2,\cdots,N$ are arbitrary constants, and $\mu(\mathrm{t})$ is the solution of equation (2.8) (it includes x arbitrary (real) constants).

Theorem 8 Suppose that functions $a(t)$, $b(t)$ satisfy condition B on Γ and equation (2.9) has $k' \neq 0$ linearly independent solutions, then problem B is solvable if and only if conditions (2.10) are satisfied. Its solution will be (2.11) too. The diversity here only is that $\mu(t)$ includes $x+k'$ arbitrary constants.

§ 3. Singular integral equation of composite type

From above discussions we find that when we consider the boundary value problems with derivatives of unknown function in its boundary condition for some complex elliptic equations, we can reduce such a problem to corresponding problem for analytic functions by using T operator. This method depends on the condition that a homogeneous two-dimensional singular integral equation must possess Fredholm properties. Under these hypotheses, such a problem in general, can be reduced to an one-dimensional singular integral equation. Consequently we have achieved our purpose to solve the considered problems. In this case we know that it is important to introduce one kind of special singular integral equations, which is said to be singular integral equation of composite type. As in section 1, if function $w(z)$ is represented in the form:

$$W(z) = \Phi(z) - \frac{1}{\pi}\iint_G \frac{f(\zeta)}{\zeta - z}\mathrm{d}\zeta = \frac{1}{\pi \mathrm{i}}\int_\Gamma \frac{\mu(t)}{t-z}\mathrm{d}t + \mathrm{i}C - \frac{1}{\pi}\iint_G \frac{f(\zeta)}{\zeta - z}\mathrm{d}\zeta,$$

then substituting it into the boundary condition (1.1), we can get the following integral equation

$$\alpha(t)\mu(t) + \mathrm{Re}\left\{\frac{\lambda(t)}{\pi \mathrm{i}}\int_\Gamma \frac{\mu(t)}{\tau - t}\mathrm{d}\tau\right\} + \beta(t)\int_{\Gamma_p}\mu(\tau)\mathrm{d}s +$$

$$\mathrm{Re}\left\{h(t)f(t) - \lambda(t)\,\frac{1}{\pi}\iint_G \frac{f(\zeta)}{\zeta - t}\mathrm{d}\zeta\right\} = \gamma(t), \qquad (3.1)$$

where $\alpha(t)-i\beta(t)=\lambda(t)$, $f(t)$ is the boundary value of function $f(z)$ on Γ. Obviously, it is a singular integral equation of composite type. As soon as we get the solution $\{u(t), f(z)\}$ of system

$$h(z)f(z)-\lambda(z)\frac{1}{\pi}\iint_G\frac{f(\zeta)}{\zeta-z}\mathrm{d}\tau=0,$$

$$K\mu(t)\equiv\alpha(t)\mu(t)+\mathrm{Re}\left\{\frac{\lambda(t)}{\pi i}\int_\Gamma\frac{\mu(t)}{\tau-t}\mathrm{d}\tau\right\}+\beta(t)\int_{\Gamma_p}\mu(t)\mathrm{d}s=\gamma(t),$$

when $|h(z)|>0$, $z\in G$, then by (1.2) we can also get the solution of problem A.

In general, we consider the following singular integral equation of composite type:

$$N(\mu(t),\omega(z))\equiv K\mu(t)+A_G\omega(z)=\gamma(t), \tag{3.2}$$

where

$$K\mu(t)\equiv A(t)\mu(t)+\frac{B(t)}{\pi i}\int_\Gamma\frac{\mu(\tau)}{\tau-t}\mathrm{d}\tau+\int_\Gamma K(\tau,t)\mu(\tau)\mathrm{d}\tau,$$

$$A_G\omega(z)\equiv A_G^O\omega(z)+\widetilde{K}\omega(z),$$

$$A_G^O\omega(z)\equiv a(z)\omega+b(z)\bar{\omega}+c(z)S_G\omega+d(z)\overline{S_G\omega},$$

$$\widetilde{K}\omega(z)\equiv\iint_G[\widetilde{K}_1(\zeta,z)\omega(\zeta)+\widetilde{K}_2(\zeta,z)\overline{\omega(\zeta)}]\mathrm{d}\zeta,$$

$$S_G\omega\equiv-\frac{1}{\pi}\iint_G\frac{\omega(\zeta)}{\zeta-z}\mathrm{d}\zeta,$$

kernels $K(\tau,t)$, $\widetilde{K}_1(\zeta,z)$, $\widetilde{K}_2(\zeta,z)$ only have the weak singularities.

By the results in[3] we know that equation $A_G\omega(z)=0$ must possess the Fredholm properties, if the following conditions:

(1) $\Delta_{10}(z)\equiv|a(z)+c(z)|^2-|b(z)+d(z)|^2\neq0$,

(2) $$\frac{|a(z)|^2+|b(z)|^2-|c(z)|^2-|d(z)|^2}{\Delta_{10}(z)}>\frac{2}{\Delta_{10}(z)}|a(z)b(z)-c(z)d(z)|,$$

(3) $|a(z)+c(z)|>|b(z)+d(z)|$

are satisfied in the closed domain $G+\Gamma$. These conditions will be said to be condition Q.

Equation (3.2) includes two unknown functions $\mu(t)$ and $\omega(z)$. Obviously, if $\{\mu(t),\omega(z)\}$ are solutions of system

$$\begin{aligned} K\mu(t) &= \gamma(t), \\ A_G\omega(z) &= 0, \end{aligned} \tag{3.3}$$

then they must be solutions of equation (3.2) too. Inversely, if $\{\mu(t), \omega(z)\}$ are solution of equation (3.2), then

$$K\mu(t) - \gamma(t) = -A_G\omega(z), \tag{3.2$'$}$$

the right hand side of equation (3.2)$'$ depends on z, and its left hand side depends on t only. So that it must be a constant. Then $\{\mu(t), \omega(z)\}$ satisfy the system:

$$\begin{aligned} K\mu(t) &= \gamma(t) + C, \\ A_G\omega(z) &= -C. \end{aligned} \tag{3.4}$$

Because our purpose is to solve equation (3.2), so that for simplicity instead of (3.4) we can solve system (3.3).

Using the suitable representation of analytic functions, let $z \to t$ and substitute them into the boundary condition, thus we can get one singular integral equation of composite type:

$$N(\mu(t), \omega(t)) = K\mu(t) + A_G\omega(t) = \gamma(t), \tag{3.5}$$

where

$$\begin{aligned} A_G\omega(t) \equiv\ & a(t)\omega(t) + b(t)\overline{\omega(t)} - \\ & c(t)\frac{1}{\pi}\iint_G \frac{\omega(\zeta)}{(\zeta - t)^2}\mathrm{d}\zeta - d(t)\overline{\frac{1}{\pi}\iint_G \frac{\omega(\zeta)}{(\zeta - t)^2}\mathrm{d}\zeta} + \\ & \iint_G [\widetilde{K}_1(\zeta, t)\omega(\zeta) + \widetilde{K}_2(\zeta, t)\overline{\omega(\zeta)}]\mathrm{d}\zeta, \end{aligned}$$

and $a(t)$, $b(t)$, $c(t)$, $d(t)$ are boundary values of functions $a(z)$, $b(z)$, $c(z)$, $d(z)$ on Γ respectively.

For singular integral equation of composite type (3.5) the following theorem holds:

Theorem 9 Given on Γ continuous functions $a(t)$, $b(t)$, $c(t)$, $d(t)$ there are continuous functions $a(z)$, $b(z)$, $c(z)$, $d(z)$ in G which take boundary values $a(t)$, $b(t)$, $c(t)$, $d(t)$ on Γ respectively, suppose they satisfy condition Q, then the equation $A_G\omega(z)=0$ has exactly N (a finite number) linearly independent solutions $\omega_j(z)$, $j=1,2,\cdots,N$. Consequently equation $K\mu(t)=\gamma(t)$ and equation (3.5) are solvable simultaneously.

Finally, we can also get the conditions of solvability for singular integral equations of composite type by the theory of one and two-dimensional singular integral equations.[2][3]

References

[1] Vekua I N. Generalized analytic functions. Reading, 1962. (Fizmatgiz 1959 in Russian)

[2] Muskhelishvifi N I. Singular integral equations. Noordhoff, Groningen, 1953.

[3] Dzhuraev A D. Methods of singular integral equations. Nauk, 1987. (in Russian)

[4] Bojarski B. Dokl. Akad. Nauk SSSR, 1955, 102(2). (in Russian).

[5] Danilyuk I I. Dokl. Akad. Nauk SSSR. 1958, 122, 1(2). (in Russian).

Beijing Mathematics,
1996,2(1):131-137.

一类二阶复偏微分方程的 Schwarz 问题①

Schwarz's Problem for Some Complex Partial Differential Equations of Second Order

Abstract In this paper Problem S (Schwarz's Problem), Problem A for bianalytic functions and Problem S_1 (Schwarz's Problem) for complex harmonic functions are investigated. Some theorems of their solvability are also obtained.

Keywords bianalytic functions; complex harmonic functions; Schwarz's problem; problem A.

§ 1. Introduction

In [1] has suggested the concept of bianalytic functions and some their properties are considered. The basic boundary value problems for them are also constructed. In paper [2] some further properties of bianalytic functions and the Dirichlet's problems for biharmonic functions are considered and some theorems of uniqueness and existence have been proved. In paper [3] the Riemann-Hilbert's problem for bianalytic functions is considered and theorems of solvability are obtained. In this pa-

① Supported by the National Natural Science Foundation of China.

per Schwarz's problem and Problem A for bianalytic and complex harmonic functions will be investigated and the theorems of their solvability are also obtained.

Suppose G is a bounded simply connected domain On the plane, ∂G is the boundary of G. Without loss of generality, we can assume, that the origin belongs to G.

We know that [1] a bianalytic function $W(z)$ in G can be represented by formula:

$$W(z)=\bar{z}\varphi_1(z)+\varphi_2(z),\ z\in G, \tag{1.1}$$

where $\varphi_i(z)$, $i=1,2$ are arbitrary analytic functions in G, and will say $\varphi_1(z)$ and $\varphi_2(z)$ is analytic factor and analytic addition of $W(z)$ respectively.

Should be noted that in (1.1) instead of $\bar{z}$ can use expression $\overline{z-a}$, here a is a certain exterior point of given domain G. [2]

For simplicity in the later if lost sight of analytic addition $\varphi_2(z)$, we also will say that $W(z)$ is a bianalytic function and denote it by $W^*(z)=(\overline{z-a})\varphi_1(z)\in D_2^*(G)$, here $D_2^*(G)$ consists of all bianalytic functions which is represented in form $W^*(z)=(\overline{z-a})\varphi_1(z)$.

But for convenience in applications we sometimes for analytic factor $\varphi_1(z)$ and analytic addition $\varphi_2(z)$ require one more condition that $\varphi_1(0)=0$ and $\varphi_2(0)=0$ respectively. [4]

Problem S To define the solution of complex equation:

$$\frac{\partial^2 W}{\partial \bar{z}^2}=0,\ z\in G,\qquad \frac{\partial}{\partial \bar{z}}=\frac{1}{2}\left(\frac{\partial}{\partial x}+\mathrm{i}\frac{\partial}{\partial y}\right), \tag{1.2}$$

which satisfies the boundary conditions as follows:①

$$\mathrm{Re}\left[(z-a)^{-1}\frac{\partial W}{\partial \bar{z}}\right]_{\partial G}=\gamma_1(t),$$

$$\mathrm{Re}[W(z)]_{\partial G}=\mathrm{Re}[(\overline{z-a})\varphi_1(z)+\varphi_2(z)]_{\partial G}=\gamma_2(t),\ t\in\partial G, \tag{1.3}$$

① Obviously, instead of the first condition of (1.3) maybe use $\mathrm{Re}\left[\frac{\partial W}{\partial \bar{z}}\right]\partial G=\gamma_1(t)$, but for convenience we still use the condition as here.

where $\gamma_i(t)$, $i=1, 2$ are given continuous functions and a is definite exterior point of G.

Obviously, a bianalytic function $W(z)$ corresponds a pair of analytic functions $\varphi_1(z)$—analytic factor and $\varphi_2(z)$—analytic addition.

Conversely, a pair of analytic functions $(\varphi_1(z), \varphi_2(z))$ corresponds a bianalytic function $W(z)$ which analytic factor and analytic addition is $\varphi_1(z)$ and $\varphi_2(z)$ respectively. Therefore to define a bianalytic function equivalents to define a pair of analytic functions. We set up the following problem:

Problem S* To define a pair of analytic functions $(\varphi_1(z), \varphi_2(z))$, $z \in G$, they are continuous in the closed domain $G+\partial G$ and satisfy the following boundary conditions:

$$\operatorname{Re}\left[\begin{pmatrix} (z-a)^{-1} & 0 \\ \overline{(z-a)} & 1 \end{pmatrix}\begin{pmatrix} \varphi_1(z) \\ \varphi_2(z) \end{pmatrix}\right]_{\partial G} = \begin{pmatrix} \gamma_1(t) \\ \gamma_2(t) \end{pmatrix}, \ t \in \partial G, \qquad (1.4)$$

where $\gamma_i(t)$, $i=1, 2$ are given continuous functions and a is definite exterior point of G.

§ 2. Solution of Problem S and Problem S*

Suppose G is a bounded simply connected domain and at first we consider the Problem S^*.

We have the following theorem:

Theorem 1 If condition $\varphi_1(0)=\varphi_2(0)=0$ (i. e. $W(0)=0$) is satisfied, then problem S^* has just unique solution-a pair of analytic functions $\{\varphi_1(z), \varphi_2(z)\}$.

$$\varphi_1(z) = (z-a)(\mathfrak{F}_{\gamma_1(t)} + i\beta_1^0), \ \varphi_2(z) = \mathfrak{F}_{\gamma_2(t)} - \mathfrak{F}_{\{|t-a|^2\gamma_1(t)\}} + i\beta_2^0, \qquad (2.1)$$

here β_i^0, $i=1,2$ is definite constant a $\gamma_1(t)$ and $\gamma(t)=\gamma_2(t)-|t-a|^2\gamma_1(t)$ are all well-known real continuous functions.

Proof From the first condition of (1. 3) $\operatorname{Re}[(t-a)^{-1}\varphi_1(t)] = \gamma_1(t)$ we have $\varphi_1(z)=(z-a)(\mathfrak{F}_{\gamma_1(t)}+i\beta_1)$, here $\mathfrak{F}$ is Schwarz's operator,[5] β_1 is arbitrary real constant, as soon as require $\varphi_1(0)=0$, we can get unique analytic function $\varphi_1(z)$ (analytic factor).

Furthermore from the second condition of (1.4) we know:

$$\mathrm{Re}[W(t)] = \mathrm{Re}[(\overline{t-a})\varphi_1(t)] + \mathrm{Re}[\varphi_2(t)] = \gamma_2(t),\ t \in \partial G,$$

or $\mathrm{Re}[\varphi_2(t)]=\gamma_2(t)-|t-a|^2\gamma_1(t)=\gamma(t)$, $t\in\partial G$, here $\gamma(t)=\gamma_2(t)-|t-a|^2\gamma_1(t)$ is a well-known continuous function.

We have $\varphi_2(z)=\mathfrak{F}_{\gamma(t)}+\mathrm{i}\beta_2=\mathfrak{F}_{\gamma_2(t)}-\mathfrak{F}_{\{|t-a|^2\gamma_1(t)\}}+\mathrm{i}\beta_2$, here β_2 is arbitrary real constant, as soon as require $W(0)=0$, consequently, $\varphi_2(0)=0$, we can get unique analytic function $\varphi_2(z)$ (analytic addition).

Theorem 2 Problem S and Problem S* are equivalent in the following sense, i. e. if $W(z)$ is solution of Problem S, then its analytic factor $\varphi_1(z)$ and analytic addition $\varphi_2(z)$ will be solution of Problem S*; conversely, if a pair of analytic functions $(\varphi_1(z), \varphi_2(z))$ is solution of Problem S*, then $W(z)=(\overline{z-a})\varphi_1(z)+\varphi_2(z)$ will be solution of Problem S.

Theorem 3 If condition $\varphi_1(0)=\varphi_2(0)=0$ (i. e. $W(0)=0$) is satisfied, then Problem S has just unique solution:

$$\mathfrak{F}(z) = (\overline{z-a})\varphi_1(z) + \varphi_2(z), \tag{2.2}$$

here $\varphi_1(z)$, $\varphi_2(z)$ are solutions of Problem S*. (see formula (2.1))

If want to define the solutions of Schwarz's problem in class of bianalytic functions which has poles at some point, then we will have some other results. At first we suppose domain G is unit disc, and denote the class of all bianalytic functions which possesses the n-order pole at origin fly $\mathfrak{D}(G)$. Obviously, any function $W(z)\in\mathfrak{D}(G)$ can be represented in form $W(z)=(\overline{z-a})\varphi_1(z)+\varphi_2(z)$, where $\varphi_i(z), i=1,2$, is analytic function in $G-\{0\}$ and at first $\varphi_1(z)$ possesses n-order pole at origin, concerning $\varphi_2(z)$ we assume it maybe has poles no more than n-order at origin.

Problem A To define bianalytic function $W(z)\in\mathfrak{D}(G)$, it satisfies the boundary condition as following:

$$\mathrm{Re}\left[(z-a)^{-1}\frac{\partial W}{\partial \bar{z}}\right]_{\partial G} = \gamma_1(t),$$

$$\operatorname{Re}[W(z)]_{\partial G}=\operatorname{Re}[(\overline{z-a})\varphi_1(z)+\varphi_2(z)]_{\partial G}=\gamma_2(t),\ t\in\partial G,\tag{2.3}$$

where $\gamma_i(t)$, $i=1,2$, are arbitrary given continuous functions; when $\gamma_i(t)\equiv 0$, $i=1,2$, the Problem A will be called Problem A^0.

§ 3. Solution of Problem A

At first, we assume that G is unit disc and start to discuss Problem A^0; The following theorem holds:

Theorem 4 If $\varphi_1(0)=\varphi_2(0)=0$ (i. e. $W(0)=0$), then Problem A^0 has the solution:

$$Q(z)=(\overline{z-a})(z-a)Q_1(z)+Q_2(z),$$

Where $Q_i(z)$, $i=1,2$, is represented by formula (3.2) (3.4) (or (3.1) (3.3) when G is unit disc), and β_0^i, $i=1,2$, are definite constants.

Proof At that time for function $Q_1(z)=(z-a)^{-1}\dfrac{\partial W}{\partial \bar z}$ we have the Laurent-series expansion of $(z-a)^{-1}\dfrac{\partial W}{\partial \bar z}=(z-a)^{-1}\varphi_1(z)=\sum\limits_{k=-n}^{+\infty}c_k^1 z^k$.

From the first condition of (2.3) ($\gamma_i(t)\equiv 0, i=1,2$) we have

$\operatorname{Re}\sum\limits_{k=-n}^{+\infty}c_k^1 e^{iks}=0$; if assume $c_k^1=\alpha_k^1+i\beta_k^1$, then we get:

$$\sum_{k=-n}^{+\infty}(\alpha_k^1\cos ks-\beta_k^1\sin ks)=0.$$

or the basis of uniqueness of Fourier-series expansion, we have:

$$\alpha_0^1=0,\ -\alpha_k^1=\alpha_{-k}^1,\ \beta_{-k}^1=\beta_k^1,\ k=1,2,\cdots,n,$$

$$\alpha_k^1=\beta_k^1=0,\ k=n+1,n+2,\cdots,$$

consequently, $c_0^1=i\beta_0^1$, $c_{-k}^1=-\bar c_k^1$, $(k=1,2,\cdots,n)$, $c_k^1=0,(k>n)$. Therefore we get formula:

$$Q_1(z)=(z-a)^{-1}\varphi_1(z)=i\beta_0^1+\sum_{k=1}^{n}(c_k^1 z^k-\bar c_k^1 z^{-k}),\tag{3.1}$$

so that analytic factor $\varphi_1(z)$ equal to $(z-a)Q_1(z)$ in unit disc here β_0^1 is definite as soon as $\varphi_1(0)=0$ is required.

If G is arbitrary bounded simply connected domain, using the conformal mapping we can get the analytic factor of Problem A^0. Suppose

$\omega=\omega(z)$ conformally maps domain G onto unit disc on plane ω it maps point z_0 to origin of plane ω, and $\omega'(z_0)>0$.

Consequently, we get formula

$$Q_1(z)=\mathrm{i}\beta_0^1+\sum_{k=1}^{n}\{c_k^1[\omega(z)]^k-\bar{c}_k^1[\omega(z)]^{-k}\}, \tag{3.2}$$

and analytic factor $\varphi_1(z)$ equal to $(z-a)Q_1(z)$ in domain G.

About the analytic addition $\varphi_2(z)$ we know that by the second condition of (2.2) ($\gamma_i(t)\equiv 0$, $i=1,2$), we have:

$$\mathrm{Re}[W(z)]|_{\partial G}=\mathrm{Re}[(\overline{t-a})\varphi_1(t)]+\mathrm{Re}[\varphi_2(t)]=0,\ t\in\partial G,$$

or

$$\mathrm{Re}[\varphi_2(t)]=-\mathrm{Re}[(\overline{t-a})\varphi_1(t)]=|t-a|^2\mathrm{Re}[(t-a)^{-1}\varphi_1(t)]=0.$$

Analogously, if we assume $c_k^2=\alpha_k^2+\mathrm{i}\beta_k^2$, then on the basis of uniqueness of Fourier-series expansion, we have:

$$\alpha_0^2=0,\ -\alpha_k^2=\alpha_{-k}^2,\ \beta_{-k}^2=\beta_k^2,\ k=1,2,\cdots,m,\ m\leqslant n,$$

$$\alpha_k^2=\beta_k^2=0,\ k=m+1,m+2,\cdots,$$

consequently,

$$c_0^2=\mathrm{i}\beta_0^2,\ c_{-k}^2=-\bar{c}_k^2,\ (k=1,2,\cdots,m),\ c_k^2=0,\ (k>m,\ m\leqslant n).$$

Therefore we get the following representation:[5]

$$Q_2(z)=\varphi_2(z)+\mathrm{i}\beta_0^2+\sum_{k=1}^{m}(c_k^2z^k-\bar{c}_k^2z^{-k}), \tag{3.3}$$

here β_0^2 is definite as soon as $\varphi_2(0)=0$ is required.

If G is arbitrary bounded simply connected domain, using the following mapping that $\omega=\omega(z)$ conformally maps domain G onto unit disc on plane ω, it maps point z_0 to origin of plane ω, and $\omega'(z_0)>0$.

We will have

$$Q_2(z)=\mathrm{i}\beta_0^2+\sum_{k=1}^{m}\{c_k^2[\omega(z)]^k-\bar{c}_k^2[\omega(z)]^{-k}\}, \tag{3.4}$$

Now we start to discuss Problem A. Suppose $\mathfrak{F}(z)=(\overline{z-a})\varphi_1(z)+\varphi_2(z)$ (see (2.2)) is solution of Problem S, it satisfies boundary condition (1.3), if $W(z)$ is any solution of Problem A, then difference $W(z)-\mathfrak{F}(z)$ will satisfy boundary condition of Problem A^0, therefore, we have

$$W(z)-\mathfrak{F}(z)=Q(z),\quad \text{or}\quad W(z)=\mathfrak{F}(z)+Q(z), \tag{3.5}$$

here $\mathfrak{F}(z)$ is solution of Problem S and $Q(z)=(\overline{z-a})(z-a)Q_1(z)+Q_2(z)$ is solution of Problem A^0.

Finally we get:

Theorem 5 If $\varphi_1(0)=\varphi_2(0)=0$ (i. e. $W(0)=0$), then Problem A has the solution (3. 5).

We know: general solutions of Problem A^0 and Problem A linearly contain, at most, 4n arbitrary real constants α_k^i, $\beta_k^i(c_k^i=\alpha_k^i+i\beta_k^i)$, $i=1$, 2, $k=1,2,\cdots,n$.

Especially, if bianalytic function $W(z)$ does not have pole (i. e. in the case $n=0$), then formula (3. 5) will be changed into $\mathfrak{F}(z)$ (see Theorem 3).

Remark If condition (2. 2) has changed into the follows:

$$\operatorname{Im}\left[(z-a)^{-1}\frac{\partial W}{\partial \bar{z}}\right]_{\partial G}=\gamma_1(t),$$

$$\begin{aligned}\operatorname{Im}[W(z)]_{\partial G}&=\operatorname{Im}[(\overline{z-a})\varphi_1(z)+\varphi_2(z)]_{\partial G}\\&=\gamma_2(t),\ t\in\partial G,\end{aligned}\qquad(3.6)$$

then it solution will be $W(z)=\mathrm{i}\mathfrak{F}(z)+\mathrm{i}Q(z)$.

§ 4. Schwarz's problem for complex harmonic functions

At first we consider some properties of complex harmonic functions. We know that[1] function $W(z)$ is a complex harmonic function, if it satisfies the following complex equation:

$$\Delta W(z)\equiv\frac{\partial^2 W(z)}{\partial z\partial\bar{z}}=\frac{\partial^2 W(z)}{\partial\bar{z}\partial z}=0,\ z\in G,\qquad(4.1)$$

here $\frac{\partial}{\partial\bar{z}}=\frac{1}{2}\left(\frac{\partial}{\partial x}+\mathrm{i}\frac{\partial}{\partial y}\right)$, $\frac{\partial}{\partial z}=\frac{1}{2}\left(\frac{\partial}{\partial x}-\mathrm{i}\frac{\partial}{\partial y}\right)$.

By equation (4. 1) we know $\frac{\partial W}{\partial\bar{z}}=f(z)$, here $f(z)=f_1(x,y)+\mathrm{i}f_2(x,y)$ is an arbitrary analytic function in G; (or $\frac{\partial W}{\partial\bar{z}}=g(z)$, where $g(z)=g_1(x,y)+\mathrm{i}g_2(x,y)$ is an arbitrary analytic function in G). Consequently, we get the first representation of complex harmonic functions:

$$W(z)=-\frac{1}{\pi}\iint_G\frac{\overline{f(\zeta)}}{\zeta-z}\mathrm{d}\zeta\mathrm{d}\eta+\Phi(z), \tag{4.2}$$

or

$$W(z)=-\frac{1}{\pi}\iint_G\frac{g(\zeta)}{\zeta-\bar{z}}\mathrm{d}\zeta\mathrm{d}\eta+\overline{\Psi(z)}, \tag{4.3}$$

here $\zeta=\xi+\mathrm{i}\eta\in G$, and $\Phi(z)$, $\Psi(z)$ are arbitrary analytic functions in G.

The second representation of complex harmonic function $W(z)$ is the following:

Theorem 6 If $W(z)$ is a complex harmonic function in G, then we have the following representation: (the second representation)

$$W(z)=\overline{\varphi_1(z)}+\varphi_2(z), \tag{4.4}$$

here $\varphi_i(z)$, $i=1,2$, is arbitrary analytic function in G, and $\overline{\varphi_1'(z)}=\overline{f(z)}$ (see (4.2)); (or $\varphi_2'(z)=g(z)$, see (4.3)).

In the later, functions $\overline{\varphi_1'(z)}$ and $\varphi_2'(z)$ will be called the conjugate analytic and analytic addition of complex harmonic function $W(z)$ respectively.

Proof We consider function $W_1(z)=\overline{\varphi_1(z)}$, obviously, will be have $\frac{\partial W_1(z)}{\partial\bar{z}}=\overline{\varphi_1'(z)}$, and if $\varphi_1'(z)=f(z)$, then from (4.2) have $\frac{\partial(W-W_1)}{\partial\bar{z}}=0$, or $W(z)-W_1(z)=\varphi_2(z)$, consequently, (4.4) has proved.

Analogously, we can prove representation (4.4) from (4.3).

Remark It should be noted that the density of representations (4.2) and (4.3) ($\overline{f(z)}$ and $g(z)$) is just $\overline{\varphi_1'(z)}$ and $\varphi_2'(z)$ respectively.

Problem S_1 (**Schwarz's problem**) To define function $W(z)$ satisfying equation (4.1), by the following boundary conditions:①

$$\operatorname{Re}\left.\frac{\partial W}{\partial\bar{z}}\right|_{\partial G}=\gamma_1(t),\ t\in\partial G,$$

$$\operatorname{Re}W_{\partial G}=\operatorname{Re}\overline{\varphi_1}\,|_{\partial G}+\operatorname{Re}\overline{\varphi_2}\,|_{\partial G}=\gamma_2(t), \tag{4.5}$$

① Obviously, here instead of the first condition of (4.5) maybe use $\operatorname{Re}\frac{\partial W}{\partial z}|_{\partial G}=\gamma_1(t)$.

here $\gamma_i(t)$, $i=1, 2$ is arbitrary given (real) continuous function.

Theorem 7 Problem S_1 always has unique solution. If $W(0)=0$, $\varphi_2(0)=\varphi_1'(0)=0$.

Proof Using the second representation (4. 4) and the first condition of (4. 5), we have:

$$\operatorname{Re}\overline{\varphi'_1(z)}\,|_{\partial G}=\operatorname{Re}\varphi'_1(z)\,|_{\partial G}=\gamma_1(t),\ t\in\partial G,$$

consequently, $\varphi_1'(z)=\mathfrak{F}_{\gamma_1(t)}+\mathrm{i}\beta_1$, as soon as require $\varphi_1'(0)=0$, we can get unique function $\varphi_1'(z)$ and $\varphi_1(z)=\int_0^z(\mathfrak{F}_{\gamma_1(t)}+\mathrm{i}\beta_1)\mathrm{d}z$ is completely definite.

Furthermore, by the second condition of (4. 5) we have $\operatorname{Re}\varphi_2(z)\,|_{\partial G}=\gamma_2(t)-\operatorname{Re}\overline{\varphi_1(z)}\,|_{\partial G}=\gamma(t)$, here $\gamma(t)$ is a will-known function. Finally we get:

$$W(z)=\int_0^z(\mathfrak{F}_{\gamma_1(t)}+\mathrm{i}\beta_1)\mathrm{d}z+\mathfrak{F}_{\gamma(t)}+\mathrm{i}\beta_2,\ z\in G, \tag{4. 6}$$

as soon as require $W(0)=0$, $\varphi_2(0)=\varphi_1'(0)=0$, we can get unique solution $W(z)$ of Problem S_1 by formula (4. 6).

If we define the solution of Problem S_1 in class of complex harmonic functions which has n-order poles at $z=0$, (for example $\varphi_1(z)$ or $\varphi_2(z)$ has n-order poles at $z=0$) we also can get analogical results as in section 3.

References

[1] Zhao Zhen. Bianalytic functions, complex harmonic functions and their basic boundary value problems. Journal of Beijing normal university (Natural sci.), 1995,31(2):175-179. (in Chinese)

[2] Zhao Zhen. Bianalytic function and its applications. Proceedings of the second Asian mathematical conference, Thailand, 1995.

[3] Zhao Zhen. Riemann-Hilbert's problem for bianaiytic functions. Journal of Beijing normal university, (Natural sci.), 1996,32(3):316-320.

[4] Lu Jianke. Complex methods of plane elasticity. Wuhan: Publ house of Wuhan university press, 1986. (in Chinese)

[5] Zhao Zhen. Singular integral equations. Publ house of Beijing normal university, 1984. (in Chinese)

北京师范大学学报(自然科学版),
1996,32(3):316-320.

双解析函数的 Riemann-Hilbert 问题[①]

Riemann-Hilbert's Problem for Bianalytic Functions

Abstract RH Problem (Riemann-Hilbert's problem) for bianalytic functions are investigated. Some theorems of solvability are also obtained.

Keywords bianlytic functions; Riemann-Hilbert's problem; theorem of solvability.

§ 0. Introduction

In [1] has suggested the concept of bianalytic functions and some their properties are considered. The basic boundary value problems for them are also constructed. In paper [2] some further properties of bianalytic functions and the Dirichlet's problems for biharmonic functions are considered and some theorem of uniqueness and existence have been obtained. In this paper the Riemann-Hilbert's problem (RH problem) will be investigated and the theorems of its solvability are also obtained.

Suppose G is a domain on the plane, ∂G is the boundary of G. In

① Supported by the National Natural Science Foundation of China.
Received: 1995-11-25.

general, G maybe is a multiply connected domain, in this case ∂G is composed by $p+1$ mutually disjoint closed Liapunov's curves L^0, L^1, $\cdots$, L^p (here L^0 contains inside all the others). Without loss of generality, we can assume, that the origin belongs to G.

We know that[1] a bianalytic function $W(z)$ can be represented by formula

$$W(z) = \bar{z}\varphi_1(z) + \varphi_2(z), \tag{0.1}$$

where $\varphi_i(z)$, $i=1, 2$ is arbitrary analytic function.

Should be noted that in (0.1) instead of $\bar{z}$ can be use expression $(\overline{z-a})$, here a is a certain exterior point of given domain G. (in [2])

For simplicity in the later if lost sight of analytic addition $\varphi_2(z)$, we also will say that $W(z)$ is a bianalytic function and denote it by $W^*(z)=(\overline{z-a})\varphi_1(z)\in D_2^*(G)$, here $D_2^*(G)$ consists all bianalytic function which is replesented in form $W^*(z)=(\overline{z-a})\varphi_1(z)$.

But for convenience in applications we often for analytic factor $\varphi_1(z)$ and analytic addition $\varphi_2(z)$ require one more condition that $\varphi_1(0)=0$ and $\varphi_2(0)=0$ respectively[3].

RH Problem To define the solutions of equation

$$\frac{\partial^2 W}{\partial \bar{z}^2} = 0, \ z \in G, \qquad \frac{\partial}{\partial \bar{z}} = \frac{1}{2}\left(\frac{\partial}{\partial x} + \mathrm{i}\frac{\partial}{\partial y}\right), \tag{0.2}$$

which satisfies the boundary conditions as follows:

$$\begin{aligned} \operatorname{Re}[\lambda(t)W^*(t)] &= \operatorname{Re}[\lambda(t)(\overline{t-a})\varphi_1(t)] = \gamma_1(t), \\ \operatorname{Re}[\lambda(t)W(t)] &= \operatorname{Re}[\lambda(t)W^*(t)] + \operatorname{Re}[\lambda(t)\varphi_2(t)] \\ &= \gamma_2(t), \ t \in \partial G, \end{aligned} \tag{0.3}$$

where $\lambda(t)$, $\gamma_i(t)$, $i=1, 2$ are given functions belonging to class H, $\lambda(t)\neq 0$, without loss of generality we can assume $|\lambda(t)|=1$. The integer $\kappa=(\frac{1}{2\pi})[\arg\overline{\lambda(t)}]_{\partial G}$ will be called the index of RH problem.

§ 1. Solution of RH problem in the case of simply connected domain

Suppose G is a simply connected domain and at first we consider the homogeneous problem (i. e. $\gamma_i(t)<0$, $i=1, 2$).

(1) If index $\kappa \geqslant 0$, then homogeneous RH problem for analytic factor $\varphi_1(z)$ (its index is also κ) has just $2\kappa+1$ linearly independent solutions. We denote the complete system of linearly independent solutions by $\varphi_{1,1}(z), \varphi_{1,2}(z), \cdots, \varphi_{1,2\kappa+1}(z)$. Furthermore nonhomogeneous RH problem for analytic factor $\varphi_1(z)$: $\operatorname{Re}[\lambda(t)(\overline{t-a})\varphi_1(t)]=\gamma_1(t)$ is always solvable for arbitrary $\gamma_1(t)$ belonging to class H (its general solution contains $2\kappa+1$ arbitrary constants).

From the second condition of (0.2) ($\lambda_i(t)<0$, $i=1, 2$) we know:

$$\begin{aligned}\operatorname{Re}[\lambda(t)W(t)] &= \operatorname{Re}[\lambda(t)(\overline{t-a})\varphi_1(t)]+\operatorname{Re}[\lambda(t)\varphi_2(t)]\\ &= \operatorname{Re}[\lambda(t)\varphi_2(t)] = 0,\ t \in \partial G.\end{aligned}$$

This is a homogeneous RH problem for analytic addition $\varphi_2(z)$, (its index is also κ) it also has $2\kappa+1$ linearly independent solutions. We denote the complete system of linearly independent solutions by $\varphi_{2,1}(z)$, $\varphi_{2,2}(z), \cdots, \varphi_{2,2\kappa+1}(z)$.

Furthermore nonhomogeneous RH problem for analytic addition $\varphi_2(z)$: $\operatorname{Re}[\lambda(t)\varphi_2(t)]=\gamma_2(t)-\gamma_1(t)=\gamma_3(t)$ is always solvable for arbitrary $\gamma_2(t)$ belonging to class H (its general solution contains $2\kappa+1$ arbitrary constants).

Theorem 1 If index $\kappa \geqslant 0$, then homogeneous RH problem for bianalytic functions $W(z)$:

$$\begin{aligned}&\operatorname{Re}[\lambda(t)W^*(t)]=\operatorname{Re}[\lambda(t)(\overline{t-a})\varphi_1(t)]=0,\\ &\operatorname{Re}[\lambda(t)W(t)]=\operatorname{Re}[\lambda(t)(\overline{t-a})\varphi_1(t)]+\operatorname{Re}[\lambda(t)\varphi_2(t)]\\ &\qquad=\operatorname{Re}[\lambda(t)\varphi_2(t)]=0,\end{aligned} \tag{1.1}$$

has just $4\kappa+2$ linearly independent solutions. Its complete system of linearly independent solutions is

$$\begin{aligned}&(\overline{z-a})\varphi_{1,1}(z),\ (\overline{z-a})\varphi_{1,2}(z),\ \cdots, (\overline{z-a})\varphi_{1,2k+1}(z),\\ &\varphi_{2,1}(z), \varphi_{2,2}(z), \cdots, \varphi_{2,2k+1}(z),\end{aligned} \tag{1.2}$$

where $\varphi_{1,1}(z), \varphi_{1,2}(z), \cdots, \varphi_{1,2k+1}(z)$ and $\varphi_{2,1}(z)$, $\varphi_{2,2}(z)$, $\cdots, \varphi_{2,2\kappa+1}(z)$, is complete system of linearly independent solutions of (1.1) for $\varphi_1(z)$ and $\varphi_2(z)$ respectively.

Proof Obviously every function of (1.2) is a solution of problem (1.1) and functions (1. 2) are linearly independent. If $\sum_{j=1}^{2\kappa+1} c_j(\overline{z-a})\varphi_{1,j}(z) + \sum_{k=1}^{2\kappa+1} b_k\varphi_{2,k}(z) = 0$, then operating two sides of the last equation by $\frac{\partial}{\partial \bar{z}}$, we have $\sum_{j=1}^{2\kappa+1} c_j\varphi_{1,j}(z) = 0$, consequently, $c_j=0$, $j=1, 2, \cdots, 2\kappa+1$, hence $b_k=0$, $k=1, 2, \cdots, 2\kappa+1$. In addition we prove any solution of homogeneous RH problem $W(z)=W^*(z)+\varphi_2(z)=(\overline{z-a})\varphi_1(z)+\varphi_2(z)$ can be linearly represented by system (1. 2) indeed any solution has its analytic factor $\varphi_1(z)$ and analytic addition $\varphi_2(z)$ which satisfy condition (1. 1), obviously, they can be represented by $\sum_{j=1}^{2\kappa+1} c_j\varphi_{1,j}(z)$ and $\sum_{k=1}^{2\kappa+1} b_k\varphi_{2,k}(z)$ respectively. □

Using the conclusions about RH problem for analytic functions, the following theorem holds:

Theorem 2 If index $\kappa \geqslant 0$, then nonhomogeneous RH problem (0. 3) for any given $\gamma_i(t)$, $i=1, 2$ always is solvable, and its general solution contains $4\kappa+2$ arbitrary constants.

(2) If index $\kappa<0$, then homogeneous RH problem for analytic factor $\varphi_1(z)$ (its index is also κ) has only trivial solution.

Furthermore nonhomogeneous RH problem for analytic factor $\varphi_i(z)$: $\mathrm{Re}[\lambda(t)(\overline{t-a})\varphi_1(t)]=\gamma_1(t)$ is solvable, iff $\gamma_1(t)$ belonging to class H satisfies $-2\kappa-1$ conditions as follows:

$$\int_{\partial G} \gamma_1(t)\,\overline{\lambda(t)(t-a)}^{-1}\Phi_j^+(t)\mathrm{d}t = 0,\ j = 1,2,\cdots,-2\kappa-1, \quad (1.3)$$

where $\Phi_j(z)$ is the solution of so-called conjugate problem for analytic functions, which satisfies the conjugate boundary condition $\mathrm{Re}[\mathrm{i}t'(s)\overline{\lambda(t)}(t-a)\varphi(t)]=0$, and $\Phi_j^+(t)$ is the boundary value of $\Phi_j(z)$. (in [4])

From the second condition of (1. 1) $\gamma_i(t)<0$, $i=1, 2$) we know:

$$\begin{aligned}\mathrm{Re}[\lambda(t)W(t)] &= \mathrm{Re}[\lambda(t)(\overline{t-a})\varphi_1(t)] + \mathrm{Re}[\lambda(t)\varphi_2(t)] \\ &= \mathrm{Re}[\lambda(t)\varphi_2(t)] = 0,\ t \in \partial G.\end{aligned}$$

This is a homogeneous RH problem for analytic addition $\varphi_2(z)$ (its index is also κ), it has only trivial solution.

Furthermore nonhomogeneous RH problem for analytic addition $\varphi_2(z)$: $\mathrm{Re}[\lambda(t)\varphi_2(t)]=\gamma_2(t)-\gamma_1(t)$ is solvable, iff given function $\gamma_2(t)-\gamma_1(t)$ belonging to class H satisfies $-2\kappa-1$ conditions as follows:

$$\int_{\partial G}[\gamma_2(t)-\gamma_1(t)]\,\overline{\lambda(t)}\Psi_j^+(t)\mathrm{d}t=0,\ j=1,2,\cdots,-2\kappa-1, \tag{1.4}$$

where $\Psi_j(z)$ is the solution of so -called conjugate problem for analytic function, which satisfies the conjugate boundary condition $\mathrm{Re}[\mathrm{i}\,t'(s)\overline{\lambda(t)}\psi(t)]=0$, and $\Psi_j^+(t)$ is the boundary value of $\Psi_j(z)$. (in [4])

Theorem 3 If index $\kappa<0$, then homogeneous RH problem (1. 1) for bianalytic functions $W(z)$, has only trivial solution.

Using the conclusion about RH problem for analytic function, the following theorem holds:

Theorem 4 If index $\kappa<0$, then nonhomogeneous RH problem (0. 3) for any given $\gamma_i(t)$, $i=1$, 2 is solvable, iff they satisfy $-4\kappa-2$ conditions (1. 3) and (1. 4).

When problem (0. 3) is solvable, it will has unique solution.

§ 2. Solution of RH problem in the case of multiply connected domain

Suppose G is a $(p+1)$-multiply connected domain. We restrict to consider those cases: the index $\kappa>p-1$ and $\kappa<0$. Concerning the cases $0\leqslant\kappa\leqslant p-1$ will be not considered.

(1) If index $\kappa<0$, then homogeneous RH problem for analytic factor $\varphi_1(z)$ (its index is also κ) has only trivial solution.

Furthermore nonhomogeneous RH problem for analytic factor $\varphi_1(z)$: $\mathrm{Re}[\lambda(t)(\overline{t-a})\varphi_1(t)]=\gamma_1(t)$ is solvable, iff $\gamma_1(t)$ belonging to class H satisfies $-2\kappa+p-1$ conditions as follows:

$$\int_{\partial G} \gamma_1(t)\,\overline{\lambda(t)}\,\overline{(t-a)}^{-1}\Phi_j^{+}(t)\mathrm{d}t = 0,\ j = 1,2,\cdots,-2\kappa+p-1, \tag{2.1}$$

where $\Phi_j(z)$ is the solution of so-called conjugate problem for analytic functions, which satisfies the conjugate boundary condition $\mathrm{Re}[it'(s)\overline{\lambda(t)}(t-a)\varphi(t)]=0$, and $\Phi_j^{+}(t)$ is the boundary value of $\Phi_j(z)$. (in [4])

From the second condition of (1.1) ($\gamma_i(t)<0$, $i=1$, 2) we know:

$$\begin{aligned}\mathrm{Re}[\lambda(t)W(t)] &= \mathrm{Re}[\lambda t(t)(\overline{t-a})\varphi_1(t)] + \mathrm{Re}\,[\lambda(t)\varphi_2(t)] \\ &= \mathrm{Re}[\lambda(t)\varphi_2(t)] = 0,\ t\in\partial G.\end{aligned}$$

This is a homogeneous RH problem for analytic addition $\varphi_2(z)$ (its index is also κ), it has only trivial solution.

Furthermore nonhomogeneous RH problem for analytic addition $\varphi_2(z)$: $\mathrm{Re}[\lambda(t)\varphi_2(t)]=\gamma_2(t)-\gamma_1(t)$ is solvable, iff given function $\gamma_2(t)-\gamma_1(t)$ belonging to class H satisfies $-2\kappa+p-1$ conditions as follows:

$$\int_{\partial G}[\gamma_2(t)-\gamma_1(t)]\,\overline{\lambda(t)}\Psi_j^{+}(t)\mathrm{d}t = 0,\ j = 1,2,\cdots,-2\kappa+p-1, \tag{2.2}$$

where $\Psi_j(z)$ is the solution of so-called conjugate problem for analytic functions, which satisfies the boundary condition $\mathrm{Re}[it'(s)\overline{\lambda(t)}\psi(t)]=0$, and $\Psi_j^{+}(t)$ is the boundary value of $\Psi_j(z)$. (in [4])

Theorem 5 If index $\kappa<0$, then homogeneous RH problem (1. 1) for bianalytic functions $W(z)$, has only trivial solution.

Using the conclusions about RH problem for analytic functions, the following theorem holds:

Theorem 6 If index $\kappa<0$, then nonhomogeneous RH problem (0. 3) for any given $\gamma_i(t)$, $i=1$, 2 is solvable, iff they satisfy $-4\kappa+2p-2$ conditions (2. 1) (2. 2).

When problem (0. 3) is solvable, it will has unique solution.

(2) If index $\kappa>p-1$, then, homogeneous RH problem. for analytic factor $\varphi_1(z)$ (its index is also κ) has just $2\kappa-p+1$ linearly independent,

solutions. We denote the complete system of linearly independent solutions by $\varphi_{1,1}(z)$, $\varphi_{1,2}(z),\cdots,\varphi_{1,2\kappa-p+1}(z)$. Furthermore nonhomogeneous RH problem for analytic factor $\varphi_1(z)$: $\mathrm{Re}[\lambda(t)(t-a)\varphi_1(t)]=\gamma_1(t)$ is always solvable for arbitrary $\gamma_1(t)$ belonging to class H (its general solution contains $2\kappa-p+1$ arbitrary constants).

From the second condition of (0.3) ($\gamma_i(t)<0$, $i=1, 2$) we know:

$$\begin{aligned}\mathrm{Re}[\lambda(t)W(t)] &= \mathrm{Re}[\lambda(t)(\overline{t-a})\varphi_1(t)]+\mathrm{Re}[\lambda(t)\varphi_2(t)]\\ &= \mathrm{Re}[\lambda(t)\varphi_2(t)]=0,\ t\in\partial G.\end{aligned}$$

This is a homogeneous RH problem for analytic addition $\varphi_2(z)$ (its index is also κ), it also has $2\kappa-p+1$ linearly independent solutions. We denote the complete system of linearly independent solutions by $\varphi_{2,1}(z)$, $\varphi_{2,2}(z),\cdots,\varphi_{2,2\kappa-p+1}(z)$.

Furthermore nonhomogeneous RH problem for analytic addition $\varphi_2(z)$: $\mathrm{Re}[\lambda(t)\varphi_2(t)]=\gamma_2(t)-\gamma_1(t)=\gamma_3(t)$ is always solvable for arbitrary $\gamma_3(t)$ belonging to class H (its general solution contains $2\kappa-p+1$ arbitrary constants).

Theorem 7 If index $\kappa>p-1$, then homogeneous RH problem (1.1) for bianalytic functions $W(z)$, has just $4\kappa-2p+2$ linearly independent solutions. Its complete system of linearly independent solutions is

$$\begin{aligned}&(\overline{z-a})\varphi_{1,1}(z),\ (\overline{z-a})\varphi_{1,2}(z),\cdots,\ (\overline{z-a})\varphi_{1,2\kappa-p+1}(z),\\ &\varphi_{2,1}(z),\varphi_{2,2}(z),\cdots,\varphi_{2,2\kappa-p+1}(z),\end{aligned}\qquad(2.3)$$

where $\varphi_{1,1}(z),\varphi_{1,2}(z),\cdots,\varphi_{1,2\kappa-p+1}(z)$ and $\varphi_{2,1}(z)$, $\varphi_{2,2}(z),\cdots,\varphi_{2,2\kappa-p+1}(z)$ is a complete system of linearly independent solutions of (1.1) for $\varphi_1(z)$ and $\varphi_2(z)$ respectively.

Using the condition about RH problem for analytic functions, the following theorem holds:

Theorem 8 If index $\kappa>p-1$, then nonhomogeneous RH problem (0.3) for any given $\gamma_i(t)$, $i=1, 2$ always is solution, and its general solution contains $4\kappa-2p+2$ arbitrary constants.

References

[1] Zhao Zhen. Bianalytic functions, complex harmonic function and their basic boundary value problems. Journal of Beijing normal university (Natural sci.), 1995, 31(2): 175－179.

[2] Zhao Zhen. Bianalytic function and its applications. Proceedings of the second Asian mathematical conference, Thailand, 1995:223.

[3] Lu Jianke. Complex methods of plane elasticity. Wuhan: Publ house of Wuhan university press, 1986.

[4] Zhao Zhen. Singular integral equations. Beijing: Publ house of Beijing normal university, 1984.

摘要 讨论了双解析函数的黎曼一希尔伯特问题(RH 问题):寻求方程$\frac{\partial^2 W}{\partial \bar{z}^2}=0, z\in G$的解,要求它满足边界条件

$$\mathrm{Re}\,[\overline{\lambda(t)}W^*(t)]=\mathrm{Re}\,[\overline{\lambda(t)}\,\overline{(t-a)}\varphi_1(t)]=\gamma_1(t),$$

$$\mathrm{Re}\,[\overline{\lambda(t)}W(t)]=\mathrm{Re}\,[\overline{\lambda(t)}W^*(t)]+\mathrm{Re}\,[\overline{\lambda(t)}\varphi_2(t)]=\gamma_2(t),\ t\in\partial G,$$

这里,$\lambda(t)$,$\gamma_i(t)$,$i=1,2$ 是给定的 H 类函数,$\lambda(t)=0$,不失一般性,可以认为[$\lambda(t)=1$]. 对于指数 κ 的不同情况我们分别得到了双解析函数黎曼-希尔伯特问题的可解性定理.

关键词 双解析函数;黎曼-希尔伯特问题;可解性定理.

北京师范大学学报(自然科学版),
1998,34(2):174-178.

双调和函数的 Dirichlet 问题①

Dirichlet's Problems for Biharmonic Functions

Abstract Some properties and mechanic illustration of bianalytic functions are considered; Dirichlet's problem and Dirichlet's problem in the changed form for biharmonic function are introduced, theorems of solvability for them are also obtained. Dirichlet's problem for bianalytic functions is also considered and the conclusion of its solvability is obtained.

Keywords bianalytic functions; biharmonic functions; Dirichlet's problem; Dirichlet's problem in the changed form.

§ 1. Some further properties of bianalytic functions

Due to their wonderful properties, analytic functions possess a lot of applications in mechanics and physics-mathematics. But if discuss the physical fields with sources or curls, then this important tool of analytic functions will be haven no use.

In [1] has suggested the concept of semi-analytic functions and con-

① Project 19571010 supported by NSFC.
本文与陈方权合作.
Received: 1997-09-20.

sidered some their properties. But from definition of semi-analytic functions, we know there for one equation of C-R equations system have no any requirements, thus in applications meet a large number, of difficuties on problem of uniqueness. So that we try to add some. control conditions to semi-analytic functions.

In paper [2] some properties of bianalytic functions are considered and basic boundary value problems are also established. In this paper some further properties and mechanic illustration of bianalytic functions are considered; Dirichlet's problem and Dirichlet's problem in the changed form are introduced; the theorems of uniqueness and existence for them are also obtained.

A bianalytic function $W(z)$ giving in a domain G, can be represented by formula[2]:

$$W(z) = \bar{z}\varphi_1(z) + \varphi_2(z),\ z \in G, \tag{1.1}$$

where $\varphi_i(z)$, $i=1, 2$ is arbitrary analytic function.

It should be noted that in (1.1) instead of $\bar{z}$ can be use expression $(\overline{z-a})$, here a is a certain exterior point of given domain G.

From the representation formula we can get a lot of properties of bianalytic functions, for example, properties of zeroes and singularities, expansion in generalized Taylor's series and generalized Laurent's series, the real and imaginary part are all biharmohic functions etc[2].

Under some control conditions with the help of bianalytic functions and so-called complex harmonic function we also can get some results solving the problem of physical fields with sources or curls[2].

From the definition of bianalytic functions directly we have:

Theorem 1 Suppose $W(z)$ is a bianalytic function in G, then $\frac{\partial W}{\partial z}$ and for any n-order derivative $\frac{\partial^n W}{\partial z^n}$ on z will be also bianalytic functions, and $\Delta W(z)$ is an analytic function.

In addition for any n we have the following representation:

$$\frac{\partial^n W(z)}{\partial z^n} = -\frac{1}{\pi}\iint_G \frac{\varphi_1^{(n)}(\zeta)}{\zeta - z} d\zeta d\eta + \varphi_2^{(n)}(z),\ z \in G. \tag{1.2}$$

where $\varphi_i^{(n)}(z)$ is the n-order derivative of function $\varphi_i(z)$, $i=1, 2$.

Definition A function $P(z)=\bar{z}p_1(z)+p_2(z)$ will be called primitive function on z of a bianalytic function

$$W(z)=\bar{z}\varphi_1(z)+\varphi_2(z),\ \text{if}\ p_1(z)=\int_0^z\varphi_1(z)\mathrm{d}z,\ p_2(z)=\int_0^z\varphi_2(z)\mathrm{d}z.$$

Therefore $\frac{\partial p}{\partial z}=W(z)$.

Theorem 2 If $W(z)$ is a bianalytic function, then its primitive function on z is also bianalytic function.

But for convenience in applications we often for analytic factor $\varphi_1(z)$ and analytic addition $\varphi_2(z)$ require one more condition that $\varphi_1(0)=0$ and $\varphi_2(0)=0$ respectively[3] (i. e. $W(0)=0$). We know that if we discuss the strain state on plane and denote the components of stress and shear stress by σ_x, σ_y and τ_{xy} respectively, then we will have

$$\sigma_x+\sigma_y=4\mathrm{Re}\{\varphi'_1(z)\},\ \sigma_x-\sigma_y+2\mathrm{i}\tau_{xy}=2[\bar{z}\varphi''_1(z)+\varphi''_2(z)]. \tag{1.3}$$

It means that analytic factor $\varphi_1(z)$ and analytic addition $\varphi_2(z)$ of a bianalytic function have their direct mechanic illustrations in the theory of elasticity, i. e. using $\varphi_1(z)$ and $\varphi_2(z)$ we can get the Kolosov's function and complex Airy's function.

When we use the methods of function theory to solve problems of elasticity on plane, the so-called stress function $U(x, y)$ being a biharmonic function, will play very important function. We can successfully represent the stress function $U(x, y)$ as the real part of a certain bianalytic function, i. e. $U(x, y)=\mathrm{Re}[W(z)]$, here $W(z)$ is a certain bianalytic function. It means that investigations bianalytic functions possess important factual sense.

We very well know that if a harmonic function $u(x, y)$ is given in a simply connected domain G, then can get an analytic function $\Phi(z)$, $z\in G$, whose real part is just $u(x, y)$. For bianalytic function $W(z)$ we have the similar property, i. e. if a biharmonic function $u(x, y)$ is given in a simply connected domain G, then can constitute a biharmonic function $v(x, y)$, such that $W(z)=u(x, y)+\mathrm{i}\,v(x, y)$, $z\in G$ will be a bianalytic function.

Indeed, if $u(x, y)$ is given in G, then $\frac{\partial u}{\partial x}$, $\frac{\partial u}{\partial y}$, $\Delta u=\zeta(x, y)$ are all definite in G. Consequently, we can get an analytic function $\zeta(x, y)+\mathrm{i}\eta(x, y)=\Delta u+\mathrm{i}\Delta v=\Delta W(z)=\varphi(z)$, here $\eta(x, y)=\Delta v$ is the conjugate harmonic function of Δu.

If we take an analytic function $\frac{\partial W}{\partial \bar{z}}=\int_0^z \varphi(z)\mathrm{d}z$, then obviously it is a definite analytic function in G. On the other hand, we have

$$\frac{\partial W}{\partial \bar{z}}=\frac{1}{2}\left[\left(\frac{\partial u}{\partial x}-\frac{\partial v}{\partial y}\right)+\mathrm{i}\left(\frac{\partial u}{\partial y}+\frac{\partial v}{\partial x}\right)\right] \tag{1.4}$$

and

$$\frac{\partial v}{\partial x}=2\mathrm{Im}\,\frac{\partial W}{\partial \bar{z}}-\frac{\partial u}{\partial y},\ -\frac{\partial v}{\partial y}=2\mathrm{Re}\,\frac{\partial W}{\partial \bar{z}}-\frac{\partial u}{\partial x},\ z=x+\mathrm{i}y\in G. \tag{1.5}$$

Using the formula

$$v(x,y)=\int_{(0,0)}^{(x,y)}\frac{\partial v}{\partial x}\mathrm{d}x+\frac{\partial v}{\partial y}\mathrm{d}y, \tag{1.6}$$

we can get the conjugate biharmonic function of $u(x, y)$ as follows:

$$v(x, y)=\int_{(0,0)}^{(x,y)}\left(2\mathrm{Im}\int_0^z\varphi(z)-\frac{\partial u}{\partial y}\right)\mathrm{d}x-\left(2\mathrm{Re}\int_0^z\varphi(z)\mathrm{d}z-\frac{\partial u}{\partial x}\right)\mathrm{d}y, \tag{1.7}$$

where $\varphi(z)=\Delta u+\mathrm{i}\Delta v$ is a definite analytic function and $\eta(x, y)=\Delta v$ is the conjugate harmonic function of Δu.

Furthermore if we define $\Delta W(0)=0$ and $W(0)=0$, then $v(x, y)$ will be unique.

§ 2. Dirichlet's problem (Problem D)

In this paragraph we consider the Dirichlet's problem for biharmonic functions and obtain some theorems of its solvability.

Suppose G is a simply connected domain and ∂G is the boundary of G.

Problem D (Dirichlet's problem) To define the solution of the following equation:

$$\Delta^2 u(x,y)=0,\ z=x+\mathrm{i}y\in G, \tag{2.1}$$

which satisfies the boundary value conditions as follows:

$$\Delta u|_{\partial G} = \gamma_1(t),\ u|_{\partial G} = \gamma_2(t),\ t \in \partial G, \tag{2.2}$$

where $\gamma_1(t)$, $\gamma_2(t)$ are given functions belonging to class $C(\partial G)$ (in general they are not constants). In the later we shall denote problem D, when $\gamma_i(t) \equiv 0$, $i=1, 2$ by problem D^0.

Theorem 3 Problem D^0 has only trivial solution (zero-solution).

Proof Obviously, from the first condition of (2.2) (when $\gamma_1(t) \equiv 0$) we get $\Delta u \equiv 0$ in G. Furthermore, by the second condition of (2.2) (when $\gamma_2(t) \equiv 0$) at once get $u(x, y) \equiv 0$.

Theorem 4 Problem D always has unique solution.

Proof From the first condition of (2.2) we get an unique definite harmonic function $\zeta(x, y) = \Delta u(x, y)$. Furthermore, by the Poisson's formula and the second condition of (2.2) we can reduce problem D to a corresponding problem for harmonic function $u^*(x, y)$, in fact

$$u|_{\partial G} = u_0(x,y)|_{\partial G} + u^*(x,y)|_{\partial G} = \gamma_2(t),$$

where $u_0(x, y)|_{\partial G}$ is a completely definite function, which will be denoted by $\gamma_3(t)$. Thus we get a Dirichlet's problem $u^*(x, y)|_{\partial G} = \gamma_2(t) - \gamma_3(t) = \gamma(t)$ for harmonic function $u^*(x, y)$.

Finally we get unique biharmonic function $u(x,y) = u_0(x, y) + u^*(x, y)$, which satisfies conditions (2.2).

Using the above results, obviously, we can also set up a Dirichlet's problem for bianalytic functions:

Problem D^* To define the solution of the following equation:

$$\frac{\partial^2 W}{\partial \bar{z}^2} = -0,\ W(z) = u(x, y) + iv(x, y),\ z = x + iy \in G, \tag{2.3}$$

which satisfies the boundary value conditions as follows:

$$\mathrm{Re}[\Delta W]|_{\partial G} = \Delta u|_{\partial G} = \gamma_1(t),$$

$$\mathrm{Re}[W]|_{\partial G} = u|_{\partial G} = \gamma_2(t),\ t \in \partial G, \tag{2.4}$$

where $\gamma_1(t), \gamma_2(t)$ are given functions belonging to class $C(\partial G)$ (in general they are not constants).

If we assume $\Delta W(0) = 0$ and $W(0) = 0$, then Problem D^* always has unique solution.

§ 3. Dirichlet's problem in the changed form (Problem B)

We know that if $W(z)$ is a bianalytic function, then, it's real and imaginary part are all biharmonic functions. But a biharmonic function $u(x, y)$ may not be the real part of a certain bianalytic function.

By formula (1. 1) we have: $\Delta W = \frac{\partial^2 W}{\partial z \partial \bar{z}} = \varphi_1'(z)$, hence, Δu, Δv will be a pair of conjugate harmonic functions. Consequently, using the Cauchy-Riemann equations system:

$$\frac{\partial \Delta u}{\partial x} - \frac{\partial \Delta v}{\partial y} = 0, \quad \frac{\partial \Delta u}{\partial y} + \frac{\partial \Delta v}{\partial x} = 0,$$

function $\Delta W(z) = \varphi_1'(z)$ can be completely defined by its real part $\Delta u(x, y)$ in a simply connected domain (at least in a neighbourhood of one point). Thus as soon as $\varphi_1'(z)$ is defined, we can get the analytic factor by formula $\varphi_1(z) = \int_0^z \varphi_1'(z)\mathrm{d}z$. In the above paragraph we suppose G is a simply connected domain. But if G is a multiply connected domain, then in general we shall get multivalued function $\varphi_i(z)$ $i=1$, 2. Now we discuss a problem in some changed form, which will be called problem B.

Suppose G is a $(p+1)$-connected domain, its boundary ∂G is composed of $p+1$ mutually disjoint closed Liapunov's curves L^0, L^1, $\cdots$, L^p (here L_0 contains inside all the other).

Problem B (Dirichlet's problem in the changed form) To define a biharmonic function $u(x, y)$, being the real part of a certain bianalytic function $W(z)$ in G, and on the boundary ∂G it satisfies the following conditions:

$$\begin{aligned} &\operatorname{Re} \Delta W\,|_{\partial G} = \Delta u\,|_{\partial G} = \gamma_1(t) + C^1(t), \\ &\operatorname{Re} W\,|_{\partial G} = u\,|_{\partial G} = \gamma_2(t) + C^2(t),\ t \in \partial G, \end{aligned} \tag{3.1}$$

where $\gamma_i(t)$ $i=1$, 2 are given functions belonging to class $C(\partial G)$. (in general they are not constants), and $C^i(t) = c_j^i$, $c_0^i = 0$, $i=1$, 2, when $t \in L^j$, $j=0, 1, 2, \cdots, p$, all these real constants c_j^i (except c_0^i, $i=1$, 2) are not given previously, but they will be completely defined by conditions of given problem.

Theorem 5　Problem B always has unique solution.

Proof　Because ΔW is analytic function in G, then from the first condition of (3.1) we can get an unique defined harmonic function $\xi(x, y)=\Delta u(x, y)$ (in [4]). Obviously, $u(x, y)= u_0(x, y)+u^*(x, y)$ will be solution of above Poisson's equation, here $u_0(x, y)$ is a completely definite solution by Poisson's formula, and $u^*(x, y)$ is harmonic function. Furthermore, by the second condition of (3.1) we can reduce problem B to a corresponding problem for harmonic function $u^*(x, y)$ in fact

$$\operatorname{Re} W\,|_{\partial G} = u\,|_{\partial G} = u_0\,|_{\partial G} + u^*\,|_{\partial G} = \gamma_2(t) + C^2(t),$$

where $u_0|_{\partial G}$ is a completely definite function, which will be denoted by $\gamma_3(t)$. Thus we will get a Dirichlet's problem in the changed form $u^*|_{\partial G}=\gamma_2(t)-\gamma_3(t)+C^2(t)=\gamma(t)+C^2(t)$ for harmonic function $u^*(x, y)$ (in [4]). Finally we get unique biharmonic function $u(x, y)=u_0(x, y)+u^*(x, y)$, which satisfies conditions (3.1).

References

[1] Wang Jianding. Semi-analytic and conjugate analytic functions. Beijing: Publ house of Beijing university of technology, 1988.

[2] Zhao Zhen. Bianalytic functions, complex harmonic functions and their basic boundary value problems. Journal of Beijing normal university (Natural sci.), 1995, 31(2): 175－179.

[3] Lu Jianke. Complex methods of plane elasticity. Wuhan: Publ house of Wuhan university press, 1986.

[4] Mushkelishvili N I. Singular integral equations. Noordhoof: Groningen, 1953.

[5] Vekua I N. Generalized analytic functions. Moscow: [s. n.], 1959.

摘要　讨论了根据给定的双调和函数可以确定一个双解析函数的重要性质(类似于解析函数所具有的性质),还讨论了双调和函数的Dirichlet问题和变形的Dirichlet问题,并得到了相应的可解性定理.对于双解析函数的Dirichlet问题也得到了相应的可解性结论.

关键词　双调和函数;双解析函数;Dirichlet问题;变形的Dirichlet问题.

Begehr H. G. W. et al. (eds.), Partial Differential and Integral Equations, Kluwer Academic Publishers, Netherlands, 1999: 211-218.

Cauchy 公式，Cauchy 型积分和双解析函数的 Hilbert 问题①

Cauchy Formula, Integral of Cauchy Type and Hilbert Problem for Bianalytic Functions

Abstract In this paper the Cauchy formula, the integral of Cauchy type and the Hilbert problem for bianalytic functions are investigated.

§ 1. Introduction

Due to their wonderful properties, analytic functions possess a lot of applications in mechanics and mathematical physics. For example, when we study a physical field in the plane without source and curl, the theory of analytic functions display its strength. But if we discuss a physical field with sources or curls, then this important tool of analytic functions will be of no use.

In [1] we have suggested the concept of bianalytic functions and have considered some of their properties. Furthermore some boundary value problems (i. e. Dirichlet's problem, Dirichlet's problem in a changed form, Riemann Hilbert's problem, Schwarz's problem, Problem A, etc.) are also considered, see [1]～[4]. In this paper we will in-

① Supported by the National Natural Science Foundation of China(19571010)，其中“1”是学科代码，“95”是年份．编辑注.

vestigate the Cauchy formula and the integral of Cauchy type for bianalytic functions and prove some useful theorems for them. Finally, we also will investigate the Hilbert problem for bianalytic functions, and prove the theorems of its solvability.

The complex form of the C-R equations is

$$\frac{\partial W}{\partial \bar{z}} = 0, \tag{1.1}$$

where $\frac{\partial}{\partial \bar{z}}=\frac{1}{2}\left(\frac{\partial}{\partial x}+\frac{\mathrm{i}\partial}{\partial y}\right)$, $W=u+\mathrm{i}v$, $z=x+\mathrm{i}y$.

In addition we also define the derivatives:

$$\frac{\partial}{\partial z}=\frac{1}{2}\left(\frac{\partial}{\partial x}-\frac{\mathrm{i}\partial}{\partial y}\right) \text{ and } \Delta\equiv\frac{\partial^2}{\partial \bar{z}\partial z}=\frac{\partial^2}{\partial z\partial \bar{z}}.$$

Definition 1 Suppose G is a region in the complex plane and define in G a complex function $W(z)$, for which the second order derivative $\frac{\partial^2 W}{\partial \bar{z}^2}$ exists, Then we will call W a bianalytic function, if $W(z)$ satisfies the partial differential equation

$$\frac{\partial^2 W}{\partial \bar{z}^2} = 0, \quad z \in G. \tag{1.2}$$

We denote the set of all bianalytic functions by $D_2(G)$.

We know, see [1], that a bianalytic function can be represented as

$$W(z) = \bar{z}\varphi_1(z) + \varphi_2(z), \tag{1.3}$$

where $\varphi_1(z)$, $\varphi_2(z)$ are analytic functions in G. It should be noted that in (1.3) instead of $\bar{z}$ one can use $\overline{(z-a)}$, where a is a certain exterior point of the domain G. For simplicity we will call $\varphi_1(z)$ and $\varphi_2(z)$ the analytic factor and analytic addition of the bianalytic function W, respectively. Later sometimes the analytic addition $\varphi_2(z)$ will be neglected (i. e. $\varphi_2(z)\equiv 0$). In general we denote, all functions in the form $w^*(z)=\bar{z}\varphi(z)$, where $\varphi(z)$ is an arbitrary analytic function, by: $D_2^*(G) \subset D_2(G)$.

When we use the methods of function theory to solve problems of elasticity in the plane, the stress function $U(x, y)$ which is biharmonic, plays a very important role. We can successfully represent the stress function $U(x,y)$ as the real part of a certain bianalytic function. i. e.

$U(x, y)=\mathrm{Re}[W(z)]$, where $W(z)$ is bianalytic. This means that investigation of bianalytic functions is of important practical interest.

§ 2. Cauchy Formula for Bianalytic Functions

It is well-known that the Cauchy formula plays a very important role in the theorem of analytic functions. By this formula an analytic function is given as soon as its boundary values are known. In the theory of bianalytic functions we also can obtain an analogue formula — the Cauchy formula for bianalytic functions. The following theorem holds.

Theorem 1 Suppose G is a bounded region in the plane, L is its boundary and $W(z)$ is a bianalytic function in G, the derivative $\frac{\partial W}{\partial \bar{z}}$ of which is continuous on $G \cup L$. Then we have the Cauchy formula

$$W(z)=\frac{1}{2\pi \mathrm{i}}\int_L \frac{W(t)}{t-z}\mathrm{d}t-\frac{1}{2\pi \mathrm{i}}\int_L \frac{\overline{t-z}}{t-z}\frac{\partial W}{\partial \bar{t}}\mathrm{d}t, \quad z \in G. \qquad (2.1)$$

Proof Suppose $L_\varepsilon=\{|\zeta-z|=\varepsilon\}, G_g=G\backslash\{|\zeta-z|\leqslant\varepsilon\}$, and the boundary of G_g is $L\backslash L_g$. We have

$$\frac{\partial}{\partial \bar{\zeta}}\left(\frac{W(\zeta)}{\zeta-z}\right)=\frac{1}{\zeta-z}\left(\frac{\partial W(\zeta)}{\partial \bar{\zeta}}\right)=\frac{\partial}{\partial \bar{\zeta}}\left(\frac{\overline{\zeta-z}}{\zeta-z}\frac{\partial W}{\partial \bar{\zeta}}\right), \quad \zeta \in G_\varepsilon.$$

Then

$$\frac{\partial}{\partial \bar{\zeta}}\left(\frac{W(z)}{\zeta-z}-\frac{\overline{\zeta-z}}{\zeta-z}\frac{\partial W}{\partial \bar{\zeta}}\right)=0, \quad z \in G_\varepsilon. \qquad (2.2)$$

or, $\left(\frac{W(z)}{\zeta-z}-\frac{\overline{\zeta-z}}{\zeta-z}\frac{\partial W}{\partial \bar{\zeta}}\right)=\Phi(\zeta)$ is an analytic function in G_ε. By the Cauchy theorem we have

$$\frac{1}{2\pi \mathrm{i}}\int_{L-L_\varepsilon}\left(\frac{W(\zeta)}{\zeta-z}-\frac{\overline{\zeta-z}}{\zeta-z}\frac{\partial W}{\partial \bar{\zeta}}\right)\mathrm{d}\zeta=0, \ z \in G. \qquad (2.3)$$

Consequently,

$$\frac{1}{2\pi \mathrm{i}}\int_L\left(\frac{W(\zeta)}{\zeta-z}-\frac{\overline{\zeta-z}}{\zeta-z}\frac{\partial W}{\partial \bar{\zeta}}\right)\mathrm{d}\zeta$$

$$=\frac{1}{2\pi \mathrm{i}}\int_{L_\varepsilon}\frac{W(\zeta)}{\zeta-z}d\zeta-\int_{L_\varepsilon}\frac{\overline{\zeta-z}}{\zeta-z}\frac{\partial W}{\partial \bar{\zeta}}\mathrm{d}\zeta, \ z \in G. \qquad (2.4)$$

Without any difficulty we can prove that the first and the second integral of the right-hand side of (2.4) will just tend to $W(z)$ and zero respectively, when $\varepsilon \to 0$.

Theorem 1 is proved. Obviously, by formula (2.1) we can get a bianalytic function as soon as the boundary values $W(t)$ and $\frac{\partial W}{\partial \bar{t}}$ are known. (This just is the solution of the basic boundary value problem, see[1].)

§ 3. Integral of Cauchy Type for Bianalytic Function

In formula (2.1) two boundary values $W(t)$ and $\frac{\partial W}{\partial \bar{t}}$ are contained. This is the difference from the Cauchy formula for analytic functions. Later for convenience we will simplify the integral of Cauchy type and consider it only for the bianalytic functions belonging to the class $D_2^*(G)$.

Definition 2 The integral of Cauchy type for bianalytic functions $w^*(z)$ belonging to the class $D_2^*(G)$ will be

$$w^*(z) = \frac{\bar{z}}{2\pi i}\int_L \frac{\varphi(t)}{t-z}dt = \frac{1}{2\pi i}\int_L \frac{\bar{t}\varphi(t)}{t-z}dt - \frac{1}{2\pi i}\int_L \frac{\overline{t-z}}{t-z}\varphi(t)dt$$

$$= \frac{1}{2\pi i}\int_L \frac{\omega(t)}{t-z}dt - \frac{1}{2\pi i}\int_L \frac{\overline{t-z}}{\bar{t}(t-z)}\omega(t)d(t), \quad z \notin L, \qquad (3.1)$$

where $\varphi(t)$ is an arbitrary continuous function and $w(t) = \bar{t}\varphi(t)$.

The first and the second term on the right-hand side of (3.1) are an analytic function and a bianalytic function, respectively. Consequently, we have $\frac{\partial^2 w}{\partial \bar{z}^2} = 0$, for $z \notin L$. Obviously, $w^*(z)$ is a bianalytic function belonging to $D_2^*(G)$, if $z \notin L$.

Theorem 2 If $W(z)$ is a bianalytic function in a bounded domain G, then $W(z) = w^*(z) + \Phi(z)$, $z \in G$, where $w^*(z)$, is a function as in (3.1) and $\Phi(z)$ is an arbitrary analytic function.

Proof We know that $W(z) = \bar{z}\varphi_1(z) + \varphi_2(z)$, see [1]. As soon as we take

$$\varphi_1(z) = \frac{1}{2\pi i}\int_L \frac{\varphi(t)}{t-z}dt, \quad \Phi(z) = \varphi_2(z),$$

we at once get Theorem 2.

Definition 3 A piecewise bianalytic function with jump curve L will be a function $\Psi(z)$ which is bianalytic in any bounded domain not containing the curve L, having a finite k-th order at infinity and being continuously extendable from left and right to L. Here L is a closed smooth curve.

Definition 4 A piecewise bianalyic function $w^*(z)$ is called characteristic bianalytjc, if it belongs to $D_2^*(G)$ in any bounded domain D not containing the curve L, has a finite k-th order at infinity and is continuously extendable from left and right to L. Here L is a closed smooth curve.

Theorem 3 Suppose L is a closed smooth curve, and $w(t)$ belongs to class H,(i. e. class of Hölder continuous functions). Then, the integral of Cauchy type $w^*(z)$ (3. 1) is a piecewise bianalytic function, which is bounded at infinity and L is the jump curve.

Using the properties of the Cauchy type integral for analytic functions, see [6], we can easily prove Theorem 3.

Furthermore, the following theorem holds.

Theorem 4 Suppose L is a closed smooth curve, and $w(t)=\bar{t}\varphi(t)$ belongs to the class H (i. e. to the class of Hölder continuous functions); then for the left boundary value $w^{*+}(t)$ and the right boundary value $w^{*-}(t)$, $t\in L$, of the integral of Cauchy type (3. 1) (for characteristic bianalytic functions) $w^*(z)$, we have the following formula.

Plemelj formulas

$$w^{*+}(t)=\frac{1}{2}\omega(t)+\frac{1}{2\pi i}\int_L\frac{\omega(\tau)}{\tau-t}d\tau-\frac{1}{2\pi i}\int_L\frac{\overline{\tau-t}}{\bar{\tau}(\tau-t)}\omega(\tau)d\tau,$$

$$w^{*-}(t)=-\frac{1}{2}\omega(t)+\frac{1}{2\pi i}\int_L\frac{\omega(\tau)}{\tau-t}d\tau-\frac{1}{2\pi i}\int_L\frac{\overline{\tau-t}}{\bar{\tau}(\tau-t)}\omega(\tau)d\tau,t\notin L. \tag{3. 2}$$

The Plemelj formulas (3. 2) can be written in the form

$$\begin{aligned}&w^{*+}(t)+w^{*-}(t)=\frac{1}{\pi i}\int_L\frac{\omega(\tau)}{\tau-t}d\tau-\frac{1}{\pi i}\int_L\frac{\overline{\tau-t}}{\bar{\tau}(\tau-t)}\omega(\tau)d\tau,\\&w^{*+}(t)-w^{*-}(t)=\omega(t),\ t\in L.\end{aligned} \tag{3. 3}$$

§ 4. Hilbert Problem for Bianalytic Functions

Suppose S^+ is a domain in the plane, L is the boundary of S^+. In general, S^+ may be a multiply connected domain. In this case L is composed by $p+1$ mutually disjoint closed Lyapunov curves L^0, L^1, L^2, $\cdots$, L^p, $L=L^0\cup L^1\cup L^2\cup\cdots\cup L^p$, (where L^0 surrounds all the others). Without loss of generality, we can assume that the origin belongs to S^+. We set the positive orientation of L such that, S^+ is always on its left. We denote the complementary set of $S^+\cup L$ by S^- and its connected component with boundary L_j by $S_j^- = 0, 1, 2, \cdots, p$, respectively. Obviously, $S^- = S_0^-\cup S_1^-\cup\cdots\cup S_p^-$, and in general S_0^- contains the point of infinity, if L_0 exists. Moreover a_j is a definite point in $S_j^- = 1, 2, \cdots, p$.

Hilbert Problem (Problem H) Find a piecewise bianalytic function with jump curve L which satisfies the boundary conditions

$$\begin{aligned} w^{*+}(t) &= G(t)w^{*-}(t) \\ w^{+}(t) &= G(t)w^{-}(t), \end{aligned} \quad t\in L, \text{ (homogeneous problem)} \quad (4.1)$$

$$\begin{aligned} w^{*+}(t) &= G(t)w^{*-}(t)+g_1(t), \\ w^{+}(t) &= G(t)w^{-}(t)+g_2(t), \end{aligned} \quad t\in L, \text{ (nonhomogeneous problem)}$$

$$(4.2)$$

where $G(t)$, $g_i(t)(i=1,2)$ are given functions of class H and $G(t)\neq 0$, $t\in L$. We will call $G(t)$ and $g_i(t)$ the coefficient and free terms of the Hilbert problem, respectively. Moreover $w^*(z)$ is a characteristic bianalytic function.

For simplicity, at first we discuss a simple case, namely the Hilbert problem for characteristic bianalytic functions.

Problem H* Find a piecewise characteristic bianalytic function with jump curve L which satisfies the boundary conditions

$$w^{*+}(t) = G(t)w^{*-}(t) \quad t\in L, \text{ (homogeneous problem)}, \quad (4.1)$$

$$w^{*+}(t) = G(t)w^{*-}(t)+g_1(t), t\in L, \text{(nonhomogeneous problem)}, \quad (4.2)$$

where $G(t)$, $g_1(t)$ are given functions of class H and $G(t)\neq 0$, $t\in L$.

By the results of the Hilbert problem for analytic functions we know that $G(t)=\dfrac{X^{+}(t)}{X^{-}(t)}$, where $X(z)$ is the so-called standard function of the Hilbert problem for analytic functions. It has order $-\kappa$ at the point at infinity, where κ is the index of Hilbert problem, see [6]. Later we will call κ also the index of the Hilbert problem for bianalytic functions, $\kappa\sum_{j=0}^{p}\lambda_j$, $2\pi\lambda_j=[\arg G(t)]\,|_{L^j}, j=0,1,2,\cdots,p$,

$$X(z)=\begin{cases}\dfrac{1}{\prod(z)}e^{\Gamma(z)}, & z\in S^{+},\\ z^{-\kappa}e^{\Gamma(z)}, & z\in S^{-},\end{cases}$$

$$\Gamma(z)=\frac{1}{2\pi i}\int_L\frac{\ln G_0(t)}{t-z}dt,\ G_0(t)=t^{-\kappa}G(t)\prod(t),\ \prod(z)=\prod_{j=1}^{p}(z-a_j)^{\lambda_j}.$$

Substituting this representation into (4.2′) we get

$$w^{*+}(t)=\frac{X^{+}(t)}{X^{-}(t)}w^{*-}(t)+g_1(t) \tag{4.3}$$

or

$$\frac{w^{*+}(t)}{X^{+}(t)}=\frac{w^{*-}(t)}{X^{-}(t)}+\frac{g_1(t)}{X^{+}(t)}. \tag{4.4}$$

Remark that the function $\dfrac{w^{*}(z)}{X(z)}$ is also a characteristic bianalytic function.

By (4.4) we have (see 1.3)

$$\frac{w^{*}(z)}{X(z)}=\frac{1}{2\pi i}\int_L\frac{g_1(t)}{X^{+}(t)(t-z)}dt-\frac{1}{2\pi i}\int_L\frac{\overline{t-z}}{\overline{t}(t-z)}\frac{g_1(t)}{X^{+}(t)}dt.$$

This is a characteristic bianalytic function which is bounded at the point at infinity. Consequently, we have

$$w^{*}(z)=\frac{X(z)}{2\pi i}\int_L\frac{g_1(t)}{X^{+}(t)(t-z)}dt-\frac{X(z)}{2\pi i}\int_L\frac{\overline{t-z}}{\overline{t}(t-z)}\frac{g_1(t)}{X^{+}(t)}dt. \tag{4.5}$$

This is a characteristic bianalytic function which has order $-\kappa$ at infinity.

We consider the following different cases for the index κ.

(1) $\kappa=0$. In this case (4.5) gives us a piecewise characteristic

bianalytic function, which is bounded at the point at infinity.

(2) $\kappa>0$. In this case $w^*(z)$ is also a characteristic bianalytic function, which is bounded at the point at infinity, and condition (4.3) is satisfied for $w^*(z)$.

(3) $\kappa<0$. Now $w^*(z)$ has order $-\kappa>0$ at the point at infinity. In order that $w^*(z)$ is bounded at the point at infinity, we must also require:

$$\int_L \frac{g_1(t)}{X^+(t)} t^k \mathrm{d}t = 0, \quad k = 0,1,2,\cdots,-\kappa-1. \tag{4.6}$$

If conditions (4.6) are satisfied, then nonhomogeneous Hilbert problem (4.2′) has the unique solution (4.5).

Concerning the homogeneous Hilbert problem (4.1′) we have the following conclusions.

(1) $\kappa>0$. Then by the condition $\frac{w^{*+}(t)}{X^+(t)}=\frac{w^{*-}(t)}{X^-(t)}$ we have $\frac{w^*(z)}{X(z)}=$ $\bar{z}P_{k-1}(z)$, which has order κ at the point at infinity;

(2) $\kappa<0$. Then $\frac{w^*(z)}{X(z)}$ must have order $-\kappa>0$ at the point at infinity. Consequently, the homogeneous Hilbert problem has only the trivial solution.

(3) $\kappa=0$. As $w^*(z)$ is not an analytic function, the homogeneous Hilbert problem also has only the trivial solution, i. e. in this case the homogeneous Hilbert problem may only have analytic solutions. We note that (3.1) is not an analytic function.

Finally we have the following theorems.

Theorem 5 If $\kappa>0$, the homogeneous Hilbert Problem (Problem H^{*0}) has κ linearly independent solutions

$$\bar{z}X(z), \ \bar{z}zX(z), \ \cdots, \ \bar{z}z^{\kappa-1}X(z). \tag{4.7}$$

If $\kappa\leqslant 0$, then Problem H^{*0} has only the trivial solution.

Theorem 6 If $\kappa>0$, the nonhomogeneous Hilbert problem (Problem H^*) has the general solution.

$$w^*(z) = \frac{X(z)}{2\pi i}\int_L \frac{g_1\bar{t}}{X^+(t)(t-z)}dt.$$

$$-\frac{X(z)}{2\pi i}\int_L \frac{\overline{t-z}}{\bar{t}(t-z)}\frac{g_1(t)}{X^+(t)}dt + \bar{z}X(z)P_{\kappa-1}(z), \tag{4.8}$$

where $w^*(z)$ is a characteristic bianalytic function, which is bounded at the point at infinity, and condition (4.3) is satisfied for $w^*(z)$. In general the solution contains κ arbitrary constants.

If $\kappa \leqslant 0$, as soon as conditions (4.6) are satisfied, then Problem H^* has the unique solution

$$w(z) = \frac{X(z)}{2\pi i}\int_L \frac{g_1(t)}{X^+(t)(t-z)}dt - \frac{X(z)}{2\pi i}\int_L \frac{\overline{t-z}}{\bar{t}(t-z)}\frac{g_1(t)}{X^+(t)}dt. \tag{4.9}$$

By using the results of Problem H^{*0} and Problem H^*, we can solve Problem H^0 and Problem H. Indeed from (4.2) we obtain that

$$\Phi^+(t) = G(t)\Phi^-(t) + [g_2(t) - g_1(t)].$$

This is just Problem H for analytic function, it is well solved in [6]. Finally, we obtain the following theorems.

Theorem 7 If $\kappa > 0$, the homogeneous Hilbert problem (Problem H^0) has the 2κ linearly independent solutions

$$\begin{gathered}\bar{z}X(z), \bar{z}zX(z), \cdots, \bar{z}z^{\kappa-1}X(z), \\ w_0(z) + X(z), w_0(z) + zX(z), \cdots, w_o(z) + z^{\kappa-1}X(z),\end{gathered} \tag{4.10}$$

where $w_0(z)$ is the general solution of Problem H^{*0}.

If $\kappa < 0$, then Problem H^0 has only the trivial solution.

Theorem 8 If $\kappa > 0$, the nonhomogeneous Hilbert problem (Problem H) has the general solution

$$\begin{aligned}w(z) = {}& \frac{X(z)}{2\pi i}\int_L \frac{g_1(t)}{X^+(t)(t-z)}dt - \frac{X(z)}{2\pi i}\int_L \frac{\overline{t-z}}{\bar{t}(t-z)}\frac{g_1(t)}{X^+(t)}dt + \\ & \bar{z}X(z)P_{\kappa-1}(z) + \frac{X(z)}{2\pi i}\int_L \frac{g_1(t)}{X^+(t)(t-z)}dt + X(z)Q_{\kappa-1}(z)\end{aligned} \tag{4.11}$$

where $P_{\kappa-1}(z)$ and $Q_{\kappa-1}(z)$ are arbitrary polynomials of order $(\kappa-1)$, and $w(z)$ is bounded at the point at infinity.

Obviously, condition (4.2) is satisfied for $w(z)$. In general the solution contains 2κ arbitrary constants.

Theorem 9 When $\kappa<0$, and the -2κ conditions

$$\int_L \frac{g_2(t)}{X^+(t)} t^k \mathrm{d}t = 0, \quad k = 0,1,\cdots -\kappa-1, \tag{4.12}$$

and (4.6) are satisfied, the nonhomogeneous Hilbert problem (Problem H) has the unique solution

$$w(z) = \frac{X(z)}{2\pi i}\int_L \frac{g_1(t)}{X^+(t)(t-z)}\mathrm{d}t -$$

$$\frac{X(z)}{2\pi i}\int_L \frac{\overline{(t-z)}}{\overline{t}(t-z)} \frac{\overline{g_1(t)}}{\overline{X^+(t)}}\mathrm{d}t + \frac{X(z)}{2\pi i}\int_L \frac{g_2(t)}{X^+(t)(t-z)}\mathrm{d}t. \tag{4.13}$$

Obviously, condition (4.12) is satisfied for $w(z)$.

References

[1] Zhao Zhen. Bianalytic functions, complex harmonic functions and their basic boundary value problems. Journal of Beijing normal university (Natural sci.), 1995, 31(2): 175-179. (in Chinese)

[2] Zhao Zhen. Bianalytic functions and their applications. Proc. Second Asian Math. Conf., Vol. 1, Thailand, 1995: 223-230.

[3] Zhao Zhen. Riemann-Hilbert's problem for bianalytic functions. Journal of Beijing normal university (Natural sci.), 1996, 32(3): 316-320.

[4] Zhao Zhen. Schwarz's problems for complex partial differential equations of second order. Beijing mathematics, 1996, 1(2): 131-137.

[5] Lu Jianke. Complex methods of plane elasticity. Wuhan: Publ house of Wuhan university press, 1986. (in Chinese)

[6] Muskhelishvili N I. Singular integral equations. Noordhoof, Groningen, 1953.

[7] Vekua I N. Generalized analytic functions. Pergomon, Oxford, 1962.

Proceedings of the Second ISAAC Congress, Vol. 1
Kluwer Academic Publishers, 2000:223-230.

Cauchy 型积分和双解析函数的广义 Harnack 定理①

On the Integral of Cauchy Type and the Generalized Harnack Theorem for Bianalytic Functions

Abstract In this paper the integral of Cauchy type, its some properties and the generalized theorem Harnack for bianalytic functions are obtained. These results are very useful for applications, especially, for solvability of boundary value problems for bianalytic functions.

§ 1. Introduction

In [1] we have suggested the concept of bianalytic functions and considered some their properties. Furthermore, some boundary value problems for them (i. e. Dirichlet's problem, Dirichlet's problem in the changed form. Riemann-Hilbert's problem, Schwarz's problem, Problem A, etc.) are also considered [1～6]. In [7] we investigated the Cauchy formula and the integral of Cauchy type for bianalytic functions and proved some useful theorems for them; finally, the Problem Hilbert for bianalytic functions is also considered. In this paper some properties of the Cauchy type integral and the generalized theorem Harnack are obtained,

The complex form of C-R equations is

① Supported by the National Natural Science Foundation of China.

$$\frac{\partial W}{\partial \bar{z}} = 0, \tag{1.1}$$

here $\frac{\partial}{\partial \bar{z}} = \frac{1}{2}\left(\frac{\partial}{\partial x} + \mathrm{i}\frac{\partial}{\partial y}\right)$.

In addition we also define the derivatives

$$\frac{\partial}{\partial z} = \frac{1}{2}\left(\frac{\partial}{\partial x} - \mathrm{i}\frac{\partial}{\partial y}\right) \text{ and } \Delta \equiv \frac{\partial^2}{\partial \bar{z}\partial z} = \frac{\partial^2}{\partial z \partial \bar{z}}.$$

Definition 1 Suppose G is a $p+1$-connected domain on the plane, the boundary of G is $L = L^0 + L^1 + \cdots + L^p$. We say that a complex function $w(z)$ in G having the second order derivative in $\bar{z}$, i. e. $\frac{\partial^2 w}{\partial \bar{z}^2}$, is called a bianalytic function, if $w(z)$ satisfies the following partial differential equation

$$\frac{\partial^2 w}{\partial \bar{z}^2} = 0, \ z \in G. \tag{1.2}$$

We denote the set of all bianalytic functions by $D_2(G)$.

We know that a bianalytic function can be represented by formula[1]

$$w(z) = \bar{z}\varphi_1(z) + \varphi_2(z), \ z \in G. \tag{1.3}$$

where $\varphi_1(z)$, $\varphi_2(z)$ are all analytic functions in G. For simplicity we will call $\varphi_1(z)$ and $\varphi_2(z)$ the analytic factor and analytic addition of bianalytic function $w(z)$ respectively, in the later we denote all functions in the form $w^*(z) = \bar{z}\varphi(z)$, by $D_2^*(G) \in D_2(G)$, where $\varphi(z)$ is arbitrary analytic function and call it by a characteristic bianalytic function.

When we use the methods of function theory to solve problems of elasticity in the plane, the stress function $U(x, y)$ which is biharmonic, will play a very important role. We can successfully represent the stress function $U(x, y)$ as the real part of a certain bianalytic function, i. e. $U(x, y) = \mathrm{Re}[w(z)]$, where $w(z)$ is a bianalytic function. This means that investigation of bianalytic functions is of important practical interest.

§ 2. Integral of Cauchy type for bianalytic function

We know that in the formula Cauchy of bianalytic functions is contained two boundary values $w(t)$ and $\frac{\partial w}{\partial \bar{t}}$, this is the difference from the formula Cauchy of analytic functions [7]. In the later for convenience we will reduce the integral of Cauchy type only for characteristic bianalytic functions, i. e. the functions belonging to class $D_2^*(G)$, as the follows:

Definition 2 The integral of Cauchy type for the characteristic bianalytic function $w^*(z)$ belonging to the class $D_2^*(G)$ is as follows:

$$\begin{aligned} w^*(z) &= \frac{\bar{z}}{2\pi \mathrm{i}}\int_L \frac{\varphi(z)}{t-z}\mathrm{d}t \\ &= \frac{1}{2\pi \mathrm{i}}\int_L \frac{\bar{t}\varphi(t)}{t-z}\mathrm{d}t - \frac{1}{2\pi \mathrm{i}}\int_L \frac{\overline{(t-z)}\varphi(t)}{t-z}\mathrm{d}t \qquad (2.1) \\ &= \frac{1}{2\pi \mathrm{i}}\int_L \frac{w(t)}{t-z}\mathrm{d}t - \frac{1}{2\pi \mathrm{i}}\int_L \frac{\overline{(t-z)}w(t)}{\bar{t}(t-z)}\mathrm{d}t,\ t\in L, \end{aligned}$$

where $\varphi(z)$ is arbitrary continuous function and $w(t)=\bar{t}\varphi(t)$. The first and the second terms of the right-hand side of (2.1) is an analytic and a bianalytic function respectively. Consequently, we have

Theorem 1 If $w(z)$ is a bianalytic function in a bounded domain G, then $w(z)=w^*(z)+\Phi(z)$, $z\in G$, where $w^*(z)$ is a function as in (2.1) and $\Phi(z)$ is an arbitrary analytic function.

Proof We know that $w(z)=\bar{z}\varphi_1(z)+\varphi_2(z)$[1], as soon as take $\varphi_1(z)=\frac{1}{2\pi \mathrm{i}}\int_L \frac{\varphi(t)}{t-z}\mathrm{d}t$, $\Phi(z)=\varphi_2(z)$, then will immediately get Theorem 1.

Definition 3 A piecewise bianalytic function with discontinuous curve L, we will call the function $\Psi(z)$, which is bianalytic in may bounded domain D not containing curve L, has a finite k-th order at the infinity and it is continuously extendible from left-hand side and right-hand side to L, where L is a closed smooth curve.

Definition 4 A piecewise bianalytic function to $w^*(z)$ is called characteristic bianalytic, if it belongs to $D_2^*(G)$ in any bounded domain D not containing curve L, has a finite k-th order at the infinity and it is continuously extendible from left-hand side and right-hand side to L, where L is a closed smooth curve.

Theorem 2 Suppose L is a closed smooth curve, and $w(t)$ belongs to class H, (i. e. the class of Hölder continuous functions), then the integral of Cauchy type $w^*(z)$ (2.1) is a piecewise bianalytic function, which is bounded at infinity, and L is the discontinuous curve.

By using the properties of the Cauchy type integral for analytic function [10], we can easily prove theorem 2.

Theorem 3 (Plemelj's formula) Suppose L is a closed smooth curve, and $w(t)$ belongs to class H (i. e. the class of Hölder continuous functions). For the integral of Cauchy type for the characteristic bianalytic functions

$$w^*(z)=\frac{1}{2\pi i}\int_L\frac{\bar{t}\varphi(t)}{t-z}dt-\frac{1}{2\pi i}\int_L\frac{(\overline{t-z})\varphi(t)}{t-z}dt \tag{2.2}$$

$$=\frac{1}{2\pi i}\int_L\frac{w(t)}{t-z}dt-\frac{1}{2\pi i}\int_L\frac{(\overline{t-z})w(t)}{\bar{t}(t-z)}dt,\ z\notin L,$$

where $\varphi(t)$ is arbitrary continuous function and $w(t)=\bar{t}\varphi(t)$, we have the following formula:

$$w^{*+}(t)=\frac{1}{2}w(t)+\frac{1}{2\pi i}\int_L\frac{w(\tau)}{\tau-t}d\tau-\frac{1}{2\pi i}\int_L\frac{(\overline{\tau-z})w(t)}{\bar{\tau}(\tau-t)}d\tau$$

$$w^{*-}(t)=-\frac{1}{2}w(t)+\frac{1}{2\pi i}\int_L\frac{w(\tau)}{\tau-t}d\tau-\frac{1}{2\pi i}\int_L\frac{(\overline{\tau-t})w(\tau)}{\bar{\tau}(\tau-t)}d\tau,\ t\in L,$$

(2.3)

By Plemelj's formula (2.3) we know its another form is as follows:

$$\begin{aligned}&w^{*+}(t)+w^{*-}(t)=\frac{1}{\pi i}\int_L\frac{w(\tau)}{\tau-t}d\tau-\frac{1}{\pi i}\int_L\frac{(\overline{\tau-t})w(\tau)}{\bar{\tau}(\tau-t)}d\tau,\\&w^{*+}(t)-w^{*-}(t)=\omega(\tau),\ t\in L.\end{aligned} \tag{2.4}$$

§ 3. The generalized Harnack theorem

Suppose $w(t)$ is a continuous function given on the boundary $L=L^0+L^1+L^2+\cdots+L^p$, problem is that $w(t)$ can be as the boundary value of a bianalytic (characteristic) in G and continuous on $G+L$ function $w(z)$? Or what will be the conditions for that $w(t)$ can be as the boundary value of the function $w(z)$?

Firstly, we know the function $w(t)$ is the boundary value of a bianalytic (characteristic) in G and continuous on $G+L$ function $w(z)$, (when L^0 not exists, we require also $w(\infty)$ is bounded), then by the Cauchy theorem of bianalytic functions have:[7]

$$w^-(z)=\frac{1}{2\pi i}\int_L\frac{w(t)}{t-z}dt-\frac{1}{2\pi i}\int_L\frac{\overline{t-z}}{t-z}\frac{\partial w(t)}{\partial\bar{t}}dt=0,\ z\in S^-,\tag{3.1}$$

where S^- is the complementary set of $G+L$.

Inversely, if condition (3.1) is satisfied, then we consider the integral of Cauchy type:

$$w^*(z)=\frac{1}{2\pi i}\int_L\frac{w(t)}{t-z}dt-\frac{1}{2\pi i}\int_L\frac{\overline{t-z}}{(t-z)}w(t)dt,\ z\notin L.\tag{3.2}$$

By (3.1) and Plemelj's formula, obviously, we have $w^{*-}(t_0)=0$, $t_0\in L$, and

$$w(t_0)=w^{*+}(t_0)-w^{*-}(t_0)=w^{*+}(t_0),\ t_0\in L.$$

This just shows that $w(t_0)$ is the boundary value of a (characteristic) bianalytic in G and continuous on $G+L$ function $w^*(z)$. Thus we obtain the following theorem:

Theorem 4 A given on L continuous function $w(t)$ can be as the boundary value of a (characteristic) bianalytic in G and continuous on $G+L$ function $w(z)$ (when L^0 not exists, we requir also that $w(\infty)$ is bounded), if and only if the condition (3.1) is satisfied.

If $w(t)$, $t\in L$ satisfies the Hölder condition, then we can change the form of (3.1) as follows:

Let $z\to t_0$, by the Plemelj's formula we can get

$$-\frac{1}{2}w(t_0)+\frac{1}{2\pi i}\int_L\frac{w(t)}{t-t_0}dt-\frac{1}{2\pi i}\int_L\frac{\overline{t-t_0}}{\bar{t}(t-t_0)}w(t)dt=0,\ t_0\in L.\tag{3.3}$$

Furthermore, from (3.3) we can also get (3.1). In fact, by the condition (3.3) we know that the function $w^*(z)$ given by (3.2) is bianalytic in S^-, where S^- is the complementary set of $G+L$ and continuously extendible to L. Obviously, we have

$$w^-(t_0)=-\frac{1}{2}w(t_0)+\frac{1}{2\pi i}\int_L\frac{w(t)}{t-t_0}dt-\frac{1}{2\pi i}\int_L\frac{\overline{t-t_0}}{\bar{t}(t-t_0)}w(t)dt=0,$$
$$t_0\in L.$$

From the Cauchy theorem, for bianalytic function we know $w(z)\equiv 0$, $z\in S^-$, i. e. the condition (3.1) is satisfied.

Theorem 5　A function $w(t)\in H(L)$ with the boundary value of a (characteristic) bianalytic in G and continuous on $G+L$ function $w^*(z)$ (when L^0 not exists, we require that $w(\infty)$ is bounded), if and only if the condition (3.3) is satisfied.

Analogously, we have that a given on L continuous function $w(t)$ can be as the boundary value of a (characteristic) bianalytic in S^- and continuous on S^-+L function $w^*(z)$ (when L_0 exists, we require also $w(\infty)$ is bounded), if and only if the following condition is satisfied:

$$w^*(z)=\frac{1}{2\pi i}\int_L\frac{w(t)}{t-z}dt-\frac{1}{2\pi i}\int_L\frac{\overline{t-z}}{\bar{t}(t-z)}w(t)dt\equiv 0,\ z\in G.$$

If the function $w(t)$ satisfies the Hölder condition, we can get also the analogous result of Theorem 5.

Theorem 6　(The generalized Harnack theorem)　Suppose $\frac{w(t)}{\bar{t}}$ is a real function, and

$$w^*(z)=\frac{1}{2\pi i}\int_L\frac{w(t)}{t-z}dt-\frac{1}{2\pi i}\int_L\frac{\overline{t-z}}{\bar{t}(t-z)}w(t)dt\equiv 0,\ z\in G,$$

then we must have $w(t)=\bar{t}c_k$, $t\in L, k=0,1,2,\cdots,p$, where c_k is a (real) constant, and $c_0=0$. (If L_0 not exists, then $k=1,2,\cdots,p$). But if $w^*(z)\equiv 0$, $z\in S^-$, then we must have $w(t)=\bar{t}C$, where C is a (real) constant.

Proof　If $w^*(z)\equiv 0$, $z\in G$, then $w(t)$ is the boundary value of a bianalytic in S^- and continuous on S^-+L function $w^{*-}(z)\in D_2^*(S^-)$, where $w^{*-}(z)$ is the sum of functions $w_k^{*-}(z)$, which are bianalytic in S_k^-, $k=0,1,2,\cdots,p$. When L_0 exists, we require also $w^{*-}(\infty)$ is bounded. Because $\frac{w(t)}{\bar{t}}$ is a real of function and $\frac{w^{*-}(z)}{\bar{z}}$ are analytic functions which is an analytic factor of bianalytic functions $w_k^{*-}(z)$, $k=0,1,2,\cdots,p$, then the imaginary part of $\frac{w_k^{*-}(z)}{\bar{z}}$ on L_k is equal to zero. Furthermore, $\operatorname{Im}\left[\frac{w_k^{*-}(z)}{\bar{z}}\right]\equiv 0$, $z\in S_k^-$, $k=0,1,2,\cdots,p$. Finally we have $\frac{w_k^{*-}(z)}{\bar{z}}\equiv c_k$, $z\in S_k^-$, $k=0,1,2,\cdots,p$. By using the condition $\frac{w_k^{*-}(z)}{\bar{z}}=0$, if $z=\infty$, we have $c_0=0$. In the other word we have $w_k^{*-}(t)=\bar{t}c_k$, $k=0,1,2,\cdots,p$. Analogously, we can prove the second part of Theorem 6.

Remark In Theorem 6 instead of $w^*(z)\equiv 0$ we can use the condition $\operatorname{Re}\left[\frac{w^*(z)}{\bar{z}}\right]\equiv 0$, then the conclusion of Theorem 6 still is right.

In fact, we first suppose $\operatorname{Re}\left[\frac{w^*(z)}{\bar{z}}\right]\equiv 0$, $z\in G$, then $\frac{w^*(z)}{\bar{z}}=Ai$, where A is a (real) constant. Thus we have $\frac{w^{*+}(t)}{\bar{t}}=Ai$, by using the Plemelj's theorem we obtain

$$\frac{w^{*-}(t)}{\bar{t}}=\frac{w^{*+}(t)}{\bar{t}}-\frac{w(t)}{\bar{t}}=Ai-\frac{w(t)}{\bar{t}}.$$

Moreover $\operatorname{Im}\left[\frac{w^{*-}(t)}{\bar{t}}\right]=A$, $t\in L$, and $\operatorname{Im}\left[\frac{w^{*-}(z)}{\bar{z}}\right]\equiv A$, $z\in S_k^-$, $k=0,1,2,\cdots,p$. In the other word $\frac{w^{*-}(z)}{\bar{z}}=Ai-c_k$, $z\in S_k^-$. Finally we have

$$\frac{w(t)}{\bar{t}}=Ai-\frac{w^{*-}(t)}{\bar{t}}=Ai-Ai+c_k=c_k,\ t\in L_k.$$

or $w(t)=\bar{t}c_k$, $t\in L_k$, $k=0,1,2,\cdots,p$. When L^0 exists we also have $c_0=0$. Analogously. If the condition $\operatorname{Re}\left[\frac{w^*(z)}{\bar{z}}\right]\equiv 0$, $z\in S_k^-$ is satisfied, then the second conclusion of Theorem 6 still is right ($w(t)=\bar{t}C$, where C is real constant).

References

[1] Zhao Zhen. Bianalytic functions, complex harmonic functions and their basic boundary value problems. Journal of Beijing normal university (Natural sci.), 1995, 31(2):175-179.

[2] Zhao Zhen. Bianalytic functions and their applications. Proceedings of Second Asian Math. Conference, 1995. 11, Tailand, 223-230.

[3] Zhao Zhen. Riemman Hilbert's problem for bianalytic functions. Journal of Beijing normal university (Natural sci.), 1996, 32(3):316-320.

[4] Zhao Zhen. Schwarz's problems for complex partial differential equations of second order. Beijing mathematics, 1995, (2):131-137.

[5] Zhao Zhen. Some boundary problems for bianalytic and complex harmonic functions. Report on international conference of complex analyses. Ningxia university, China, 1996, 8.

[6] Zhao Zhen, Chenfang Quan. Dirichlet's problems for bianalytic functions. Journal of Beijing normal university (Natural sci.), 1998, 34(2):174-178.

[7] Zhao Zhen. Cauchy formula, integral of Cauchy type and Hilbert problem for bianalytic functions. Proceedings of ISAAC, 1997, USA, Deleware university, Kluwer academic publishers, 1998: 211-218.

[8] Zhao Zhen. The Schwarz's problem for a class of complex partial differential equation of third order. Journal of Ningxia university (Natural sci.), 1998, 19(1):6-8.

[9] Lu Jianke. Complex methods of plane elasticity. Wuhan: Publ house of Wuhan university press, 1986. (in Chinese)

[10] Mushkelishvili N I. Singular integral equation. Noordhoof, Groningen, 1953.

[11] Vekya I N. Generalized analytic functions. Moskow, 1959. (in Russian)

Integral Equation and Related Problems.
World Scientific, 2000:290-296.

关于双解析函数的几个重要性质

On Some Important Properties of Bianalytic Functions①

Abstract In this paper the derivatives of higher order, the Cauchy inequality and Liouville's theorem for bianalytic functions are obtained. These results are very important for the theory of bianalytic functions.

§ 1. Introduction

In [1] we have suggested the concept of bianalytic functions and considered some their properties. Furthermore, some boundary value problems for them (i. e. Dirichlet's problem, Dirichlet's problem in the changed form, Riemann-Hilbert's problem, Schwarz's problem, Problem A, etc.) are also considered [1～6]. In [7] we investigated the Cauchy formula and the integral of Cauchy type for bianalytic functions and proved some useful theorems for them; finally, the Hilbert problem for bianalytic functions is also considered. In this paper some properties of bianalytic functions, i. e. the formula of higher order derivatives and the Cauchy inequality and Liouville's theorem are obtained.

The complex form of C-R equations is

① Supported by the National Natural Science Foundation of China.

$$\frac{\partial w}{\partial \bar{z}} = 0, \tag{1.1}$$

here $\frac{\partial}{\partial \bar{z}}=\frac{1}{2}\left(\frac{\partial}{\partial x}+\mathrm{i}\frac{\partial}{\partial y}\right)$.

In addition we also define the derivatives

$$\frac{\partial}{\partial z}=\frac{1}{2}\left(\frac{\partial}{\partial x}-\mathrm{i}\frac{\partial}{\partial y}\right) \text{ and } \Delta\equiv\frac{\partial^2}{\partial \bar{z}\partial z}=\frac{\partial^2}{\partial z\partial \bar{z}}.$$

Definition 1 Suppose G is a $p+1$-connected domain on the plane, the boundary of G is $L=L^0+L^1+\cdots+L^p$, and define in G a complex function $w(z)$, which exists the two order derivative in $\bar{z}$, i.e. $\frac{\partial^2 w}{\partial \bar{z}^2}$, we will call it is a bianalytic function, if $w(z)$ satisfies the following partial differential equation

$$\frac{\partial^2 w}{\partial \bar{z}^2} = 0, \quad z \in G. \tag{1.2}$$

In the later we denote the set of all bianalytic functions by $D_2(G)$.

We know that a bianalytlic function can be represented by formula[1]

$$w(z) = \bar{z}\varphi_1(z) + \varphi_2(z), \quad z \in G. \tag{1.3}$$

where $\varphi_1(z)$, $\varphi_2(z)$ are all analytic functions in G. For simplicity we will call $\varphi_1(z)$ and $\varphi_2(z)$ the analytic factor and analytic addition of bianalytic function $w(z)$ respectively.

When we use the methods of function theory to solve problems of elasticity on the plane, the stress function $U(x, y)$ which is biharmonic, will play a very important role. We can successfully represent the stress function $U(x, y)$ as the real part of a certain bianalytic function, i.e. $U(x, y)=\mathrm{Re}[w(z)]$, where $w(z)$ is a bianalytic function. This means that investigation of bianalytic functions is of important practical interest.

§ 2. The derivatives of higher order for bianalytic functions

In this section we will prove a very important property of bianalytic functions, i.e. a bianalytic (in given regain G) function $w(z)$ must has its partial derivative on z of any order n: $\frac{\partial^n w}{\partial z^n}$, and the formula of the derivatives of higher order for bianalytic functions is also obtained.

Theorem 1 If function $w(z)$ is bianalytic in G, $w(z)$ and $\frac{\partial w}{\partial \bar{z}}$ are continuous on $\overline{G+L}$, then function $w(z)$ must have the partial derivatives on z of n order, and we have the following formula:

$$\frac{\partial^n w}{\partial z^n}=\frac{1}{2\pi \mathrm{i}}\int_L \frac{w(t)}{(t-z)^{n+1}}\mathrm{d}t-\frac{1}{2\pi \mathrm{i}}\int_L \frac{\overline{t+z}}{(t-z)^{n+1}}\frac{\partial w}{\partial \bar{t}}\mathrm{d}t,\ z\in G,\ n\in \mathbf{N}^*. \tag{2.1}$$

Proof Firstly we prove the following formula, i. e. the special case of $n=1$:

$$\frac{\partial w}{\partial z}=\frac{1}{2\pi \mathrm{i}}\int_L \frac{w(t)}{(t-z)^2}\mathrm{d}t-\frac{1}{2\pi \mathrm{i}}\int_L \frac{\overline{t+z}}{(t-z)^2}\frac{\partial w}{\partial \bar{t}}\mathrm{d}t,\ z\in G. \tag{2.2}$$

Obviously $\frac{\partial w}{\partial z}=\lim\limits_{z\to z_0}\frac{w(z)-w(z_0)}{z-z_0}$①, where z_0 is any definite point of G; by the Cauchy formula for bianalytic functions (2.2), we have

$$\begin{aligned}&w(z)-w(z_0)\\&=\frac{1}{2\pi \mathrm{i}}\int_L\left[\frac{1}{t-z}-\frac{1}{t-z_0}\right]w(t)\mathrm{d}t+\frac{1}{2\pi \mathrm{i}}\int_L\left[\frac{\overline{t+z_0}}{t-z_0}-\frac{\overline{t+z_0}}{t-z}\right]\frac{\partial w}{\partial \bar{t}}\mathrm{d}t\\&=\frac{z-z_0}{2\pi \mathrm{i}}\int_L\frac{w(t)}{(t-z)(t-z_0)}\mathrm{d}t-\frac{z-z_0}{2\pi \mathrm{i}}\int_L\frac{\overline{t+z_0}}{(t-z)(t-z_0)}\frac{\partial w}{\partial \bar{t}}\mathrm{d}t,\end{aligned}$$

hence,

$$\begin{aligned}&\frac{w(z)-w(z_0)}{z-z_0}-\left[\frac{1}{2\pi \mathrm{i}}\int_L\frac{w(t)}{(t-z_0)^2}\mathrm{d}t-\frac{1}{2\pi \mathrm{i}}\int_L\frac{\overline{t+z_0}}{(t-z_0)^2}\frac{\partial w}{\partial \bar{t}}\mathrm{d}t\right]\\&=\frac{1}{2\pi \mathrm{i}}\int_L\left[\frac{1}{(t-z)(t-z_0)}-\frac{1}{(t-z_0)^2}\right]w(t)\mathrm{d}t-\\&\quad\frac{1}{2\pi \mathrm{i}}\int_L\left[\frac{\overline{t+z_0}}{(t-z)(t-z_0)}-\frac{\overline{t+z_0}}{(t-z_0)^2}\right]\frac{\partial w}{\partial \bar{t}}\mathrm{d}t\\&=\frac{(z-z_0)}{2\pi \mathrm{i}}\int_L\frac{w(t)}{(t-z)(t-z_0)^2}\mathrm{d}t-\frac{(z-z_0)}{2\pi \mathrm{i}}\int_L\frac{\overline{t+z_0}}{(t-z)(t-z_0)^2}\frac{\partial w}{\partial \bar{t}}\mathrm{d}t.\end{aligned} \tag{2.3}$$

① Here we consider only the partial derivation on z, so we can suppose $\bar{z}$ is still not changed when z is changed.

Now we estimate the last integrals in (2.3) on modules: Because functions $w(z)$ and $\frac{\partial w}{\partial \bar{z}}$ are all continuous on $G+L$, and G is a bounded domain, so that must exit a positive constant M, such that

$$| w(t) | \leqslant M, \quad \left|\frac{\partial w}{\partial \bar{t}}\right| \leqslant M, \quad | t+z_0 | \leqslant M. \tag{2.4}$$

Suppose d is the distant from point z_0 to the boundary L, i. e. $d=\min\limits_{t\in L}|t-z_0|>0$, let $|z-z_0|<\frac{d}{2}$, hence

$$| t-z | \geqslant | t-z_0 | - | z_0-z | \geqslant d-\frac{d}{2}=\frac{d}{2}. \tag{2.5}$$

Finally, we have

$$\left|\int_L \frac{w(t)}{(t-z)(t-z_0)^2}\mathrm{d}t\right| \leqslant \frac{M}{\frac{d}{2}\cdot d^2}\cdot l=\frac{2Ml}{d^3},$$

$$\left|\int_L \frac{\overline{t+z_0}}{(t-z)(t-z_0)^2}\frac{\partial w}{\partial \bar{t}}\mathrm{d}t\right| \leqslant \frac{M^2}{\frac{d}{2}\cdot d^2}\cdot l=\frac{2M^2 l}{d^3},$$

where l is the length of boundary L. Consequently, we get

$$\left|\frac{w(z)-w(z_0)}{z-z_0}-\left[\frac{1}{2\pi \mathrm{i}}\int_L \frac{w(t)}{(t-z_0)^2}\mathrm{d}t-\frac{1}{2\pi \mathrm{i}}\int_L \frac{\overline{t+z_0}}{(t-z_0)^2}\frac{\partial w}{\partial \bar{t}}\mathrm{d}t\right]\right|$$
$$\leqslant \frac{Ml}{\pi d^3}(1+M) | z-z_0 |.$$

Let $z\to z_0$, where z_0 is any definite point of G, so that

$$\lim_{z\to z_0}\frac{w(z)-w(z_0)}{z-z_0}=\frac{\partial w}{\partial z}=\frac{1}{2\pi \mathrm{i}}\int_L \frac{w(t)}{(t-z_0)^2}\mathrm{d}t-\frac{1}{2\pi \mathrm{i}}\int_L \frac{\overline{t+z_0}}{(t-z_0)^2}\frac{\partial w}{\partial \bar{t}}\mathrm{d}t$$

In the following we prove Theorem 1 for any positive integer by mathematical induction.

In fact we have proved that formula (2.2), it means that Theorem 1 is true, when $n=1$. We suppose formula (2.1) holds, when $n=k$, i. e. formula

$$\frac{\partial^k w}{\partial z^k}=\frac{k!}{2\pi \mathrm{i}}\int_L \frac{w(t)}{(t-z)^{k+1}}\mathrm{d}t-\frac{k!}{2\pi \mathrm{i}}\int_L \frac{\overline{t+z}}{(t-z)^{k+1}}\frac{\partial w}{\partial \bar{t}}\mathrm{d}t, \quad z\in G \tag{2.6}$$

holds. Now we prove that formula (2.1) still holds for $n=k+1$, i. e. the following formula

$$\frac{\partial^{k+1} w}{\partial z^{k+1}} = \frac{(k+1)!}{2\pi \mathrm{i}} \int_L \frac{w(t)}{(t-z)^{k+2}} \mathrm{d}t - \frac{(k+1)!}{2\pi \mathrm{i}} \int_L \frac{\overline{t+z}}{(t-z)^{k+2}} \frac{\partial w}{\partial \bar{t}} \mathrm{d}t, \ z \in G \tag{2.7}$$

holds. Denote by $\frac{\partial w^k(z)}{\partial z^k}$ the k order partial derivative of $w_z^{(k)}(z)$ on z.

For any $z_0 \in G$ by formula (2.6) we have

$$\begin{aligned}
& w_z^{(k)}(z) - w_z^{(k)}(z_0) \\
&= \frac{k!}{2\pi \mathrm{i}} \int_L w(t) \left[\frac{1}{(t-z)^{k+1}} - \frac{1}{(t-z_0)^{k+1}} \right] \mathrm{d}t - \\
&\quad \frac{k!}{2\pi \mathrm{i}} \int_L \left[\frac{\overline{t+z_0}}{(t-z)^{k+1}} - \frac{\overline{t+z_0}}{(t-z_0)^{k+1}} \right] \frac{\partial w}{\partial \bar{t}} \mathrm{d}t \\
&= \frac{k!}{2\pi \mathrm{i}} \int_L w(t) \frac{(t-z_0)^{k+1} - (t-z)^{k+1}}{(t-z)^{k+1}(t-z_0)^{k+1}} \mathrm{d}t - \\
&\quad \frac{k!}{2\pi \mathrm{i}} \int_L \overline{(t+z_0)} \frac{(t-z_0)^{k+1} - (t-z)^{k+1}}{(t-z)^{k+1}(t-z_0)^{k+1}} \frac{\partial w}{\partial \bar{t}} \mathrm{d}t \\
&= \frac{k!}{2\pi \mathrm{i}} \int_L w(t) \frac{[(t-z_0)-(t-z)] \sum_{l=0}^{k} (t-z_0)^{k-1}(t-z)^l}{(t-z)^{k+1}(t-z_0)^{k+1}} \mathrm{d}t - \\
&\quad \frac{k!}{2\pi \mathrm{i}} \int_L \overline{(t+z_0)} \frac{[(t-z_0)-(t-z)] \sum_{l=0}^{k} (t-z)^{k-1}(t-z_0)^l}{(t-z)^{k+1}(t-z_0)^{k+1}} \frac{\partial w}{\partial \bar{t}} \mathrm{d}t.
\end{aligned} \tag{2.8}$$

Moreover we get

$$\begin{aligned}
& \frac{w_z^{(k)}(z) - w_z^{(k)}(z_0)}{z-z_0} - \left[\frac{(k+1)!}{2\pi \mathrm{i}} \int_L \frac{w(t)}{(t-z_0)^{k+2}} \mathrm{d}t - \right. \\
&\quad \left. \frac{(k+1)!}{2\pi \mathrm{i}} \int_L \frac{\overline{t+z_0}}{(t-z_0)^{k+2}} \frac{\partial w}{\partial \bar{t}} \mathrm{d}t \right] \\
&= \frac{k!}{2\pi \mathrm{i}} \int_L w(t) \left[\frac{\sum_{l=0}^{k} (t-z_0)^{k-1}(t-z)^l}{(t-z)^{k+1}(t-z_0)^{k+1}} - \frac{k+1}{(t-z_0)^{k+2}} \right] \mathrm{d}t - \\
&\quad \frac{k!}{2\pi \mathrm{i}} \int_L \overline{(t+z_0)} \left[\frac{\sum_{l=0}^{k} (t-z)^l (t-z_0)^{k-l}}{(t-z)^{k+1}(t-z_0)^{k+1}} - \frac{k+1}{(t-z_0)^{k+2}} \right] \frac{\partial w}{\partial \bar{t}} \mathrm{d}t \qquad (2.9) \\
&= \frac{k!}{2\pi \mathrm{i}} \int_L w(t) \left[\frac{\sum_{l=0}^{k} (t-z)^l (t-z_0)^{k-l+1} - (t-z)^{k+1-l}}{(t-z)^{k+1}(t-z_0)^{k+2}} \right] \mathrm{d}t -
\end{aligned}$$

$$\frac{k!}{2\pi i}\int_L \frac{\overline{(t+z_0)}\sum_{l=0}^{k}(t-z)^l(t-z_0)^{k-l+1}-(t-z)^{k+1-l}}{(t-z)^{k+1}(t-z_0)^{k+2}}\frac{\partial w}{\partial \bar{t}}dt.$$

We estimate the last integrals in (2. 9) on modules: Because function $w(z)$ and $\frac{\partial w}{\partial \bar{t}}$ are all continuous on $G+L$, and G is a bounded domain, there must exit a positive constant number M_1, such that

$$|w(t)|\leqslant M_1,\ \left|\frac{\partial w}{\partial \bar{t}}\right|\leqslant M_1,\ |t+z_0|\leqslant M_1. \tag{2.10}$$

For any l ($0\leqslant l\leqslant k$), we have

$$\begin{aligned}&|(t-z)^l|\left[(t-z_0)^{k-l+1}-(t-z)^{k-l+1}\right]|\\ \leqslant&|t-z|^l|z-z_0|\sum_{i=0}^{k-l}|(t-z_0)^{k-l-i}(t-z)^i|\\ \leqslant&(2M_1)^l|z-z_0|\sum_{l=0}^{k-l}|(2M_1)^{k-l-i}(2M_1)^i\\ \leqslant&(k-l+1)(2M_1)^k|z-z_0|\leqslant(k+1)(2M_1)^k|z-z_0|.\end{aligned} \tag{2.11}$$

Suppose $|z-z_0|<\frac{d}{2}$, using the above estimation we get

$$\left|\frac{w_z^{(k)}(z)-w_z^{(k)}(z_0)}{z-z_0}-\frac{(k+1)!}{2\pi i}\int_L\frac{w(t)}{(t-z_0)^{k+1}}dt-\frac{(k+1)!}{2\pi i}\int_L\frac{\overline{t+z_0}}{(t-z)^{k+2}}\frac{\partial w}{\partial \bar{t}}dt\right|$$
$$\leqslant\frac{(k+1)!}{2\pi d^{2k+3}}(2M_1)^{2k+1}(1+M_1)|L||z-z_0|, \tag{2.12}$$

where $|L|$ is the length of boundary L. Finally, let $z\to z_0$, we take the limit in (2. 9), then get

$$\frac{\partial^{k+1}w}{\partial z^{k+1}}=\frac{(k+1)!}{2\pi i}\int_L\frac{w(t)}{(t-z_0)^{k+2}}dt-\frac{(k+1)!}{2\pi i}\int_L\frac{\overline{t+z_0}}{(t-z)^{k+2}}\frac{\partial w}{\partial \bar{t}}dt.$$

Because z_0 is any point of G, thus Theorem 1 is proved.

Remark For any analytic function $f(z)$, we have $\frac{\partial f}{\partial z}=\frac{df}{dz}=f'(z)$, because $\frac{\partial f}{\partial \bar{z}}=0$, and $\frac{\partial \bar{z}}{\partial z}=0$. From the formula (1. 3) it follows that for any bianalytic function $w(z)$ we have[1]

$$w(z)=\bar{z}\varphi_1(z)+\varphi_2(z), \tag{1.3}$$

where $\varphi_i(z)(i=1,2)$ are analytic functions. Moreover we have

$$\frac{\partial w}{\partial z}=\bar{z}\varphi'_1(z)+\varphi'_2(z).$$

Obviously, $\frac{\partial w}{\partial z}$ is also a bianalytic function.

Analogously we can prove that $\frac{\partial^n w}{\partial z^n}$ is also a bianalytic function, $n \in \mathbf{N}^*$.

§ 3. Cauchy inequality and Liouville's theorem for bianalytic functions

Theorem 2 (Cauchy inequality) suppose function $w(z)$ is bianalytic in the circle $|z-z_0| \leqslant R$, $|w(z)|$ and $\left|\frac{\partial w}{\partial \bar{z}}\right|$ are bounded, i. e. $|w(z)| \leqslant M$, $\left|\frac{\partial w}{\partial \bar{z}}\right| \leqslant M$, then for the partial derivatives on z of n order, we have the following inequalities

$$\left|\frac{\partial^n w}{\partial z^n}\right| \leqslant \frac{n!M(1+M)}{R^n},\ n \in \mathbf{N}^* \tag{3.1}$$

Proof By formula (2.1), take $z=z_0$ for any $R_1<R$, we have

$$\frac{\partial^n}{\partial z^n} w(z_0) = \frac{n!}{2\pi \mathrm{i}} \int_{|t-z_0|=R_1} \frac{w(t)}{(t-z_0)^{n+1}} \mathrm{d}t - \frac{n!}{2\pi \mathrm{i}} \int_{|t-z_0|=R_1} \frac{\overline{t+z_0}}{(t-z_0)^{n+1}} \frac{\partial w}{\partial t} \mathrm{d}t. \tag{3.2}$$

Consequently,

$$|w_z^{(n)}(z_0)| \leqslant \frac{n!}{2\pi} \cdot \frac{M(1+M)}{R_1^{n+1}} \cdot 2\pi R_1 = \frac{M(1+M)n!}{R_1^n},\ n \in \mathbf{N}^* \tag{3.3}$$

Above inequality holds for any $R_1<R$, hence we take the limit $(R_1 \to R)$ in (3.2), the formula (3.1) is derived.

Using the Cauchy inequality we can prove the following theorem.

Theorem 3 (Liouville's theorem) If $w(z)$ is a bianalytic on the hole complex plane, $w(z)$ and $\frac{\partial w}{\partial \bar{z}}$ are bounded, i. e. $|w(z)| \leqslant M$, $\left|\frac{\partial w}{\partial \bar{z}}\right| \leqslant M$, then the function $w(z)=\bar{z}C_1+C_2$, where C_1, C_2 are (complex) constants.

Proof For any point $z_0 \in G$ the function $w(z)$ is bianalytic in the circle $|z-z_0| \leqslant R_1$ by using the Cauchy inequality we obtain

$$\left|\frac{\partial}{\partial z} w(z_0)\right| \leqslant \frac{M(1+M)}{R}.$$

Let $R \to +\infty$, we get $\frac{\partial}{\partial z} w(z_0)=0$. Consequently, we have $\frac{\partial}{\partial z} w(z) \equiv 0$.

Because $\frac{\partial w}{\partial \bar{z}}=\varphi_1(z)$ (by formula (1.3)) is bounded and analytic on the hole plane, than by the Liouville's theorem for analytic function we know that $\varphi_1(z)=C_1$, i.e. $w(z)=\bar{z}C_1+\varphi_2(z)$, hence $\frac{\partial w}{\partial z}=\varphi_2'(z)=0$, thus $\varphi_2(z)=C_2$. Finally, we get $w(z)=\bar{z}C_1+C_2$. This completes the proof of Theorem 3.

References

[1] Zhao Zhen. Bianalytic functions, complex harmonic functions and their basic boundary value problem. Journal of Beijing normal university (Natural sci.) 1995, 31(2):175-179.

[2] Zhao Zhen. Bianalytic functions and its applications. Proceedings of second Asian math. conference, Tailand, 1995, 1-5.

[3] Zhao Zhen. Riemman Hilbert's problem for bianalytic functions. Journal of Beijing normal university (Natural sci.), 1996, 32(3):316-320.

[4] Zhao Zhen. Schwarz's problems for complex partial differential equations of sec. and order. Beijing math., 1996, Part 1:131-137.

[5] Zhao Zhen. Some-boundary problems for bianalytic and complex harmornic functions. Report on international conference of complex analysis, Ningxia university, China, 1996.

[6] Zhao Zhen, Chenfang Quan. Dirichlet's problems for bianalytic functions. Journal of Beijing normal university (Natural sci.), 1998, 34(2):174-178.

[7] Zhao Zhen. Cauchy formula, integral of Cauchy type and Hilbert problem of bianalytic functions. Partial differential and intergral equations, Proceedings of ISAAC congress, 1997, USA, Kluwer academic publishers, 1998, 211-218.

[8] Zhao Zhen. The Schwarz's problem for a class of complex partial differential equation of third order. Journal of Ningxia university (Natural sci.), 1998, 19(1):6-8.

[9] Lu Jianke. Complex methods of plane elasticity. Wuhan: Publ house of Wuhan university press, 1986. (in Chinese)

[10] Mushkelishvili N I. Singular integral equation. Noordhoof, Groningen, 1953.

[11] Vekya I N. Generalized analytic functions. Moskow, 1959. (in Russian)

Complex Analysis and Its Applications.
Pitman Research Notes in Mathematics Series, 1994, 305:344-351.

复椭圆型方程的一类斜导数边值问题

A Class of Boundary Value Problems with Oblique Derivatives for Complex Elliptic Equation①

§ 1. Introduction

In book [1] systematically considered the method of two-dimensional singular integral equations and solved a lot of problems in mathematical physics.

In paper [2] with help of method singular integral equations solved two kinds of boundary value problems (Problem A and Problem B).

Method of singular integral equations used to solve boundary value problems is very effective in obtaining the conditions of solvability and the representation of solutions. [4]~[6]

In this paper using method of singular integral equations we consider another boundary value problem (Problem P), and obtain the conditions of solvability for it. In the other hands we also get the representation of solutions.

① Supported by the National Natural Science Foundation of China.

§ 2. Formulation of Problem P

2.1 Preparatory knowledge

We consider a class of functions $C_\Delta(f)$ represented by formula

$$W^*(z)=\frac{2}{\pi}\iint_G f(\zeta)\ln|\zeta-z|\,\mathrm{d}\zeta, \tag{2.1}$$

where $f(z)$ is arbitrary H-continuous function; $z\in G$, G is a simple connected domain.

We know that if $\Gamma=\partial G$ satisfies a certain smooth conditions, then for any function of class $C_\Delta(f)$ hold:[4]

$$\Delta W=\frac{1}{4}\left(\frac{\partial^2 W}{\partial\bar{z}\partial z}\right)=f,\ z\in G, \tag{2.2}$$

here $\Delta=\frac{\partial^2}{\partial x^2}+\frac{\partial^2}{\partial y^2}$, $\frac{\partial}{\partial\bar{z}}=\frac{1}{2}\left(\frac{\partial}{\partial x}+\mathrm{i}\frac{\partial}{\partial y}\right)$, $\frac{\partial}{\partial z}=\frac{1}{2}\left(\frac{\partial}{\partial x}-\mathrm{i}\frac{\partial}{\partial y}\right)$.

Further more the general solutions of equation

$$\frac{\partial^2 W}{\partial\bar{z}\partial z}=f, \tag{2.3}$$

can be represented by formula:

$$\begin{aligned}W(z)&=\Phi(z)-\frac{1}{\pi}\iint_G\frac{\varphi(\zeta)}{\zeta-\bar{z}}\mathrm{d}\zeta+\frac{2}{\pi}\iint_G f(\zeta)\ln|\zeta-z|\,\mathrm{d}\zeta\\&=\Phi(z)+\bar{T}\varphi+W^*(z).\end{aligned} \tag{2.4}$$

2.2 Formulation of Problem P

Problem P: To define the solutions of equation (2.2) such that the following condition:

$$\mathrm{Re}\left[\lambda(t)\frac{\partial^2 W}{\partial\bar{z}\partial z}+a(t)\frac{\partial^2 W}{\partial z^2}+b(t)\frac{\partial W}{\partial z}+c(t)W\right]=\gamma(t), \tag{2.5}$$

is satisfied on Γ, where $\lambda(t)$, $a(t)$, $b(t)$, $c(t)$, $\gamma(t)$ are Hölder continuous functions given on Γ, $|a(t)|\neq 0$, $|\lambda(t)|>|a(t)|$, in addition, we assume that $\lambda(t)$, $a(t)$ satisfy condition B on Γ[2].

§ 3. An auxiliary problem (Problem V)

First we consider an auxiliary problem.

Problem V: To define the analytic function $\varphi(z)$, such that the following boundary condition:

$$\operatorname{Re}\left[a(t)\varphi'(t)+b(t)\varphi(t)+c(t)\iint_G k(t,\zeta)\varphi(\zeta)\mathrm{d}\zeta\right]=\gamma(t), \tag{3.1}$$

is satisfied, where $a(t)$, $b(t)$, $c(t)$, $\gamma(t)\in H(\Gamma)$,

$$k(t,\zeta)=\frac{-1}{\bar{\zeta}-\bar{t}},$$

$\varphi'(t)$ is boundary value of function $\varphi'(z)$. (In general $k(t,\zeta)$ can have any weak singularity, for example: $k(t,\zeta)=\frac{k^*(t,\zeta)}{|t-\zeta|^\alpha}$, $0\leqslant\alpha<2$, where $k^*(t,\zeta)$ is H-continuous function.)

We introduce representation of analytic functions:[6]

$$\varphi(z)=\int_\Gamma\mu(t)\ln\left(1-\frac{z}{t}\right)\mathrm{d}s+\int_\Gamma\mu(t)\mathrm{d}s+\mathrm{i}C, \tag{3.2}$$

where $\mu(t)$ is a real function, satisfied Hölder condition, and C is a real constant.

At the same time we introduce elementary functions:

$$\begin{aligned}N_0(z,t)&=\ln\left(1-\frac{z}{t}\right)+1,\\ N_1(z,t)&=\frac{\mathrm{d}}{\mathrm{d}z}\left[\ln\left(1-\frac{z}{t}\right)\right]=-\frac{1}{t-z},\\ &z\in G,\ t\in\Gamma.\end{aligned} \tag{3.3}$$

We understand function In $\left(1-\frac{z}{t}\right)$ is branch, which take value zero at point $z=0$. For every fixed t functions $N_j(z,t)$, $j=0,1$, are analytic functions, $z\in G$.

Fixed t, let $z\to t_0$ we get functions $N_0(t_0,t)$ and $N_1(t_0,t)$ have logarithmic and $(t-t_0)^{-1}$ type singularity respectively, when $t=t_0$. Easily we can find that

$$\begin{aligned}\varphi(t_0)&=\int_\Gamma N_0(t_0,t)\mu(t)\mathrm{d}s+\mathrm{i}C,\\ \varphi'(t_0)&=\pi\mathrm{i}\,\overline{t'_0}\mu(t_0)+\int_\Gamma N_1(t_0,t)\mu(t)\mathrm{d}s,\end{aligned} \tag{3.4}$$

and

$$\varphi(z)=\int_\Gamma N_0(z,t)\mu(t)\mathrm{d}s+\mathrm{i}C.$$

Consequently boundary condition (3.1) has the following form

$$N\mu \equiv A(t_0)\mu(t_0) + \int_\Gamma N(t_0,t)\mu(t)\mathrm{d}s = \gamma(t_0) - C\sigma(t_0), \quad (3.5)$$

where

$$A(t_0) = -\pi\mathrm{i}\mathrm{Im}\, a(t)\, \overline{t'_0} \quad (3.6)$$

$$\sigma(t_0) = \mathrm{Re}\left[\mathrm{i}b(t_0) + \mathrm{i}c(t_0)\iint_G k(t_0,\zeta)\mathrm{d}\zeta\right] \quad (3.7)$$

$$N(t_0,t) = \mathrm{Re}[b(t_0)N_0(t_0,t) + a(t_0)N_1(t_0,t) +$$

$$c(t_0)\iint_G k(t_0,\zeta)N_0(\zeta,t_0)\mathrm{d}\zeta] \quad (3.8)$$

After a simple computation we find that index of equation (3.5) is as following:

$$\chi = \frac{1}{2\pi}\left[\arg\frac{t'\,\overline{a(t)}}{\overline{t'}a(t)}\right]_\Gamma = 2(n+1),$$

where $n=\frac{1}{2\pi}[\arg\overline{a(t)}]_\Gamma$.

In the next we also call χ index of Problem V.

For further application we introduce also the transposed homogeneous equation

$$N'\upsilon \equiv A(t_0)\upsilon(t_0) + \int_\Gamma N(t,\ t_0)\upsilon(t)\mathrm{d}s = 0, \quad (3.9)$$

with the help of (3.2) we find that integral equation (3.5) is equivalent to problem (3.1).

Substituting $\mu(t)$ into (3.2) we can obtain the solution of problem (3.1).

Consequently we have

Theorem 1 Problem (3.1) is always solvable for each given $\gamma(t)$, if and only if the transposed equation (3.9) has no solutions distract from zero.

Theorem 2 If the transposed homogeneous equation (3.9) has exactly k' linear independent solutions $\upsilon_j(t)$, $j=1,2,\cdots,k'$, then the necessary and sufficient conditions of solvability for problem (3.1), will be

$$\int \upsilon_j(t)[\gamma(t) + C\sigma(t)]\mathrm{d}s = 0,\ j = 1,2,\cdots,k'. \quad (3.10)$$

In this case homogeneous problem (3.1) has exactly $k=\chi+k'$ linear independent solutions. Obviously, problem (3.1) has unique solution only if $\chi+k'=0$.

§ 4. Solution of Problem P

First we define the solution of Problem P only in class $C_\Delta(f)$, such a problem will be called Problem P*.

For function of class $C_\Delta(f)$ we have

$$
\begin{aligned}
W^*(z) &= \frac{2}{\pi}\iint_G f(\zeta)\ln|\zeta-z|\,d\zeta,\\
\frac{\partial W^*}{\partial z} &= -\frac{1}{\pi}\iint_G \frac{f(\zeta)}{\zeta-z}d\zeta,\\
\frac{\partial^2 W^*}{\partial z^2} &= \frac{1}{\pi}\iint_G \frac{f(\zeta)}{(\zeta-z)^2}d\zeta,\\
\frac{\partial^2 W^*}{\partial\bar{z}\partial z} &= f(z).
\end{aligned}
\tag{4.1}
$$

We consider two-dimensional singular integral equation

$$
\lambda(z)f(z)+a(z)\frac{1}{\pi}\iint_G\frac{f(\zeta)}{(\zeta-z)^2}d\zeta-b(z)\frac{1}{\pi}\iint_G\frac{f(\zeta)}{\zeta-z}d\zeta+
$$

$$
c(z)\frac{2}{\pi}\iint_G f(\zeta)\ln|\zeta-z|\,d\zeta=0, \tag{4.2}
$$

where $\lambda(z)$, $a(z)$, $b(z)$, $c(z)$ are continuous functions in G, they take the values $\lambda(t)$, $a(t)$, $b(t)$, $c(t)$ on Γ respectively. By assumption we know that $|\lambda(z)|>|a(z)|$, $z\in G$, (see § 2) and thus equation (4.2) is one of equations considered in Dzhuraev's book [1]. Therefore the Fredholm's Theorem holds for equation (4.2). In other words equation (4.2) has exactly N (a finite number) linearly independent solutions.

Let $z\to t$ and use the Sokhotski-Plemelj formulas we take boundary value in equation (4.2) and obtain the following:

$$
\lambda(t)f(t)+a(t)\frac{1}{\pi}\iint_G\frac{f(\zeta)}{(\zeta-t)^2}d\zeta-b(t)\frac{1}{\pi}\iint_G\frac{f(\zeta)}{\zeta-t}d\zeta+
$$

$$
c(t)\frac{2}{\pi}\iint_G f(\zeta)\ln|t-\zeta|\,d\zeta
$$

$$= \lambda(t) \frac{\partial^2 W^*}{\partial \bar{z} \partial z} + a(t) \frac{\partial^2 W^*}{\partial z^2} + b(t) \frac{\partial W^*}{\partial z} + c(t) W^* = 0.$$

Furthermore we have

$$\mathrm{Re}\left[\lambda(t)\frac{\partial^2 W^*}{\partial \bar{z} \partial z}+a(t)\frac{\partial^2 W^*}{\partial z^2}+b(t)\frac{\partial W^*}{\partial z}+c(t)W^*\right]=0. \tag{4.3}$$

The following theorem holds:

Theorem 3 If $\lambda(t)$, $a(t)$ satisfy condition B on Γ, then homogeneous Problem P* is always solution, and its solution can be represented in the form:

$$W^*(z) = \frac{2}{\pi}\iint_G f(\zeta)\ln|\zeta - z|\,\mathrm{d}\zeta,$$

where $f(z)$ is general solution of equation (4.2), or $f(z) = \sum_{j=1}^{N} c_j f_j(z)$, here $f_j(z)$, $j=1, 2, \cdots, N$ are a complete system of linear independent solutions of equation (4.2), $c_j (j=1,2,\cdots,N)$ are arbitrary constants.

Using representation (2.4), we have

$$\begin{aligned}
&\mathrm{Re}\left[\lambda(t) \frac{\partial^2 W}{\partial \bar{z} \partial z} + a(t) \frac{\partial^2 W}{\partial z^2} + b(t) \frac{\partial W}{\partial z} + c(t) W\right] \\
&= \mathrm{Re}[a(t)\Phi'' + B(t)\Phi' + c(t)\Phi] + \\
&\quad \mathrm{Re}\left[a(t)\varphi' + b(t)\varphi - c(t) \frac{1}{\pi}\int_G \frac{\varphi(\zeta)}{\zeta - \bar{z}}\mathrm{d}\zeta\right] + \\
&\quad \mathrm{Re}\left[\lambda(t) \frac{\partial^2 W^*}{\partial \bar{z} \partial z} + a(t) \frac{\partial^2 W^*}{\partial z^2} + b(t) \frac{\partial W^*}{\partial z} + c(t) W^*\right] \\
&= \gamma(t).
\end{aligned}$$

According to Theorem 1, Theorem 2 and Theorem 3, we can reduce Problem P to the problem with oblique derivatives

$$\mathrm{Re}[a(t)\Phi'' + b(t)\Phi' + c(t)\Phi] = \gamma(t), \tag{4.4}$$

for analytic functions $\Phi(z)$.

Problem (4.4) has been considered by Muskhelishvili N I. [6]

Using the integral representation of analytic functions as follows:[6]

$$\Phi(z) = \int_\Gamma \mu(t)\left(1 - \frac{z}{t}\right)\ln\left(1 - \frac{z}{t}\right)\mathrm{d}s + \int_\Gamma \mu(t)\mathrm{d}s + \mathrm{i}C, \tag{4.5}$$

where $\mu(t)$ is a (real) unknown function (Hölder continuous function), C is an arbitrary (real) constant.

Obviously, we have

$$\Phi''(z) = (-1)^2\int_\Gamma \frac{\mu(t)\,\overline{t'}}{t(t-z)}\mathrm{d}t = \frac{1}{2\pi\mathrm{i}}\int_\Gamma \frac{2\pi\mathrm{i}\,\overline{t'}\mu(t)}{t(t-z)}\mathrm{d}t. \qquad (4.6)$$

Let $z \to t$ in formulas (4.5) (4.6), use Sokhotslki-Plemelj formulas and then substitute them into (4.4), we can get the (real) singular integral equation as follows:

$$K\mu = A(t_0)\mu(t_0) + \int_\Gamma N(t_0, t)\mu(t)\mathrm{d}s = \gamma(t_0) - C\sigma(t_0), \qquad (4.7)$$

where

$$A(t_0) = \mathrm{Re}[\pi\mathrm{i}\,\overline{t'}_0 t_0^{-1} a(t_0)], \qquad (4.8)$$

$$N(t_0, t) = \mathrm{Re}[a(t_0)N_2(t_0, t) + b(t_0)N_1(t_0, t) + c(t_0)N_0(t_0, t)], \qquad (4.9)$$

$$N(t_0, t) = \left(1 - \frac{t_0}{t}\right)\ln\left(1 - \frac{t_0}{t}\right) + 1,$$

$$N_1(t_0, t) = (-1)\cdot\frac{1}{t}\left[1 + \ln\left(1 - \frac{t_0}{t}\right)\right],$$

$$N_2(t_0, t) = (-1)^2\cdot\frac{1}{t}\cdot\frac{1}{t - t_0},$$

$$\sigma(t) = -\mathrm{Im}\, c(t).$$

After a simple computation we find that index of equation (4.7) is as follows:

$$\chi^* = \frac{1}{2\pi}\left[\arg\frac{tt'\,\overline{a(t)}}{\bar{t}\,\overline{t'}\,a(t)}\right]_\Gamma = 2(n+2),$$

$$n = \frac{1}{2\pi}[\arg\overline{a(t)}]_\Gamma.$$

In the next we also call χ^* index of problem (4.4).

For further application we introduce also the transposed homogeneous equation

$$K'v \equiv A(t_0)v(t_0) + \int_\Gamma N(t, t_0)v(t)\mathrm{d}s = 0, \qquad (4.10)$$

with the help of representation (4.5) we know that integral equation (4.7) is equivalent to problem (4.4). Substituting $\mu(t)$ into (4.5) we can obtain the solution of problem (4.4).

Consequently we have

Theorem 4 Problem (4.4) is always solvable for each given $\gamma(t)$, if and only if the transposed homogeneous equation (4.10) has no solution distinct from zero.

Theorem 5 If the transposed homogeneous equation (4.10) has exactly k'_*, linearly independent solutions $v_j^*(t)$, $j=1,2,\cdots,k'_*$, then the necessary and sufficient conditions of solvability for problem (4.4) will be

$$\int_\Gamma v_j^*(t)[\gamma(t)+C\sigma(t)]\mathrm{d}s=0,\quad j=1,2,\cdots,k'_*. \qquad (4.11)$$

In this case homogeneous problem (4.4) has exactly $k^*=\chi^*+k'_*$ linear independent solutions. Obviously, problem (4.4) has unique solution only if $\chi^*+k'_*=0$.

Theorem 6 Suppose that functions $\lambda(t)$, $\alpha(t)$ satisfy the condition B on Γ, $a(t)\neq 0$ and $b(t)$, $c(t)$ satisfy conditions solvability of problem (3.1) (see (3.10)), then Problem P and problem (4.4) are solvable simultaneously. (Of course we are interested only in obtaining the solution of Problem P by the solution of problem(4.4).)

Theorem 7 Suppose conditions of Theorem 6 are satisfied and equation (4.10) has only zero-solution ($k'_*=0$), then the Problem P is always solvable for each given $\gamma(t)$, and its solutions can be represented m the form:

$$W(z)=\int_\Gamma \mu(t)\left(1-\frac{z}{t}\right)\ln\left(1-\frac{z}{t}\right)\mathrm{d}s+\int_\Gamma \mu(t)\mathrm{d}s+\mathrm{i}C-$$
$$\frac{1}{\pi}\iint_G \frac{\varphi(\zeta)}{\bar{\zeta}-\bar{z}}\mathrm{d}\zeta+\frac{2}{\pi}\iint_G f(\zeta)\ln|\zeta-z|\,\mathrm{d}\zeta, \qquad (4.12)$$

where $f(z)=\sum_{j=1}^{N}c_jf_j(z)$, $f_j(z)$, $j=1,2,\cdots,N$ are a complete system of linear independent solutions of equation (4.2), c_j, $j=1,2,\cdots,N$ are arbitrary constants. $\varphi(z)$ is the solutions of problem (3.1) and $\mu(t)$ is the solutions of equation (4.7) (it includes χ^* arbitrary (real) constants).

Theorem 8 Suppose conditions of Theorem 6 are satisfied, and equation (4.10) has $k'_* \neq 0$ linear independent solutions, then Problem P is solvable if and only if conditions (4.11) are satisfied. Its solution will be (4.12) too. The diversity here only is that $\mu(t)$ includes $\chi^* + k'_*$ arbitary (real) constant.

References

[1] Dzhuraev A D. Methods of singular integral equations. Nauka Moscow, 1987. (in Russian)

[2] Zhao Zhen. Boundary value problems for complex elliptic equations on the plane and singular integral equation of composite type. Proceedings of the international conference on integral equations and boundary value problems, World scientific, Singapore. 1991:281-288.

[3] Zhao Zhen. Some boundary value problems on the plane. Journal of Beijing normal university (Natural sci.), 1991, 27(1):42-44.

[4] Vekua I N. Generalized analytic functions. Reading, 1962.

[5] Vekua I N. On the linear Reimann boundary value problem. Bulletins mathematical institute of acad. sci., Georgian republic, Vol. 11, 1942.

[6] Muskhelishvili N I. Singular integral equations. Noordhoft, Groningen, 1953.

论文和著作目录

Bibliography of Papers and Works

论文目录

[序号] 作者. 论文题目. 杂志名称,年份,卷(期):起页-止页.

[1] Чжао Чжэн. Решение Обобшеннои Задачи Римана — Гидбберта Методом Раздожения в Ряд[Russian](Solutions of the generalized Lemann-Hilbert's problem with the method of series development). Докдады Академии Наук СССР,1959,128(2):253-256.

[2] Чжао Чжэн. Рещение Задачи Дирихде На Пдоскости Для Уравнения Эллиптического Типа Второго Порядка Методом Разложения в Рял[Russian](A solution of Dirichlet's problem on plane for second order elliptic differential equations with the method of series development). Доклады Академии Наук СССР,1960,132(4):781-784.

[3] Zhao Zhen. Solution of some classes boundary value problems for elliptic differential equations with the method of series development. 西伯利亚科学院,1960.

[4] Zhao Zhen. Solution of the Dirichlet problem on a plane for a second-order equation of elliptic type by expansion into series. 苏维埃数学,1960,1:663-666.

[5] 赵桢. 用展级数法解二阶椭圆型方程的平面 Dirichlet 问题. 北京师范大学学报(自然科学版),1962,(2):11-18.

[6] 赵桢,刘来福.关于 n 重调和方程的基本边值问题.北京师范大学学报(自然科学版),1963,(3):1-26.

[7] 赵桢.带位移的奇异积分方程的 Noether 理论.应用数学与计算数学,1979,(6):53-61.

[8] 赵桢.带两个 Carleman 位移的奇异积分方程的可解性问题.北京师范大学学报(自然科学版),1980,(2):1-18.

[9] 赵桢.带两个 Carleman 位移的奇异积分方程 Noether 可解的充分必要条件.数学年刊,1981,2A(1):91-100.

[10] 赵桢.关于带两个位移的广义 Hilbert 问题.数学研究与评论,1982,(1):97-108.

[11] 赵桢.关于带位移的奇异积分方程与边值问题.应用数学与计算数学,1982,(1):49-54.

[12] 赵桢.关于非线性奇异积分方程与边值问题.应用数学与计算数学,1984,(3):76-79.

[13] 赵桢.带位移的广义 Hilbert-Poincarè 问题.北京师范大学学报(自然科学版),1986,(1):17-27.

[14] 赵桢,参编.奇异积分方程.中国大百科全书:数学.北京:中国大百科全书出版社,1988:533-534.

[15] Begehr H, Wen Guochun, Zhao Zhen. An initial and boundary value problem for nonlinear composite type systems of three equations. Mathematic Pannonica, 1991, 2(1): 49-61; In: Wen Guochun, Zhao Zhen eds. Proceedings of the international conference on integral equations and boundary value problems. Beijing, 2-7 September 1990, World scientific, 1991:1-3.

[16] Zhao Zhen. Boundary value problems for complex elliptic equations on the plane and singular integral equation of composite type. In: Wen Guochun, Zhao Zhen eds. Proceedings of the international conference on integral equations and boundary value problems, World scientific, 1991:281-288.

[17] Zhao Zhen. Some boundary value problems for elliptic equation on the plane. 北京师范大学学报(自然科学版),1991,27(1):42-44.

[18] 赵桢. 复椭圆型方程的一类斜导数边值问题. 北京师范大学学报(自然科学版),1993,29(2):154-157.

[19] 赵桢. 我中学生代表团参加第 19 届全俄中学生数学奥林匹克载誉回国. 数学通报,1993,(7):31.

[20] 赵桢,编译. 第 19 届全俄中学生数学奥林匹克试题及其解答. 数学通报,1993,(9): 31-35.

[21] 赵桢,编译. 第 19 届全俄中学生数学奥林匹克试题及其解答. 数学通报,1993, (12):27-31.

[22] Zhao Zhen. A class of boundary value problems with oblique derivatives for complex elliptic equation. In: Yang Chungchun, Wen Guochun, Li Kinyin, Chiang Yikman. Complex analysis and its applications. Pitman research notes in mathematics series, Vol. 305,1994,344-351.

[23] 赵桢. 双解析函数的某些性质. 四川师范大学学报(自然科学版),1994,17(2):114-116.

[24] 赵桢. 双解析函数与复调和函数以及它们的基本边值问题. 北京师范大学学报(自然科学版),1995,31(2):175-179.

[25] 赵桢. 双解析函数的某些边值问题. 宁夏大学学报(自然科学版),1996,17(1):41-43.

[26] Zhao Zhen. Riemann-Hilbert's problem for bianalytis functions. 北京师范大学学报(自然科学版),1996,32(3):316-320.

[27] Zhao Zhen. Schwarz's problem for some complex partial differential equations of second order. Beijing mathematics, 1996, 2(1): 131-137.

[28] 赵桢. 一类三阶复偏微分方程的 Schwarz 问题. 宁夏大学学报(自然科学版),1998, 19(1):6-8.

[29] Zhao Zhen, Chenfang Quan. Dirichlet's problems for biharmonic functions. 北京师范大学学报(自然科学版),1998,34(2):174-178.

[30] 邓小琴,赵桢. 二阶椭圆复方程解的某些性质. 北京交通大学学报,1999,23(2):101-106.

[31] Zhao Zhen. Cauchy formula, Integral of Cauchy type and Hilbert

problem for bianalytic functions. In: Begehr H G et al eds. Partial differential and integral equations, Kluwer academic press, 1999: 211-218.

[32] Zhao Zhen. On the integral of Cauchy type and the generalized Harnack theorem for bianalytic functions. In: Heinrich G W et al eds. Proceedings of the second internation society for analysis applications and computation congress, Kluwer academic press, 2000, Vol. 1:223-230.

[33] Zhao Zhen. On some important properties of bianalytic functions. In: Lu Jianke, Wen Guochun eds. Proceedings of the international conference on boundary value problems, Integral equation and related problems. World scientific, 2000:290-296.

著作

[序号] 著者，译者. 书名. 出版地：出版社，出版年份.

[1] 赵桢，陈方权，蒋绍惠，刘来福，邝荣雨，译. 带位移的奇异积分方程与边值问题. 北京：北京师范大学出版社，1982.

[2] 赵桢. 奇异积分方程. 北京：北京师范大学出版社，1984.

[3] Wen Guochun, Zhao Zhen, ed. Proceedings of the international conference on integral equations and boundary value problems. Beijing, 2-7, September, World scientific, 1990.

[4] 赵桢，等主编. 北京数学奥校初中练习册(6 册). 北京：北京师范大学出版社，1995.

[5] 赵桢，章建跃，主编. 数学试题精编 精要 精解(上、中、下册). 北京：中国青年出版社，1996.

[6] 赵桢，等主编. 中小学数学素质同步训练题库(小学三、四、五、六年级各 1 册；初中一、二、三年级各 1 册；初中应用 1 册). 北京：中国友谊出版公司，1996.

[7] 赵桢，主编，刘来福，张秀平，副主编. 北京数学会北京数学培训学校教学丛书：

张秀平，分卷主编. 高中数学分卷 Ⅰ：北京：北京师范大学出版社，2007.

胡永建，分卷主编. 高中数学分卷 Ⅱ：北京：北京师范大学出版社，2007.

刘来福，分卷主编. 高中数学奥林匹克教程应用分卷. 北京：北京师范大学出版社，2008.

李延林，分卷主编. 高中数学奥林匹克教程提高分卷. 北京：北京师范大学出版社，2010.

刘来福，分卷主编. 高中数学奥林匹克教程建模分卷. 北京：北京师范大学出版社，2011.